D0216289

RANDOMNESS & UNDECIDABILITY IN PHYSICS

RANDOMNESS & UNDECIDABILITY IN PHYSICS

Karl Svozil

Institute for Theoretical Physics
Technical University of Vienna
Austria

Published by

World Scientific Publishing Co. Pte. Ltd.

P O Box 128, Farrer Road, Singapore 9128

USA office: Suite 1B, 1060 Main Street, River Edge, NJ 07661

UK office: 73 Lynton Mead, Totteridge, London N20 8DH

Library of Congress Cataloging-in-Publication Data
Svozil, Karl.
Randomness & undecidability in physics / by Karl Svozil.
p. cm.
Includes bibliographical references.
ISBN 981020809X
1. Determinism (Philosophy) 2. Stochastic processes. 3. Chaotic behavior in systems. 4. Mathematical physics. I. Title.
II. Title: Randomness and undecidability in physics.
QC6.4.D46S86 1993
530.1'5--dc20 92-19665
CIP

The cover drawings are by Karin Svozil.

This book was typeset using the TEX/LATEX typesetting system, in particular PostScript-based LATEX (PS-LATEX) by Mario Wolczko. The fonts used in this book include Times Roman, Helvetica, Courier, ITC Avant Garde Gothic Book, Symbol, AMS-fonts & D. E. Knuth's PUNK. The output from TEX was converted into the PostScript page description language with DVIPS by Tomas Rokicki. In this process, graphics were directly added from *Mathematica* in Encapsulated PostScript form.

Printed in Singapore by Utopia Press.

Dedicated to the memory of my father

Karl Svozil

3. 8. 1918 — 11. 5. 1991

Preface

Recent findings in the computer sciences, discrete mathematics, formal logics and metamathematics have opened up a *via regia* for the investigation of undecidability and randomness in physics. A translation of these formal and abstract concepts yields a fresh look into diverse features of physical modelling such as quantum complementarity and the measurement problem, but also stipulates questions related to the necessity of the assumption of continua.

Any physical system may be perceived as a computational process. One could even speculate that physical systems *exactly* correspond to, and indeed are, computations of a very specific kind; with a particular computer model in mind. From this point of view it is absolutely reasonable to investigate physical systems with concepts and methods developed by the computer sciences.

Conversely, any computer may be perceived as a physical system; not only in the immediate sense of the physical properties of its hardware. Computers are a medium to virtual realities. The foreseeable importance of such virtual realities stimulates the investigation of an "inner description," a "virtual physics," if you like, of these universes of computation. Indeed, one may consider our own universe as just one particular realisation of an enormous number of virtual realities, most of them awaiting discovery.

Besides these issues, the intuitive terms "rational" (human thought), "conceivable" and so on, have been made precise by the concepts of mechanic computation and recursive enumeration. The reader may find these developments sufficiently exciting to go on and study this new field.

The first part of this book introduces the fundamental concepts. Informally stated, recursive function theory is concerned with the question of whether an entity is computable in a very precisely defined way. Algorithmic information theory deals with the quantitative description of computation, in particular with the shortest program length. Coding and suitable algebraic representation of physical statements are the prerequisites for their algorithmic treatment.

One motive of this book is the recognition that what is often referred to as "randomness" in physics might actually be a signature of undecidability for systems whose evolution is computable on a step-by-step basis. Therefore the second part of the book is devoted to the investigation of undecidability.

To give a flavour of the type of questions envisaged: Consider an arbitrary algorithmic system which is computable on a step-by-step basis. — Any computer program is such a system. It is in general impossible to specify another algorithmic procedure (including itself) which, by performing experiments and successive input/output analysis on the first system, finds the deterministic law by which it is governed. But even if such a law is specified, it is in general impossible to predict the system behaviour in the "distant future.' In other words: no "speedup" or "computational shortcut" is possible. These statements are consequences of two classical theorems in recursion theory, the recursive unsolvability of the rule inference problem and of the halting problem.

Certain self-referential statements like "I am lying" are paradoxical and resemble the absurd attempt of *Freiherr von Münchhausen* to rescue himself from a swamp by dragging himself out by his own hair. Such paradoxes can only be consistently avoided by accepting restrictions to the expressive power and to the comprehension of the associated methods and systems — with undecidability and incompleteness as consequence.

Complementarity is a feature which can be modelled by experiments on certain finite automata. This is due to the fact that measurement of one observable of the automaton destroys the possibility to measure another observable of the same automaton and *vice versa.* Certain self-referential measurements pursuit a similar attempt: on the one hand they pretend to render the "true" value of an observable, while on the other hand they have to interact with the object to be measured and thereby inevitably change its state.

It is important to distinguish between the "intrinsic" view of an observer, who is entangled with and who is an inseparable part of the system, and the "extrinsic" perspective of an observer who is not entangled with the system *via* self-reference. Indeed, the recognition of the importance of intrinsic perception, of a "view from within," might be considered as a key observation towards a better understanding of undecidability and complementarity.

The third, last part of the book is dedicated to a formal definition of randomness and entropy measures based on algorithmic information theory.

Help and discussions with Matthias Baaz, Norbert Brunner, Cristian Calude, Gregory Chaitin, Anatol Dvurečenskij, Günther Krenn, Michiel van Lambalgen, Otto Rössler, Martin Schaller, Christoph Strnadl, Johann Summhammer, Walter Thirring and Anton Zeilinger are gratefully acknowledged. Nevertheless, all misinterpretations and mistakes are mine. Thanks go also to Jennifer Gan from World Scientific, who managed to guide me through the production of this book both kindly and smoothly.

Vienna, March 1993 Karl Svozil

Institut für Theoretische Physik
Technische Universität Wien
Wiedner Hauptstraße 8-10/136
A-1040 Vienna, Austria

e1360dab@awiuni11.bitnet
e1360dab@AWIUNI11.EDVZ.UniVie.AC.AT

Contents

List of symbols

$\aleph_0$	cardinality of a countable infinite set, e.g., $\mathbb{N}$ or $\mathbb{Q}$	
$\aleph_1$	cardinality of a continuum, e.g., $\mathbb{R}$ or $\mathbb{C}$	
C^n	n times continuously differentiable	29, 240
$\mathfrak{C}(\mathfrak{H})$	lattice of closed subspaces of a Hilbert space $\mathfrak{H}$	67
$\mathfrak{A}$	automaton propositional calculus	131
$L \vdash \varphi$	φ can be derived in L	119
$\varnothing$	empty set, empty input sequence	17
`False`, `F`, **0**	absurd proposition	
$H(s)$	algorithmic information of object s	91, 195
$H(s, t)$	joint algorithmic information of objects s, t	91
H_D	computational complexity	99
$h(N; M)$	information theoretic uncertainty, missing information	233
h	average algorithmic entropy per symbol	
$\bar{h}$	metric entropy	240
$\mathfrak{H}$	Hilbert space	66
$I(N; M)$	Shannon information gain after $N - M$ experiments	233
$\mathfrak{I}$	generic interface	81
$\upharpoonleft\!\downharpoonright$	interface by sybolic exchange	81
∞	infinity	
$\lvert x \rvert$	length of string or chain x, cardinal number of set x	91, 55, 154
$\lVert x \rVert$	length of chain x, norm of x	53, 65
$\log_2^* x$	iterated logarithm	89
λ	Lyapunov exponent	216, 240
$MO2$, MOn	"Chinese latern" lattice	149, 157
$\mathbb{N}$	set of integers	
$\#(s)$	code of symbol s	47
$O(f)$	of the order of f	88
ω	ordinal number infinity	194
2^ω	set of infinite binary sequences	194
Ω	halting "probability"	92
$\mathfrak{P}(\mathfrak{H})$	lattice of all projections of a Hilbert space $\mathfrak{H}$	69
$P(s)$	algorithmic "probability" to produce object s	92
$P(s, t)$	joint algorithmic "probability" to produce objects s, t	92
$\mathbb{Q}$	set of rationals	
$\mathbb{R}$	set of reals	
$\Sigma(n)$	busy beaver function	104
$\widetilde{}$, $\tilde{\mathfrak{A}}$	extrinsic entity, extrinsic automaton logic	79, 131

True, T, **1**	trivial proposition	
$U(p, s) = t$, $U(p) = t$	universal computer, program p, input s, output t	17
$v(s)$	partition of automaton states from input s	131
V	set of partitions of automaton states	131
$\vee$	logical or	133
$\wedge$	logical and	133
$\neg$	logical not	133

"*iff*" stands for "*if and only if;*" i.e., for a necessary and sufficient condition.

Part I

Algorithmic physics: the Universe as a computer

'So I wasn't dreaming, after all,' she said to herself, 'unless—unless we're all part of the same dream. Only I do hope it is *my* dream, and not the Red King's! I dont like belonging to another person's dream,' she went on in a rather complaining tone: 'I have a great mind to go and wake him, and see what happens!'
from "Through the Looking Glass" by Lewis Carroll [82]

Chapter 1

Algorithmics and recursive function theory

This chapter should be understood as a very brief review of algorithmics and the theory of recursive functions. For an elementary introduction, see, for instance, David Harel's book *Algorithmics* [227]. Comprehensive treatments of the subject are Hartley Rogers' *Theory of Recursive Functions and Effective Computability* [390] and P. Odifreddi's *Classical Recursion Theory* [342]. Other introductions are M. Davis, *Computability & Unsolvability* [135], M. L. Minsky, *Computation: Finite and Infinite Machines* [322] and part C of the *Handbook of Mathematical Logic* [27].

1.1 Algorithm and effective computability

The present concept of formal computation has been developed from the experience of practical "manual" calculation, at least up to some limited resource level. How could the intuitive concept of "manual" or "mechanic computation" be formalised? (In the computer sciences, what is called "mechanic" is often referred to as *"deterministic."* In the physical context, such a terminology may give rise to confusion with the term "deterministic system," which will be characterised by some "mechanically" computable evolution function but not necessarily "mechanically" computable parameter values.) Take, for instance, an intuitive, pragmatic understanding of "mechanic computation," put forward by Alan M. Turing [446] in the 30's: *"whatever can (in principle) be calculated on a sheet of paper by the usual rules is computable."* The question whether or not this — or any other (possibly more sophisticated) — understanding of "mechanic" computation is adequate, has (at least) two aspects: *(i)* syntax, investigated by formal logic, mathematics and the computer sciences and *(ii)* physics.

One important result of syntactic arguments is the concept of *"universal computation,"* as envisioned by K. Gödel, J. Herbrand, St. C. Kleene, A. Church and, most influentially, by A. M. Turing. Universal computation is usually developed by introducing a very precise (elementary) model of computation, e.g., the Turing machine, a Cellular Automaton *et cetera*. It is then shown that, provided it is "sufficiently complex," this model comprises all "reasonable" instances of computation. I.e., adding additional components to such a machine does not change its computational capacity qualitatively. (Yet it may change its efficiency in terms of time, storage requirements, program length *et cetera*.) Furthermore, all such "universal" models of computation turn out to be equivalent in the sense that there is a one-to-one translation of one into the other; i.e., every calculation that can be performed by any one of them can be done by all of the others as well. It is often assumed that universal computation remains unaffected by the specific physical realisation of the computing device. I.e., it does not really matter (in principle) what physical base you use for computing — doped silicon, light, nerve cells, billiard balls — as long as your devices work "mechanically" and are "sufficiently complex" to support universal computation.

Nevertheless, the *syntax* of universal computation is based on *primary intuitive concepts* of "reasonable" instances of "mechanic" computation, which refer to *physical insight;* i.e., which refer to the types of processes which can be performed in the physical world. In this sense, the level of physical comprehension sets the limits to whatever is acceptable as valid method of computation. If, for instance, we could employ an "oracle" (the term "oracle" will be specified later), our wider notion of "mechanic" computation would include oracle computation. Likewise, a computationally more restricted universe would intrinsically imply a more restricted concept of "mechanic" computation. For an early and brilliant discussion of this aspect the reader is referred to A. M. Turing's original work [446]. As D. Deutsch puts it ([139], p. 101),

> *"The reason why we find it possible to construct, say, electronic calculators, and indeed why we can perform mental arithmetic, cannot be found in mathematics or logic.* The reason is that the laws of physics 'happen to' permit the existence of physical models for the operations of arithmetic *such as addition, subtraction and multiplication. If they did not, these familiar operations would be non-computable functions. We might still know* of *them and invoke them in mathematical proofs (which would presumably be called 'non constructive') but we could not perform them."*

For another discussion of this topic, see R. Rosen [391] and M. Davis' book [135], p. 11, where the following question is asked:

> " *... how can we ever exclude the possibility of our presented, some day (perhaps by some extraterrestrial visitors), with a (perhaps extremely complex) device or "oracle" that "computes" a non computable function?"*

Indeed, the concept of *Turing degree* [446] yields an extension of universal computation to oracle computation. In a way, the syntactic aspect of computation could thus be "tuned" to wider concepts of (physically realisable) computation if that turns out to be necessary. As has been pointed out before, such a change cannot be decided by syntactic reasoning; it has to be motivated by physical arguments. Thus, at least at the level of investigation of "reasonable" instances of computation, the theory of computation is part of physics.

Consider, for example, electronic computing, which has transformed our comprehension of computation dramatically, shifting it from "number-crunching" to symbolic calculation, visualisation, automation, artificial intelligence and the creation of "virtual realities." Yet, as long as electronic circuits act as "mechanic" devices, electronics will not change our concept of universal "mechanic" computing. Thus electronic computing is not an example for the type of physical process which would make necessary a revision of concepts of computation.

The question remains whether such revisions will become necessary as our abilities to stimulate and control physical processes advance. It remains to be seen whether new capacities will become feasible by the use of the continuum in classical physics (see below) or by quantum effects [139, 140, 141, 7, 175, 309, 33, 92, 93] or more remote possibilities. It can for instance be conjectured that quantum computation will not enlarge the scope of computation because the enhanced physical capacities of quantum mechanics will inevitably be washed away by noise; i.e., by the randomness of quantum events and by the amplification of noise when information in the quantum domain gets interpreted classically. This resembles the "peaceful coexistence" between quantum theory and relativity (cf. section 17.2.3, p. 229).

We do not know whether our own self is limited by the "mechanic" computational capacities of the part of the world we presently cope with scientifically. Kurt Gödel (*Collected Works II* [206], p. 305) seemed to have believed that the mind transcends mechanical procedures. See also P. Odifreddi [342], p. 113, and the remark by H. Weyl, quoted on page 24.

Thus, if the term "mechanic" is a label referring to rules which completely determine some "reasonable" actions of computation, then whatever is considered as "reasonable" action of computation has to be specified also by physics.

The following notion of "algorithm" or "effective computation" is motivated by those types of tasks which can actually be performed in the physical world. Therefore, its features are bounded (from above), excluding computations in the limit of infinite time, storage space, program length *et cetera*.

D 1.1 (Algorithm, effective computability)

An algorithm *is an abstract recipe of finite size, prescribing a process which might be carried out "mechanically" by some computing agent, such as a human or a computer and which terminates after some finite time.*

An effective computation, *or* effective *procedure C, is some "mechanic" procedure which can be applied to any of a certain class of symbolic inputs s and which will eventually yield, for some inputs, a symbolic output t. The process is written similar to functional notation* $C(s) = t$.

Remarks:

(i) An algorithm needs not to be defined for *all* input values. This corresponds to a *partial* function, whose domain is not all of $\mathbb{N}$ (for details, see 1.2, p. 6). In such a case, the procedure might go on forever, never halting.

(ii) Although the input and output sequences as well as the size of the "scratch paper" (internal memory) have to be finite, *no* restriction is imposed on their actual sizes. The same is true for the length of the description of the algorithm as well as its execution time. This renders the definition hard to operationalise, since, for instance, relatively "small" programs may take a "large" time to execute and may produce "huge" output (for details, see 8.7, p. 104).

(iii) The above definition of *algorithm* and *effective computability* is informal. The question if there exist formal notions corresponding to this heuristic approach is nontrivial. In particular, the term "mechanic" has to be specified formally and sufficiently

comprehensive. The Church-Turing thesis (see below) gives a positive answer which, due to the principal incompatibility between the informal, intuitive ("algorithm") and the formal ("recursive function") approach, remains conjectural. To specify this connection, a brief characterisation of the term "recursive function" will be given in the next section.

(iv) Analogue computation is not *effective* in the above sense, since its feature of *continuity* does not correspond to any meaningful "mechanical" process.

(v) The input can be eliminated by substituting a constant for each input parameter, thereby converting, say, a `FORTRAN` statement such as `READ (*,*) A` into `A=···`. In that way, *input* code can be converted to *program* code, and *vice versa*. See also type-free coding in chapter 4, p. 45.

Since the present understanding of "mechanic" computation has been developed from the computations which can be actually realised, it is tautological that, indeed, there exist physical systems which are universal computers (at least up to some finite computational resources). The circularity of this argument is evident. A constructive "proof" of this conjecture is the assembly of a "universal" computer (here the quotes refer to the finiteness of the computer's resources) such as a personal computer workstation (insert your favourite brand here:) "...", Charles Babbage's *Analytical Engine* [439] *et cetera.* These computers are "universal" because it is possible to implement a program which *simulates* any other (not necessarily universal) computer such as a Turing machine (with finite tape). Their universality is only limited by the finiteness of their resources.

1.2 Recursive functions

The notion of *effective computation* will now be formalised by defining a specific model of computation which is (hopefully) comprehensive enough to correspond properly to the heuristic concept of algorithm.

D 1.2 (Partial function, total function) *The notion of* partial and total function *is defined as follows: A* partial function $f(x) = z$, *with* $x, z \in \mathbb{N}$ *is a* singe-valued *function, i.e., the functional value* z *is determined uniquely. The* domain *of* f *is* $\text{domain}(f) = \{x \mid \forall z = f(x)\} \subset \mathbb{N}$. *The term* partial *expresses the fact that the domain of* f *may not be* all of $\mathbb{N}$. *If* $\text{domain}(f) = \mathbb{N}$, f *is called a* total function. f *is* defined (convergent) *at* x *if* $x \in \text{domain}(f)$, *expressed by* $f(x)\downarrow$; *otherwise* f *is* undefined (divergent) *at* x, *expressed by* $f(x)\uparrow$.

1.2.1 Primitive recursive function

A first attempt towards a formal characterisation of effectively computable functions is motivated by the construction of the natural numbers $\mathbb{N}$ by the iterated process of "adding 1" [360]. By generalising this approach, one may consider a class of functions obtained by *recursive* definitions, i.e., an arbitrary function is itself a function of "simpler" functions. Thereby, the notion of "simpler" function is specified. Usually the "simplest" functions, among others, are defined to be the constant functions, the successor functions, *et cetera.* The corresponding class of *primitive recursive functions* can be defined as follows:

D 1.3 (Primitive recursive function)

The class of primitive recursive functions *is the smallest class* $\mathfrak{C}$ *of functions* $f, g, \ldots$ *from*

$\mathbb{N}$ *to* $\mathbb{N}$ *such that*

(i) All constant functions, $f(x_1, \ldots, x_k) = m$*, are in* $\mathfrak{C}$, $1 \leq k$, $0 \leq m$;

(ii) The successor function, $f(x) = x + 1$*, is in* $\mathfrak{C}$;

(iii) All identity functions, $f(x_1, \ldots, x_k) = x_i$*, are in* $\mathfrak{C}$, $1 \leq i \leq k$;

(iv) If f *is a function of* k *variables in* $\mathfrak{C}$*, and* $g_1, \ldots, g_k$ *are (each) functions of* m *variables in* $\mathfrak{C}$*, then the function* $h(x_1, \ldots, x_m) = f(g_1(x_1, \ldots, x_m), \ldots, g_k(x_1, \ldots, x_m))$ *is in* $\mathfrak{C}$, $1 \leq k, m$;

(v) if h *is a function of* $k+1$ *variables in* $\mathfrak{C}$*, and* g *is a function of* $k-1$ *variables in* $\mathfrak{C}$*, then the unique function* f *of* k *variables satisfying* $f(0, x_2, \ldots, x_k) = g(x_2, \ldots, x_k)$, $f(y+1, x_2, \ldots, x_k) = h(y, f(y, x_2, \ldots, x_k), x_2, \ldots, x_k)$ *is in* $\mathfrak{C}$, $1 \leq k$.

It would have been more elegant to use the λ-notation for characterising $\mathfrak{C}$ (for details, see H. Rogers [390], p. 6), but as this notation is difficult to interpret for the unfamiliar reader, the standard functional notation was used.

Example:

The following primitive recursive definition yields the *Fibonacci sequence* 0, 1, 1, 2, 3, 5, 8, ...:

$$\begin{aligned} f(0) &= 1 \\ f(1) &= 1 \\ f(x+2) &= f(x+1) + f(x) \quad . \end{aligned}$$

The corresponding algebraic expression is given by

$$f(x) = \frac{1}{\sqrt{5}} \left[\left(\frac{1+\sqrt{5}}{2} \right)^{x+1} - \left(\frac{1-\sqrt{5}}{2} \right)^{x+1} \right] .$$

Yet, primitive recursion, as introduced in 1.3, does not comprehend the full realm of effective computability. It can be shown that there exist *algorithmic* functions of the natural numbers which grow faster than any primitive recursive function. This has been, for instance, discussed in depth in Róza Péter's book *Rekursive Funktionen* [360], p. 86, by C. Calude, S. Marcus and I. Tevy [62] and by H. Rogers [390], p. 8. The first example of a recursive function which is not primitive recursive was obtained independently by W. Ackermann [3] and the Romanian of Swiss origin G. Sudan [423]. It is interesting that these examples are not just mathematical curiosities (they are very useful in compiler testing, for instance), but as has been pointed out by C. Calude [72], *most* recursive functions (see definition 1.2, p. 6) are *not* primitive recursive, just as most reals are not recursive. A particular example will be given next.

Counterexample:

One strategy for obtaining counterexamples of evidently recursive functions which are *not* primitive recursive is to try to construct functions which grow "very fast." Consider, for instance, the function $f(x) = A(x, x, x)$, where the *Ackermann generalised exponential* A is defined by the successive iteration

$$A(0, x, y) = y + x \tag{1.1}$$

$$A(1,x,y) = yx \quad (1.2)$$

$$A(2,x,y) = y^x \quad (1.3)$$

$$A(3,x,y) = \left. y^{y^{\cdot^{\cdot^{\cdot^{y}}}}} \right\} x \text{ times} \quad (1.4)$$

$$\vdots$$

$$A(z+1,x+1,y) = A(z,A(z+1,x,y),y) \quad . \quad (1.5)$$

Another recursive representation of the Ackermann function is (see [390], p. 8):

$$\begin{aligned}
A(0,0,y) &= y \\
A(0,x+1,y) &= A(0,x,y)+1 \\
A(1,0,y) &= 0 \\
A(x+2,0,y) &= 1 \\
A(z+1,x+1,y) &= A(z,A(z+1,x,y),y) \quad .
\end{aligned}$$

$A(x,x,x)$ is defined by using double nesting. Functions which grow even faster can be obtained by higher order nesting. It can be shown that for sufficiently large $x \in \mathbb{N}$, every primitive recursive function of x is less than $f(x)$ [360]. Nevertheless, the above definition of A is within the scope of a reasonable definition of algorithm.

Another, strategy is diagonalization (see below). Diagonalization is actually a very similar strategy. Consider functions $f_1(n) < f_2(n) < f_3(n) < \cdots$ and $g(n) = f_n(n)+1$. g grows faster than all f's. For, if $g(n)$ would correspond to some $f_i(n)$, then for $n = i$ the contradiction $f_n(n) = g(n) = f_n(n)+1$ (wrong) would follow.

In summary, the notion of primitive recursive functions turns out to be *too weak* to correspond to the algorithmic notion of effective computability.

1.2.2 Diagonalization

Another example of an algorithmically definable function which is not primitive recursive can be constructed by a method called *diagonalization* (see also Cantor's diagonalization method, chapter 9.2, p. 113). In principle, it is possible to list and label all primitive recursive functions by a finite algorithm. In doing so, one could, for instance, derive all strings corresponding to primitive recursive functions of length 1, 2, 3, and so on, and then label these functions by a natural number. In this way, an effectively computable one-to-one function between the natural numbers $\mathbb{N}$ and the class $\mathfrak{C}$ of primitive recursive functions can be defined. Let g_x be the xth function in this enumeration. Now define a diagonalization function h by

$$h(x) = g_x(x)+1 \quad . \quad (1.6)$$

Notice that for defining h at the argument x we have used the xth function g_x in the functional enumeration. (This "trick" is the reason why the method is called "diagonalization.") Notice further that, since the addition of one is an effectively computable task and since $g_x(x)$ can be obtained by an effective computation, h must be effectively computable.

Assume now that h were primitive recursive. If this would be the case, h would turn up in the enumeration of the primitive recursive functions somewhere; say at the yth place. Hence,

$$h = g_y \tag{1.7}$$

But then, if we took h at y, by combining equations (1.6) and (1.7), we would arrive at the contradiction

$$g_y(y) = h(y) = g_y(y) + 1 \quad \text{(wrong)} \quad . \tag{1.8}$$

Hence, h cannot be primitive recursive — diagonalization leads to the conclusion that the class of primitive recursive functions does not include all algorithmically definable functions.

One possible way to avoid this contradiction (and at the same time maintain functional enumeration) is to assume that the functions g_i are *partial* functions. Thereby, if in the above construction, $g_y(y)$ needs not be defined, the contradiction "$g_y(y) = g_y(y)+1$" needs not show up.

1.2.3 Partial recursive function

The following notion of *recursive function* was proposed by St. C. Kleene [136] (see also Roger's book [390], p. 16). Another, equivalent, definition has been proposed by A. Turing [446] on the basis of Turing machines (p. 17). Turing himself motivated its definition by arguing that it is a one-dimensional analogue of a sheet of paper on which calculations could be performed by the usual rules. Other equivalent definitions have been put forward by A. Church in the context of *lambda calculus* and by K. Gödel (see also M. Davis [135], p. 10).

D 1.4 (Partial recursive function)

Consider a set of instructions P, consisting of recursive relations of a general kind. A computation *is a finite sequence of equations, beginning with P, where each of the following equations is obtained from preceding equations by one of the three procedures:*

(i) by the substitution *of a numeral expression for a variable symbol throughout an entire equation;*

(ii) by the use of one equation *to* substitute *"equals for equals" at every occurrence in a* second equation*;*

(iii) by the evaluation of an instance of the successor function $f(x) = x + 1$.

An equation is deducible *from P if it is obtained in the course of* some *computation from P. Imagine a* standard list *of all equations deducible from P. Then given any P and any numerical input, the* principle output *is defined to be the* first *numerical output in the standard list generated from the set of instructions P. The resulting relation between input and principal output defines a* partial recursive function f_P, *which can be associated with the original set of instructions P. In this way one obtains the class of* partial recursive functions.

Remarks:

(i) The class of primitive recursive functions can be embedded into the class of recursive functions. The extension of recursive functions over primitive recursive functions

consists in the addition of the operation of seeking indefinitely through some standard list for the suitable equation.

(ii) One may suspect that the course of computation is not uniquely determined by the input and by the recursive relations P. Moreover, two different outputs may be obtained from the same input by different computations. These difficulties are circumvented by defining the value of the function to be *the first occurrence* in an effectively computable standard list.

(iii) The *type-free* representation of the recursive functions by *combinatory logic* and, more specifically, by the *lambda calculus*. In this representation, no difference is made between a "function" and its "argument" and "value(s)."

(iv) There "exist" non recursive functions which grow too fast to be effectively computable. One example is the *busy beaver function* $\Sigma_U(n)$ which, roughly speaking, is defined to be the greatest number producible on a computer U by programs of program length less than or equal to n. (For a precise definition, see 8.7, p. 104.) $\Sigma_U(n)$ is a non recursive function; nevertheless it can also be used for upper bounds on the runtime of such programs; see 8.9, p. 105. That this function grows "very fast" indeed can be seen from a program evaluating the *Ackermann generalised exponential* $A(n,n,n)$; with $\Sigma_U(n) \gg A(n,n,n)$. A computer implementation of A requires programs of "small" length, yet, for large n, $A(n,n,n)$ is considerably "large." This applies to computer programs which are permitted to have sufficient size to implement the Ackermann generalised exponential and other "fast growing" functions. (On the contrary, a restriction to "very small size" program codes of only a view bits length would not allow the implementation of $A(n,n,n)$ or similar, fast growing algorithms.) Indeed, if the busy beaver function would be "growing slowly," one would be able to solve all halting problems of finite-size programs (p. 105).

1.2.4 Enumeration of partial recursive functions

The set of primitive recursive functions is one-to-one enumerable. In the same way, it is possible to enumerate all recursive functions, or, equivalently, the sets of instructions P. This statement will not be proved here; informally it may be seen as a direct consequence of an effective enumeration of all sets of instructions P. I.e., every integer x can be associated with a partial recursive function ψ_x (corresponding to a set of instructions P_x) which is at the $(x+1)$'st place of this enumeration.

D 1.5 (Gödel number of recursive function)

P_x is the set of instructions associated with the integer x in the fixed enumeration of all sets of instructions. $x = \#(P_x)$ is called index *or* Gödel number *of P_x. # is the associated recursive Gödel function.*

We shall use the method of Gödel numbers to derive the cardinality of the class of recusive functions.

T 1.6 *The following statements hold:*

(i) There are exactly $\aleph_0$ (a countable *infinity of) partial recursive functions. The partially recursive functions can be recursively enumerated.*

(ii) There are exactly $\aleph_0$ total recursive functions. The total recursive functions cannot be recursively enumerated.

Informally stated, the non recursive enumerability of the total recursive functions originates in the impossibility to recursively determine which function is a partial function and which one is a total function. This is a consequence of the recursive unsolvability of the halting problem (cf. section 9.3, p. 114).

Proof:

Since all constant functions (with values in $\mathbb{N}$) are recursive and $|\mathbb{N}| = \aleph_0$, there are *at least* $\aleph_0$ constant functions. The Gödel numbering shows that there are *at most* $\aleph_0$ partial recursive functions. By a similar argument as in the section dealing with *diagonalization* (p. 8), the assumption of the enumeration of all total recursive functions yields a contradiction.

A probably more intuitive, algorithmic, proof is the enumeration of all Turing machines. (For a definition of a *Turing machine*, see 1.18, p. 17.) Such an explicit enumeration of Turing machines is, for instance, outlined in M. Davis' [135], chapter 4.)

Remarks:

(i) Unlike the primitive recursive functions (which are *total* functions), the class of recursive functions contains *partial* functions. Therefore, the argument against the enumeration of all algorithmically definable functions using *diagonalization* (p. 8) does not apply here. Moreover, as will be seen from the *recursive unsolvability of the halting problem* (see chapter 9.3, p. 114), it is in general undecidable if a recursive function converges or diverges at some particular argument.

(ii) The inverse of the Gödel function # may be partial or total. In the above definition, $\#^{-1}$ is total. For an example of a construction of the Gödel number $\#(t)$ corresponding to a term t, see definition 4.6, p. 49. This definition can be used to enumerate all recursive function with a slight change in notation, i.e., whenever we have to code "+" or some other algebraic operation allowed in a recursion we use the code for "~" or some other symbol which we dont use in this context. In this example, $\#^{-1}$ is a *partial* function.

1.2.5 Existence of uncomputable reals

One major goal of this book is the presentation of mathematical and physical entities which are non recursive. Such entities may be non recursive functions, numbers, sequences and problems which are unsolvable by recursive functions.

A somewhat related subject are mathematical proof methods which do not have any "constructive" or algorithmic flavour, for instance the principle of the excluded middle. One example is a proof of the following theorem: *"There exist irrational numbers $x, y \in \mathbb{R}-\mathbb{Q}$ with $x^y \in \mathbb{Q}$."* *Proof:* case 1: $\sqrt{2}^{\sqrt{2}} \in \mathbb{Q}$; case 2: $\sqrt{2}^{\sqrt{2}} \not\in \mathbb{Q}$, then $\sqrt{2}^{\sqrt{2}^{\sqrt{2}}} = 2 \in \mathbb{Q}$. The question which one of the two cases is correct; i.e., which number is rational, remains unsolved in the context of the proof.

Informally, recursive reals may be described as the real numbers whose expressions as a decimal are calculable "bit by bit" by finite means, i.e., by effective computations (c.f., [446]). This sloppy definition may give rise to confusions; in particular the term "bit by bit" needs an explanation: One might, for instance, conclude correctly that all rationals, all rational powers of rationals (e.g., $\sqrt{2}$), Euler's number e, the number π *et cetera* are computable. But then, one might be tempted to wrongly infer that the halting probability Ω (p. 92) is computable too, since G. Chaitin has published an algorithm for computing Ω in the limit of infinite time [110]. Yet, one necessary condition for a real to be computable is that it *converges effectively* (i.e., that there exists a computable radius of convergence). — This is what is meant by the phrase "bit by bit." This criterion (effectively computable radius of convergence) is not satisfied by any computation for Ω. For a detailed definition of recursive reals, see definition 1.25, p. 28, as well as M. B. Pour-El and J. I. Richards, *Computability in Analysis and Physics* [373], as well as the forthcoming book *Computability --- A Mathematical Sketchbook* by D. S. Bridges [53], among others.

T 1.7 *The following statements hold:*

(i) There are exactly $\aleph_0$ (a countable *infinity of) recursive reals.*

(ii) The set of recursive reals is not recursively enumerable. I.e., there does not exist an effective enumeration of all (countable many) recursive reals.

(iii) "Almost all" reals are non recursive.

Proof:

ad (i) Since all constant functions (with values in $\mathbb{N}$) are recursive and $|\mathbb{N}| = \aleph_0$, there are *at least* $\aleph_0$ recursive reals. By theorem 1.6, there are *at most* denumerable many (i.e., $\aleph_0$) partial recursive functions (one may think of an enumeration of all Turing machines).

ad (ii) Proof by diagonalization: assume that there exists an effectively computable enumeration of the decimal reals in the interval [0, 1] of the form

$$\begin{aligned} r_1 &= 0.r_{11}r_{12}r_{13}r_{14}\cdots \\ r_2 &= 0.r_{21}r_{22}r_{23}r_{24}\cdots \\ r_3 &= 0.r_{31}r_{32}r_{33}r_{34}\cdots \\ r_4 &= 0.r_{41}r_{42}r_{43}r_{44}\cdots \\ &\vdots \end{aligned} \tag{1.9}$$

Consider the real number formed by the diagonal elements $0.r_{11}r_{22}r_{33}\cdots$. Now change each of these digits, avoiding zero and nine. (This is necessary because reals with different digit sequences are identified if one of them ends with an infinite sequence of nines and the other with zeros, for example 0.0999 ... = 0.1000 ....) The result is a real $r' = 0.r_1'r_2'r_3'\cdots$ with $r_n' \neq r_{nn}$ which thus differs from each of the original numbers in at least one (i.e., the "diagonal") position. Therefore there exists at least one real which is not contained in the original enumeration. Despite the assumption of the recursive enumerability of the recursive reals, all the operations involved in the argument are straightforwardly computable. Therefore, if this assumption were correct, r' should be computable as well. Therefore, r' should show up somewhere in the original enumeration. The fact that r' is *not* contained therein and the resulting contradiction allows only one consistent consequence: an effectively computable enumeration of the computable reals does not exist; any algorithmic attempt to enumerate the recursively enumerable reals is incomplete.

In contradistinction, there exists an effectively computable enumeration of the rational numbers $\frac{i}{j}$, for instance by taking the counter diagonals in

$$\begin{array}{cccc} \frac{1}{1} & \frac{1}{2} & \frac{1}{3} & \cdots \\ \frac{2}{1} & \frac{2}{2} & \frac{2}{3} & \cdots \\ \frac{3}{1} & \frac{3}{2} & \frac{3}{3} & \cdots \\ \frac{4}{1} & \frac{4}{2} & \frac{4}{3} & \cdots \\ \vdots & \vdots & \vdots & \ddots \end{array}$$

starting from the top left.

ad (iii) By theorem 1.7*(i)*, the set of recursive reals is countable. The measure of a countable set of points on the real line is zero. Therefore, the measure of the set of all uncomputable reals is one. I.e., in the measure theoretic sense, "almost all" reals are uncomputable. This can be demonstrated by the following argument: Let $M = \{r_i\}$ be an infinite point set (i.e., M is a set of points r_i) which is denumerable and which is the subset of a dense set. Then, for instance, every $r_i \in M$ can be enclosed in the interval

$$I(i, \delta) = [r_i - 2^{-i-1}\delta, r_i + 2^{-i-1}\delta] \quad , \tag{1.10}$$

where δ may be arbitrary small (we choose δ to be small enough that all intervals are disjoint). Since M is denumerable, the measure μ of these intervals can be summed up, yielding

$$\sum_i \mu(I(i, \delta)) = \delta \sum_{i=1}^{\infty} 2^{-i} = \delta \quad . \tag{1.11}$$

From $\delta \to 0$ follows $\mu(M) = 0$.

Other denumerable point sets of reals are the *rationals* $\mathbb{Q}$ and the *algebraic reals.* (Algebraic reals x satisfy some equation $a_0x^n + a_1x^{n-1} + \cdots + a_n = 0$, where $a_i \in \mathbb{N}$ and not all a_i's vanish.) Consequently, their measures vanish. The complement sets of *irrationals* $\mathbb{R} - \mathbb{Q}$ and *transcendentals* (non algebraic reals) are thus of measure one [226].

Despite such vague and probabilistic ideas as to "pull out" an arbitrary element of the "continuum urn" (you might have to assume the axiom of choice for doing that) which is then non recursive with probability one, it is hard to come up with a concrete elementary example of a provable non recursive real. One such example is the *algorithmic halting probability* Ω, which has been introduced by G. Chaitin [109, 110] and which is even provable random, has been defined by equation (7.10), p. 92 (see also 14.1.2, p. 196). Indeed, the algorithmic probability $P(S)$, defined by equation (7.9), p. 92, associated with an arbitrary recursively enumerable set S is provable non recursive.

Another "example" is a (binary) real $r = 0.r_1r_2r_3\cdots$ generated by infinitely many *coin tosses* of a fair coin and by identifying $r_i = 0$ or 1 for the outcome "head" or "tail" of the i'th coin toss, respectively. r is non recursive with probability one and recursive with vanishing probability.

1.2.6 Church-Turing thesis

Since the 1930's, other formal characterisations of functions corresponding to computational processes have been proposed (see again Rogers' book [390], p. 18 for details).

Probably the most prominent definition of recursive functions has been given by Alan Turing, who explicitly constructed a machine model (the Turing machine) to formalise the notion of computation. So far, all these definitions have turned out to be equivalent (i.e., there exist "translations" from one functional characterisation to another). We may therefore suspect that the class of the recursive functions defined above is inclusive enough to be a suitable characterisation of "mechanical" procedures; at least syntactically.

Likewise, a wide variety of functions, each agreed to be intuitively algorithmic, have been studied. Each one of these functions turned out to be a partial recursive function. The conjectural evidence of a connection between the informal, heuristic notion of *effective computability* and the precise notion of *partial recursive function* culminates in the Church-Turing thesis (sometimes referred to as "Church thesis"):

C 1.8 (Church-Turing thesis)

Any algorithm corresponds to a partial recursive function. In other words: whatever is felt to be effectively computable can be brought within the scope of the class of partial recursive functions.

(Vice versa, by our understanding of "mechanic" computation, every partial recursive function corresponds to an algorithm. In other words: a (partial) recursive function is a mathematical entity which can be defined by an algorithm.

There is an equivalence between effectively computable algorithms *and* partial recursive functions.*) The terms* effectively computable, mechanically computable, computable *and* recursively enumerable *will be used synonymously.*

Remarks:

(i) The technical (methodological) importance of this equivalence is that proofs (sometimes referred to as *"proofs by Church's thesis"*) can be carried through by convenient informal methods associated with algorithmics. At the same time, they have a concrete mathematical meaning, since, by the Church-Turing thesis, they refer to the class of partial recursive functions. Thereby, the formalised models of computation (such as the Turing machine) may remain "on the shelf." Indeed, most proofs in this and later chapters will be *"proofs by Church's thesis."*

(ii) The nontrivial claim of the Church-Turing thesis is that *every* conceivable effectively computable process corresponds to a partial recursive function. The input corresponds to the function argument and the output corresponds to the function value. The converse statement relating recursive functions to effective computations is trivial, given our understanding of "mechanic" effective computation. In summary,

$$\text{recursive} \underset{\Leftarrow}{\overset{\overset{\text{trivial}}{\Rightarrow}}{\text{nontrivial}}} \text{effectively computable} \quad .$$

(iii) As has been already emphasized, the Church-Turing thesis includes a physical as well as a syntactic claim. In particular, it specifies which types of computations are physically realisable. As physical statements may change with time, so may our concept of effective computation.

1.2.7 Computation = polynomial equation

This section reviews a result by M. Davis, H. Putnam and J. Robinson [134], which has been strengthened by J. P. Jones and Y. V. Matijasevic [247]. This result, given the Church-Turing thesis, can be informally stated as *"any computation can be encoded as polynomial."* More precisely, any computation can be encoded in an exponential diophantine equation.

First we define the notion of *diophantine equation*:

D 1.9 (Diophantine equation)
A(n) polynomial (exponential) diophantine equation $L(x_1, \ldots, x_n) = R(x_1, \ldots, x_n)$ *is build up from non-negative integer variables* $x_1, \ldots, x_n$ *and from non-integer constants by using the operation of addition* $A + B$*, multiplication* $A \cdot B$ *(and exponentiation* A^B*).*
An example for an exponential diophantine equation is $a^n + b^n = c^n$ with $a, b, c, n \in \mathbb{N}$.

D 1.10 *A* predicate $P(a_1, \ldots, a_n)$ *is called* recursively enumerable *if there is an algorithm which, given the non-negative integers* $a_1, \ldots, a_n$*, will eventually, e.g., by generating a list of all n-touples satisfying P, discover that these numbers have the property P. P is* recursive *if, in addition to that, an algorithm exists which will eventually discover that these numbers do* not *have the property P.*

The predicate $P(a_1, \ldots, a_n)$ *is called* exponential / polynomial diophantine *if P holds if and only if there exist non-negative integers* $x_1, \ldots, x_m$ *such that*

$$L(a_1, \ldots, a_n, x_1, \ldots, x_m) = R(a_1, \ldots, a_n, x_1, \ldots, x_m) \quad .$$

The following result is mentioned without proof; for a proof, see J. P. Jones and Y. V. Matijasevic [247], which is reviewed G. Chaitin [110].

T 1.11 *A predicate is polynomial / exponential diophantine if and only if it is recursively enumerable.*
By the Church-Turing thesis, the following equivalence holds:

C 1.12 "diophantine equation" ≡ "effectively computable process."

1.2.8 Recursively enumerable ≠ recursive

The predicates *recursively enumerable* and *recursive* will be defined for sets:

D 1.13 *A set* $A \subset \mathbb{N}$ *is called* recursively enumerable *if* $A = \emptyset$ *or A is the range of a (partial) recursively enumerable function.*

A set $A \subset \mathbb{N}$ *is called* recursive *if* both *A as well as its complement* $\mathbb{N}-A$ *are recursively enumerable.*
We only mention here one important result of recursion theory.

T 1.14 *There exists a set* $A \subset \mathbb{N}$ *which is recursively enumerable but not recursive.*
An example is the set of theorems of a "sufficiently complex" formal system arithmetic, or, equivalently, the outputs of "sufficiently complex" infinite computations. See also chapter 9.5 on page 118. Furthermore,

T 1.15 (Recursive inseparability) *There exists a* recursively inseparable *pair of sets; i.e.,* $A, B \subset \mathbb{N}$ *such that*

(i) A and B are recursively enumerable;

(ii) $A \cap B = \emptyset$;

(iii) there is no recursive set C with $A \subset C$ *and* $B \subset \mathbb{N} - C$.

1.3 Universal computers

In this section certain classes of automata will be specified which have become important for historic and theoretic reasons. If not stated otherwise, the terms *"computing agent," "computer"* and *"automaton"* are synonyms. *Universal computers* will be introduced as the class of automata on which it is possible to implement recursive functions. *Turing machines* and *Cellular Automata* are examples of this class. Indeed, these computer models (e.g., the Turing machine model) can be used for a *definition* of recursive functions. They provide one of the alternative definitions of the partial recursive functions (c.f. section 1.2.3, p. 9).

In section 1.5, p. 24, the "mechanic" (in the computer sciences termed "deterministic") devices are followed by a specific, conceptually important non deterministic device called *oracle*. Oracles are capable of solving problems, such as the halting problem (see 9.3, p. 114), which are recursively unsolvable. A "model" for an oracle problem solver will be given which, due to "Zeno squeezing" of its intrinsic time scale, may perform computations in the infinite time limit.

As has already been mentioned before, universal computation is invariant under changes of "sufficiently complex" machine models. In J. von Neumann's words (see *Theory of Self-Reproducing Automata*, ed. by A. W. Burks [340], p. 50):

> *The importance of Turing's research is just this: that if you construct an automaton [[A]] right, then any additional requirements about the automaton can be handled by sufficiently elaborated instructions. This is only true if A is sufficiently complicated, if it has reached a certain minimum level of complexity. In other words, a simpler thing will never perform certain operations, no matter what instructions you give it; but there is a very definite finite point where an automaton of this complexity can, when given suitable instructions, do anything that can be done by automata at all.*

Consider a set of instructions P of the type introduced for the definition of recursive functions. Assume further some particular *listing* of the set of instructions. Let P_x be the set of instructions associated with the Gödel number $x = \#(P_x)$ (for details, see 1.5, p. 10). Let φ_x stand for the function associated with P_x.

The following consideration yields an algorithmic definition of a *universal algorithm* and, by the Church-Turing thesis, of a *universal recursive function* $u(x, y)$: given any two numbers $x, y \in \mathbb{N}$, find P_x, for instance by going through the enumeration of all sets of instructions until the $(x + 1)$st place; then apply P_x to input y to calculate $\varphi_x(y)$. If $\varphi_x(y)$ gets an output, take this output value for $u(x, y)$. Having done so, we have obtained an effectively computable algorithm (associated with u) which yields $\varphi_x(y)$ on inputs x, y. By this definition of u, $u(x, y)$ effectively *imitates* any $\varphi_x(y)$. By the Church-Turing thesis, there exists a $z \in \mathbb{N}$, such that the algorithm u corresponds to the partial recursive function

$\varphi_z(x, y)$. This can be stated as follows:

T 1.16 (Existence of universal function)

Assume arbitrary recursive functions φ_x. Then there exists an index z and an associated universal partial function $\varphi_z(x, y) = u(x, y)$ *such that for all x and y*

$$\varphi_z(x, y) = u(x, y) = \begin{cases} \varphi_x(y) & \text{if } \varphi_x(y) \downarrow \text{ (convergent)} \\ \text{divergent} & \text{if } \varphi_x(y) \uparrow \text{ (divergent)} \end{cases} . \tag{1.12}$$

One immediate consequence of this theorem is that there exists a critical degree of "computational strength" of the P's beyond which all further complexity can be absorbed into increased program size (i.e., algorithmic information), increased use of memory and increased computation time (i.e., computational complexity).

D 1.17 (Universal Computer)

Any physical system or any device on which a universal function $u = \varphi_z$ can be implemented is called universal computer, *or* universal machine. *A universal computer can compute all computable functions.*

The notation $U(p, s) = t$ will be used for a *universal computer U* with program p, input (string) s and output (string) t. $\emptyset$ denotes the *empty* input or output (string). Furthermore, $U(p, \emptyset) = U(p)$. Concatenation of programs and input/output strings is allowed. E.g., given two input strings $s_1 = s_{11}s_{12}s_{13}\cdots s_{1i}$ and $s_2 = s_{21}s_{22}s_{23}\cdots s_{2j}$, a term denoted by "$s_1, s_2$," or "$s_1 s_2$" is treated by the automaton as the string $s_{11}s_{12}s_{13}\cdots s_{1i}s_{21}s_{22}s_{23}\cdots s_{2j}$.

Examples

1.3.1 Turing Machine

There exist several examples of universal computers, all being fictious because no finite machine can be universal. The most prominent device is the *Turing machine* [239, 136, 390] consisting of a finite *memory unit* and an infinite *tape* (which is tessellated into a sequence of squares) on which information is read or written. As has been noted before, A. Turing motivated the Turing Machine by the one-dimensional version of a sheet of paper on which arbitrary calculations can be performed according to "usual rules."

D 1.18 (Turing machine) *Assume discrete time cycles, labelled by 0,1,2,.... A* Turing machine *is an agency or automaton with the following features:*

(i) a finite *number of* internal states $a_0, \dots, a_k$, *which form a set $A = \{a_i\}$; a_0 is called the* passive state, *the other states $a_1, \dots, a_k$ are called* active states;

(ii) a linear, infinite tape, *consisting of* squares *and extending to the right; the tape can be moved backward and forward through the machine so that at each moment only one square is scanned;*

(iii) each square is capable of having read or printed on it any one of a finite *number of* tape symbols $s_0, s_1, \dots, s_l$; *where s_0 is the* blank symbol; *so there are $l+1$ possible square conditions;*

At each moment of time, the machine situation *is defined by (i) a particular machine state a_i, by (ii) a particular position of the tape in the machine (i.e., which square is scanned), and (iii) by a particular printing of the whole tape.*

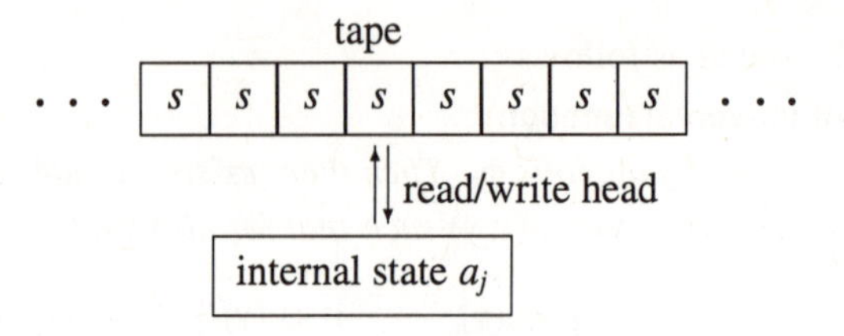

Figure 1.1: Scheme of a Turing machine.

In the active *state, the Turing machine performs an act between the time t and* $t+1$, *consisting of three possible parts: (i) the writing of a tape symbol* $s_{i'}$ *as the state of the square scanned at time t; subsequently accompanied by (ii) a* tape shift, *so that at the time* $t+1$ *the machine is either one square left of, or in the same, or one square right of the position scanned at t, denoted by L, C, R, respectively, and (iii) the choice of a new internal state* $a_{j'}$ *for the next time cycle* $t+1$. *Any machine act can therefore be described by the triple* $s_{i'}\left\{\begin{array}{c} L \\ C \\ R \end{array}\right\} a_{j'}$ *, or* $i'\left\{\begin{array}{c} L \\ C \\ R \end{array}\right\} j'$.

The pair $s_i a_j$, *or ij is called the machine* configuration. *The machine configuration determines the next act. If the configuration calls for a left move L but the scanned square is already the leftmost square of the tape, the machine goes into the passive state* a_0. *There are* $(l+1)k$ *active configurations. A particular Turing Machine is therefore specified by a* machine table, *showing for each active machine configuration which act has to be performed. (A scheme of a Turing machine is drawn in Fig. 1.1.)*

Example:

The following example is taken from St. C. Kleene's review article [267]. Consider the Turing machine specified by the machine table 1.1 with 11 internal states and two square symbols 0 for blank and 1 for s_1, respectively. At time $t=0$, the Turing machine is in internal state 1 and scans the 3rd position of the tape, whose squares are in the states 0110 $\cdots$ ("$\cdots$" denotes blank states here). The time evolution of this machine is enumerated in table 1.2.

1.3.2 Cellular Automata

Cellular Automata (CA) have been introduced by the late J. von Neumann [340] to achieve universal computation and the (re-) production of automata by automata. Informally speaking, Cellular Automata are a "large (usually an infinite) collection" of computationally interconnected finite automata. Each finite automaton is being thought of as a *cell*. The interconnections between the automaton cells are *locally* organised in D-dimensional space. For practical, mostly graphical, reasons $D=1$ or 2 is often chosen. Conversely, a CA can be perceived as the *tessellated plane, or* $\mathbb{R}^D$, with *locally interconnected processing units*. More precisely,

D 1.19 (Cellular Automaton)

Assume discrete time cycles, labelled by 0, 1, 2, ... *A* Cellular Automaton *(CA) is an infi-*

a_i	0	1
0	HALT	HALT
1	0C0	1R2
2	0R3	1R9
3	1L4	1R3
4	0L5	1L4
5	0L5	1L6
6	0R2	1R7
7	0R8	0R7
8	0R8	1R3
9	1R9	1L10
10	0C0	0R11
11	1C0	1R11

Table 1.1: Table specifying a Turing machine with 11 internal states and two tape symbols 0 (for blank) and 1.

nite collection of locally connected finite automaton (cells) of the tessellated $\mathbb{Z}^D$, $D \in \mathbb{N}$. Let $a_{i,\vec{n}}(t)$ stand for the i'th internal state of the $\vec{n}$'th automaton (cell), $\vec{n} = (n_1, n_2, \ldots, n_D)$, at time t. Let $\{\vec{n}\}$ characterise the D-dimensional neighbourhood of $\vec{n}$, including $\vec{n}$, and let $a_{i,\{\vec{n}\}}(t)$ stand for the internal states of the automaton cells around $\vec{n}$, including the origin $\vec{n}$. The internal state $a_{i,\vec{n}}(t+1)$ of the automaton at $\vec{n}$ at time $t+1$ is given by a computable function ϕ of the state and its neighbourhood states at time t, i.e.,

$$a_{i,\vec{n}}(t+1) = \phi(a_{i,\{\vec{n}\}}(t)) \quad .$$

Remarks:

(i) Cellular Automata feature *parallel processing,* as opposed to the sequential Turing machine concept. Instead of the infinite tape, they operate with an infinite number of finite automata.

(ii) There exist several neighbourhood declarations, the most prominent ones being the *von Neumann neighbourhood* and the *Moore neighbourhood,* with $1+2D$ neighbours (including the centre) for $D > 0$ and $1 + 2D + 2^D$ neighbours (including the centre) for $D > 1$, respectively. For $D = 2$, they are drawn in Fig. 1.2. Another neighbourhood declaration is the *Margolus neighbourhood,* which is obtained by partitioning the cell array into finite, disjoint and uniformly arranged pieces (blocks), by block rules updating the block as a whole; and by a change of the partition with every time step (for details, see T. Toffoli and N. Margolus, *Cellular Automata Machines* [445], chapter 12).

(iii) Universality of a specific CA in the sense of the Church-Turing thesis can be proved by *embedding* a Turing machine into a suitable Cellular Automaton structure [340, 61, 122]. One particular "fancy" example, E. Fredkin's *Billiard Ball Model* [308], mimics universal computation *via* the "elastic scattering" of CA "billard balls".

(iv) CA's perform optimally for physical configurations which can be decomposed into locally connected parallel entities. For more information and for recent developments,

time / tape symbol$^{\text{internal state}}$	1	2	3	4	5	6	7	...
0	0	1	1^{1}	0	0	0	0	...
1	0	1	1	0^{2}	0	0	0	...
2	0	1	1	0	0^{3}	0	0	...
3	0	1	1	0^{4}	1	0	0	...
4	0	1	1^{5}	0	1	0	0	...
5	0	1^{6}	1	0	1	0	0	...
6	0	1	1^{7}	0	1	0	0	...
7	0	1	0	0^{7}	1	0	0	...
8	0	1	0	0	1^{8}	0	0	...
9	0	1	0	0	1	0^{3}	0	...
10	0	1	0	0	1^{4}	1	0	...
11	0	1	0	0^{4}	1	1	0	...
12	0	1	0^{5}	0	1	1	0	...
13	0	1^{5}	0	0	1	1	0	...
14	0^{6}	1	0	0	1	1	0	...
15	0	1^{2}	0	0	1	1	0	...
16	0	1	0^{9}	0	1	1	0	...
17	0	1	1	0^{9}	1	1	0	...
18	0	1	1	1	1^{9}	1	0	...
19	0	1	1	1^{10}	1	1	0	...
20	0	1	1	0	1^{11}	1	0	...
21	0	1	1	0	1	1^{11}	0	...
22	0	1	1	0	1	1	0^{11}	...
23	0	1	1	0	1	1	1^{0}	...

Table 1.2: Evolution of a Turing machine.

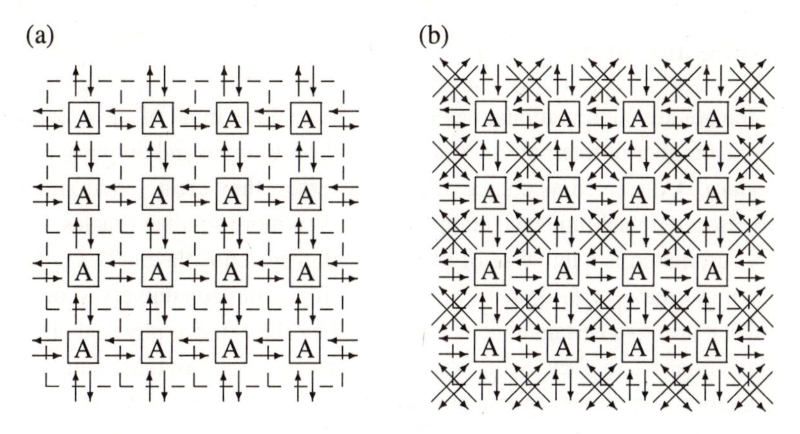

Figure 1.2: Von Neumann (a) and Moore (b) neighborhoods for twodimensional CA. A symbolizes a finite automaton.

see T. Toffoli and N. Margolus, *Cellular Automata Machines* [445], St. Wolfram's *Theory and Application of Cellular Automata* [454] as well as references [453, 323, 221]. For technical realisations, see, among others, references [445, 398, 233].

1.3.3 Register machines

Register machines (also called *counter machines*) use a finite number of registers, each containing a non negative integer. Registers are addressable and arithmetical operations can be performed on them. In this model there is no need to arithmetize words over an alphabet of symbols. For details, see for instance J. P. Jones & Y. V. Matijasevic [247] and G. Chaitin [110].

1.3.4 Digital computers are physical systems which are universal up to finite complexities

The statement that commercially available general-purpose computers are in some sense "universal computers" is quite obvious [186]. Otherwise they would not be very helpful for general algorithmic tasks, such as commercial applications, evaluation of particular functions (e.g., solutions of differential equations) and so on. It is, for instance, possible to simulate an arbitrary Turing machine with *finite* tape on them.

As available general-purpose computers are *finite automata*, i.e., as they are limited by the finiteness of memory size, tape length, runtime and virtually everything one can think of, they cannot compute *all* effectively computable algorithms. In terms of the Church-Turing thesis, the corresponding functional class is a subset of the recursive functions.

It is evident that tasks which are uncomputable on universal computers (such as the *halting problem*, see 9.3 on p. 114) cannot be solved on available general-purpose machines either. In this respect, they share some features with the fictious but truly universal abstractions, in particular with respect to *undecidability*.

1.4 Finite automata

Finite automata are computational devices which are finite in all of their features. They have a finite number of internal states, a finite number of input and output symbols, finite tape *et cetera.* Conceived as computational universes, they create environments which are more restricted than universal computers or, even more so, than oracles.

Every automaton is a universe of its own, with a specific "flavour," if you like. Programmers may create *"cyberspaces"* (a synonym for automaton universes) of their imagination which are, to a certain extend, not limited by the exterior physical laws to which they and their hardware obey. Seen as isolated universes, some of these animations might have nothing in common with our physical world. Yet, others may serve as excellent playgrounds for the physicist.

Why should physicists investigate finite automata? There is a modest and a radical answer to this question. The modest finite automata thesis is this: There exist algorithmic features of automaton universes such as complementarity (see chapter 10, p. 127) or undecidability (see chapter 12, p. 181) which translate into physics and which are difficult to analyse by non algorithmic means.

The radical finite automata thesis is this: Because the physical universe actually *is* a finite automaton. This viewpoint has probably been most consequently put forward by E. Fredkin [186] in the context of Cellular Automata. (Although a universal Cellular Automaton is no finite machine — it has infinite extension — its cell space is discrete, leaving room to only a finite, although "very large," number of cells per unit volume element of whatever is considered as the fundamental scale.) The radical finite automaton thesis contains two claims, each of which should in principle be testable, at least to some extend: *(i)* the "laws of nature" are mechanistic; i.e., computable in the usual Church-Turing sense (although intrinsically there might not exist any effective procedure to find these laws and although forecasts corresponding to halting problems in general may not be computable); and *(ii)* under certain "mild" assumptions, the computational capacities of physical systems are finite. See also chapter 3, p. 37 for a discussion of related topics.

1.4.1 Definition

In the following, an automaton shall be characterised algebraically. For a more detailed treatment, see, among others, E. Moore's original article [328], as well as the books by J. H. Conway [125], J. E. Hopcroft & J. D. Ullman [240], J. R. Büchi [59] and W. Brauer [28]. Readers less interested in formal definitions may skip these sections. For them, it suffices to keep in mind that a (i,k,n)-automaton has i internal states, k input and n output symbols; it is characterised by its transition and output functions δ and o. Two states are called *distinguishable* if there exists at least one experiment which distinguishes between them; i.e., which yields non identical output. Two automata are *isomorphic* if there exists a one-to-one translation between them and if corresponding output symbols are identical.

D 1.20 (*k*-algebra, (i,k,n)-automaton) *A* k-algebra $A = (A, e, I, \delta)$ *is a system consisting of*

(i) a set $A = \{a_1, \dots a_i\}$, *corresponding to the* internal states *(finite automata have a finite number of internal states),*

(ii) an element $e \in A$, corresponding to the initial state,

(iii) a set $I = \{s_1, s_2, s_3, \ldots, s_k\}$, corresponding to the k input symbols,

(iv) a transition function $\delta : A \times I \to A$, *corresponding to a set of operators $\{\delta_1, \delta_2, \delta_3, \ldots, \delta_k\}$, called* transition operators.

A (i,k,n)-automaton $A = (A, e, I, O, \delta, o)$ *is a k-algebra* $A = (A, e, I, \delta)$ *with*

(v) a set $O = \{t_1, t_2, t_3, \ldots, t_n\}$, corresponding to the n output symbols, *and*

(vi) an output function $o : A \to O$, *mapping internal states into* output symbols. *O^* and I^* denote the set of all output and input sequences.*

This type of an automaton is often called *Moore automaton*, since its output function o only depends on the internal state of the automaton. Another, more general automaton definition is the *Mealy automaton*, which is defined similarly to the Moore automaton, except that the output function $o : A \times I \to O$ depends also on some particular input (instead of merely the internal automaton state). Whereas for the Moore automaton type one gets "the first output free," any output from the Mealy automaton is obtained only after investment of one input symbol. If one discards the first output symbol from a Moore automaton, the class of Moore automata and the class of Mealy automata are equivalent; i.e., any Moore automaton can be translated into an isomorphic Mealy automaton and *vice versa*.

Any finite k-algebra can be represented by a *transition table*, with the present internal states enumerated in the first row, the input symbols enumerated in the first column and the future internal states in the remaining matrix. A finite (i,k,n)-automaton is additionally characterised by the output function. This function may be represented by an *output table,* listing all internal states (and the input symbols for Mealy-type automata) and their corresponding output symbol(s). A *transition graph* is obtained by drawing circles for every internal state and by drawing directed arrows to indicate the transition function on some input. For examples, see chapter 10, p. 127.

Transitions are indicated in the following way: let $v \in A$, then the transition from v with input i is $v\delta_i$. This notation translates into the standard notation. Let U denote a universal computer and p_A stand for the program which simulates a (i,k,n)-automaton on U, such that initially the (i,k,n)-automaton is in internal state e. Then $U(p_A, s) = t$, with the input sequences s generated by concatenation of input symbols $s_1 \cdots s_n$, and with the output sequences t generated by concatenation of output symbols $[t_0]t_1 \cdots t_n$ ($[t_0]$ is optional for Moore-type automata).

The "halting state" can be simulated by the introduction of a special state, say h. If the automaton is in this state, no input will cause it to leave that state.

1.4.2 Distinguishability of states

D 1.21 (Distinguishable internal states) *Let $w = s_1 s_2 \cdots s_k$ be an input sequence (of length k), then let*

$$o(a_{i,w}) = o(a_i)o(\delta_{s_1}(a_i))o(\delta_{s_2}(\delta_{s_1}(a_i)))\cdots o(\delta_{s_k}(\cdots\delta_{s_1}(a_i)\cdots)) \quad .$$

A pair of states $\{a_i, a_j\}$ *of an automaton A is called* (pairwise) distinguishable *if and only if there exists at least one experiment for which the outcome depends on which one of these states the automaton was in at the beginning of the experiment.*

An arbitrary number of states $\{a_1, a_2 \ldots, a_i\}$ *are called* distinguishable *if and only if there exists at least one experiment performed on A of which the outcome depends on which one of these states the automaton was in at the beginning of the experiment.*

An internal state a_i *of an automaton A is called* indistinguishable *from a state* a_j *of A if and only if every experiment performed on A starting in the state* a_i *produces the same output as if it would start in state* a_j*. I.e., two states* a_i *and* a_j *are* indistinguishable *if and only if*

$$o(a_{i,w}) = o(a_{j,w})$$

for all $w \in I^*$.

1.5 Oracles

1.5.1 Definition

D 1.22 (Oracle (Turing [446])) *An* oracle *is some agent, or "black box," which, upon being consulted, supplies the true (correct) answers about mathematical or algorithmic or physical entities.*

The "magical insight" characterises the *non deterministic* feature of oracles. The corresponding problem may or may not be decidable by any effective computation, the latter case being more interesting. An example for such a problem is the halting problem (for details see 9.3, p. 114).

1.5.2 Zeno squeezing

A model for an *oracle* problem solver will be introduced next. This machine exceeds the capacity of any presently realisable, finite machine and also of any universal computer such as the Turing machine. Its design is based on a universal computer with a "squeezed" time of computation. The only difference between universal computation and this type of oracle computation is the speed of execution. Zeno squeezed oracle computation performs computations *in the limit of infinite time of computation.* In order to achieve this limit, two time scales are introduced: the *intrinsic time scale of the process of computation* which approaches infinity in finite *extrinsic or proper time of some outside observer.* By the introduction of two time scales it is possible to produce an otherwise uncomputable and random output even for finite proper times.

A very similar setup has been introduced in Hermann Weyl's book *Philosophy of Mathematics and Natural Science* [452], which has been discussed by A. Grünbaum in *Philosophical Problems of Space and Time* [219], p. 630. H. Weyl raised the question whether it is kinematically feasible for a machine to carry out an *infinite* sequence of operations in *finite* time. H. Weyl writes ([452], p. 42):

> ... *Yet, if the segment of length 1 really consists of infinitely many subsegments of length 1/2, 1/4, 1/8, ..., as of 'chopped-off' wholes, then it is*

> *incompatible with the character of the infinite as the 'incompletable' that Achilles should have been able to traverse them all. If one admits this possibility, then there is no reason why a machine should not be capable of completing an infinite sequence of distinct acts of decision within a finite amount of time; say, by supplying the first result after 1/2 minute, the second after another 1/4 minute, the third 1/8 minute later than the second, etc. In this way it would be possible, provided the receptive power of the brain would function similarly, to achieve a traversal of all natural numbers and thereby a sure yes-or-no decision regarding any existential question about natural numbers! ...*

The considerations below concentrate on algorithmics aspects. For paradoxical usages of Zeno squeezed oracles, see section 9.4, p. 117.

The argument is similar to a paradox which is often referred to as *"Achilles and the Tortoise,"* or *"Achilles and Hector,"* ascribed to Zeno of Elea (5th century B.C.). For a detailed treatment, see H. D. P. Lee's *Zeno of Elea* [294], as well as G. S. Kirk's and J. E. Raven's *The Presocratic Philosophers* [265], p. 292. It describes a race between Hector and Achilles as follows: for simplicity, assume that Hector is one unit of distance ahead of Achilles, and that Achilles runs twice as fast as Hector. Will Achilles ever catch up with and overtake Hector? Obviously not, if one argues as follows: Achilles runs to Hector's new position; however in the time that it took Achilles to run one unit of distance, Hector has advanced 1/2 unit of distance. Achilles runs to Hector's new position; however in the time that it took Achilles to run 1/2 unit of distance, Hector has advanced 1/4 unit of distance. Achilles runs to Hector's new position; however in the time that it took Achilles to run 1/4 unit of distance, Hector has advanced 1/8 unit of distance. ... *ad infinitum.* According to Aristotle in *Physics Z9*, Zeno seems to have argued that, by assuming infinite divisibility of space and time, one arrives at the conclusion that Achilles never catches up with Hector, which, given every-day experience that a faster body overtakes a slower one, is an absurd conclusion; Hence, if Zeno is interpreted correctly, he seems to have argued that infinite divisibility of space and time contradicts experience.

Zeno's argument has been formalised and interpreted in terms of undecidability by E. G. K. López-Escobar [304] (see also E. W. Beth, *The Foundations of Metamathematics* [41], p. 492). In a sense, Zeno may have anticipated Cantor's method of diagonalization: E. G. K. López-Escobar constructs an expression for the points successively reached by Achilles; then he constructs a new point, the "meeting point" of Achilles and Hector, for which this expression does not hold and which is nevertheless reached by Achilles. This contradicts the assumption that the expression describes the relative motion of Achilles and Hector *completely.* A sketch of the formal argument goes as follows: Let $\texttt{POS}(r_A(t), r_H(t))$ stand for the binary predicate that represents the simultaneous positions of Achilles $r_A(t)$ and Hector $r_H(t)$ at time t. The time will be measured *discretely*, i.e., $t = 0, 1, 2, \ldots$. As introduced here, the time parameter t is *not* Achilles' and Hector's "proper time," but rather a "squeezed" time parameter. Intuitively speaking, the higher is t, the smaller is the amount of "proper time" corresponding to one unit of t-time (from t to $t + 1$). In the limit of $t \to \infty$, one unit of t-time corresponds to a "proper time" of measure zero. (For more details, see the oracle problem solver described below.)

Assume again that Hector is one unit of distance ahead of Achilles and that Achilles runs twice as fast as Hector. From the "axiom"

$$\mathtt{POS}(0, 1) \tag{1.13}$$

and the "rule of inference"

$$\mathtt{POS}(r_A(t) = r, r_H(t) = r + s) \Rightarrow \\ \mathtt{POS}(r_A(t+1) = r + s, r_H(t+1) = r + s + \frac{s}{2}) \quad , \tag{1.14}$$

one can derive by Zeno's argument that, since $s > 0$, $\mathtt{POS}(r_A(t), r_H(t))$ only for $r_A(t) < r_H(t)$. On the other hand, by summation of the infinite series of distances resulting from Achilles' moves towards Hector, one obtains that Achilles will reach Hector at

$$r_A(\infty) = r_H(\infty) = 1 + \frac{1}{2} + \frac{1}{4} + \frac{1}{8} + \cdots = \frac{1}{1 - \frac{1}{2}} = 2 \quad . \tag{1.15}$$

From this argument follows that $\mathtt{POS}(2, 2)$ is a "true" predicate. (And, from another argument using calculus or from practical experience, it follows that $\mathtt{POS}(r_A, r_H)$ with $r_A > r_H$ may be "true.") But it is also clear that $\mathtt{POS}(2, 2)$ cannot be derived by Zeno's argument. Therefore, the whole method of reasoning of Zeno may be considered *incomplete* with respect to other forms of reasoning (e.g., using summation techniques for infinite geometric series). In other words, Zeno's formalism cannot decide and therefore is *inappropriate* for statements of the form "$\mathtt{POS}(r_A, r_H)$" if $r_A \geq r_A$.

Let us come back to the original goal: the construction of a *"Zeno squeezed oracle,"* or, in A. Grünbaum's terminology, of an "infinity machine." It can be conceived as follows: Consider two time scales τ and t.

(i) The *proper time* τ measures the physical system time by clocks in a way similar to the usual operationalisations; whereas

(ii) a discrete *cycle time* $t \in \mathbb{N}$ characterises a sort of "intrinsic" time scale for a process running on an otherwise universal machine.

For some unspecified reason we assume that this machine would allow us to "squeeze" its intrinsic time t with respect to the proper time τ by a geometric progression. Let $k < 1$, then any time cycle of t, if measured in terms of τ, is squeezed by a factor of k with respect to the foregoing time cycle (see Fig. 1.3), i.e.,

$$\tau_0 = 0 \tag{1.16}$$

$$\tau_{t>0} = \sum_{n=1}^{t} k^n = \frac{k^t - 1}{k - 1} \quad . \tag{1.17}$$

In the limit of infinite cycle time $t \to \infty$, the proper time $\tau_\infty = 1/(1-k)$ remains finite.

With such a device, countable many intractable and uncomputable problems, such as the halting problem, would become solvable. One could, for instance "effectively compute" the algorithmic halting probability Ω in the limit from below [cf. ref. [109] and (14.2), p. 196] by identifying the time of computation with t. In the limit $t \to \infty$, for $k < 1$ one obtains $\lim_{t\to\infty} \omega_t = \Omega$ at the finite proper time $\tau < \infty$. With such a

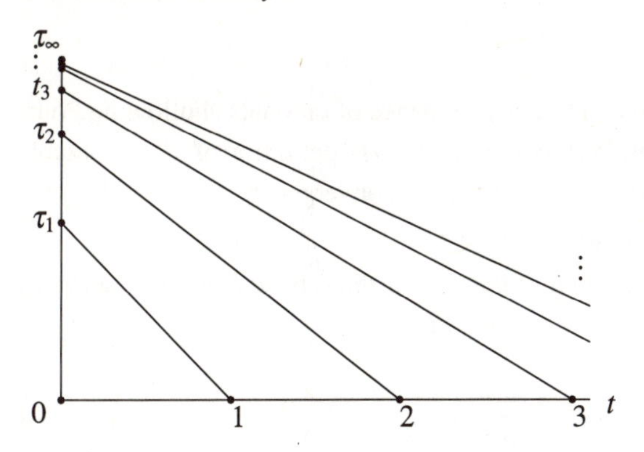

Figure 1.3: Zeno squeezed time cycles of a universal computer constituting an oracle problem solver.

Zeno squeezed oracle, many other problems would be within the reach of a "constructive solution." Take, for instance, Fermat's conjecture, which could be "(dis)proved" by the following algorithm (in FORTRAN-style pseudocode):

```
DO LABEL1 a=1, 1, ∞;
DO LABEL1 b=1, 1, ∞;
DO LABEL1 c=1, 1, ∞;
DO LABEL1 n=3, 1, ∞;
IF a**n+b**n=c**n THEN PRINT (*,*) a,b,c,n;
                   PRINT (*,*) 'FERMAT WAS WRONG!!!'; STOP 1;
        ELSE GOTO LABEL1;
LABEL1: CONTINUE;
PRINT (*,*) 'FERMAT WAS RIGHT!!!'; STOP 2;
END;
```

In the spirit of Zeno, the non existence of oracle problem solvers could be viewed as a further indication of the non existence of infinite divisibility. Indeed, it is not evident why such an oracle problem solver can be excluded in classical physics, e.g., in continuum mechanics. [This is related to infinite (measurement) precision in classical physics.] It is not completely unreasonable to put forward the provocative statement that, at least in principle, oracles of the above type are allowed in classical physics, and that classically the Church-Turing thesis 1.8, p. 14, could be falsified by the actual construction of an oracle problem solver. (One might speculate that a simple process corresponding to diagonalization might cause the inconsistency of such a classical universe. The choice seems to be unlimited computational capacity *versus* consistency; cf. section 9.4, p. 117).

1.6 Recursive analysis

After the definition of effective computability, a flavour of the applications to analysis is given next. A much more detailed account can be found in M. B. Pour-El and J. I. Richards' book *Computability in Analysis and Physics* [373]. (See also O. Aberth's book *Computable Analysis* [2].) A review of this research by D. S. Bridges is in [45] and by

G. Kreisel is in [278].

A real number can be represented as a Cauchy sequence of rationals. Following Pour-El & Richards, a translation into recursion theory requires *(i)* a definition of a "computable sequence of rationals," and *(ii)* a definition of "effective convergence."

D 1.23 (Computable sequence of rationals)
A sequence $\{r_n\}$ of rationals is computable if there exist three recursive functions $a(n)$, $b(n)$, $c(n)$ from $\mathbb{N}$ to $\mathbb{N}$ such that $b(n) \neq 0$ for all n and

$$r_n = (-1)^{c(n)} \frac{a(n)}{b(n)} \quad .$$

D 1.24 (Effective convergence) *A sequence $\{r_n\}$ of rationals converges effectively to a real number x if there exists a recursive function $\varepsilon : \mathbb{N} \to \mathbb{N}$ such that for all $n \in \mathbb{N}$*

$$k \geq \varepsilon(n) \quad \text{implies} \quad |r_k - x| \leq 2^{-n} \quad .$$

D 1.25 (Computable real) *A real number x is computable if there exists a computable sequence of rationals which converges effectively to x.*

Remarks:

(i) A complex number is computable if both its real and imaginary parts are computable; A vector is computable if each of its components is computable; a sequence $\{x_k\}$ of real numbers is computable if all reals x_k are computable.

(ii) Let $\{x_k\}$ and $\{y_k\}$ be computable sequences of real numbers. Then the following sequences are computable: $x_k \pm y_k$, $x_k y_k$, x_k/y_k $(y_k \neq 0 \forall k)$, $\min(x_k, y_k)$, $\max(x_k, y_k)$, e^{x_k}, $\sin x_k$, $\log x_k$ $(x_k > 0 \forall k)$, ... These and similar results can be proven by Taylor series or other expansions which converge effectively.

(iii) Differentiation and integration can be properly reformulated in terms of recursion theory (for details see [373], p. 33).

T 1.26 (Non recursive enumerability of the waiting time)
Let $a : \mathbb{N} \to A \subset \mathbb{N}$ be a one-to-one recursive function generating a recursively enumerable but non recursive set A. Let $w(n)$ denote the waiting time, *defined by*

$$w(n) = \max\{m \mid a(m) \leq n\} \quad .$$

Then there is no recursive function $c(n)$ such that $w(n) \leq c(n)$ for all n.

Proof by contradiction: Assume such a $c(n)$ would exist. In order to decide whether or not some $k \leq n$ is in A, we could evaluate all $a(i)$ associated with $i \leq w(n) \leq c(n)$. If k does not turn up, then, according to our definition of $w(n)$, it will never turn up. But this renders an effective decision procedure whether or not some natural number is in A, which is a contradiction to the assumption that A is a non recursive set. (The non recursive enumerability of the waiting time is similar to the non recursive enumerability of the busy beaver function; see 8.7, p. 104.) The following result is cited without proof (for details, see [373], p. 16).

T 1.27 *Let $a : \mathbb{N} \rightarrow A \subset \mathbb{N}$ be a one-to-one recursive function generating a recursively enumerable but non recursive set A. Consider the series*

$$s_k = \sum_{m=0}^{k} 2^{-a(m)}$$

and let $x = \lim_{k\to\infty} s_k$. Then the waiting time $w(n)$ is the smallest integer such that

$$k \geq w(n) \quad \text{implies} \quad x - s_k \leq 2^{-n} \quad .$$

It follows that the real number x defined above is a computable monotone sequence of rationals $\{s_k\}$ which converges, but does *not* converge effectively.

"Pedestrian versions" of this statement and other important findings whose proofs are rather involved are listed below:

(i) There exist non computable reals corresponding to convergent sequences of computable rationals with *non effective convergence.* (C.f. G. Chaitin's Ω [110]; see also 14.1.2, p. 196.)

(ii) The maximum (minimum) value of a computable function is computable, but the point(s) where the maximum occurs need(s) not be (see, for instance, [373], p. 42, and [276]). A necessary condition for this to occur is that there are *infinitely* many maximum (minimum) points. If a computable function takes on a local maximum (minimum) at an *isolated* point, then this point is computable.

(iii) Differential equations may have *uncomputable* solutions from *computable* initial values, but these solutions must be "weak solutions" in the sense of distributions. For the wave equation in 3+1 dimensions, that means that weak solutions are not even in C^1, the class of differentiable functions whose derivations are continuous ([373], p. 73). Differential equations with computable initial values may also have uncomputable *non unique solutions*; see [369] and G. Kreisel [276].

(iv) Bounded linear operators in Banach space preserve computability, and unbounded operators do not ([373], chapter 3).

(v) Under mild side conditions, a self-adjoint operator has computable eigenvalues, although the sequence of eigenvalues need not be computable ([373], chapter 4).

The question still remains if "pathologies" such as non recursive solutions from evolution equations with recursive initial values have a physical meaning. Personally, the author believes that they seem to indicate the necessity for further restrictions of the class of solutions by physical assumptions, such as uniqueness, finiteness and continuity. This could be done on operational, on practical, on aesthetic and, in a broader sense, on metaphysical grounds. Yet, if the non recursive solutions are interpreted as physically meaningful, these solutions would correspond to physical events which would have been generated by a computer agent "much more" powerful than any universal computational device.

1.7 Formal systems correspond to computable processes

The (recorded) formalisation of theory began with Euclid and evolved to the concept of a *formal (axiom) system*, which will be introduced next. The goal is that, after specifying certain theoretical *symbols*, all relevant theoretical entities have to be expressed by *axioms* and *rules of deduction.* Then it should be possible to "mechanically" derive *theorems* by string processing or by other purely syntactic means. (In its extreme consequence, the program of formalisation can be understood as the "elimination of the necessity of meaning and intuition.")

D 1.28 (Formal system) *A* formal system *L is a system of symbols together with rules for employing them. The individual symbols are elements of an* alphabet. *Formulas are sequences of symbols. There shall be defined a class of formulas called* well-formed formulas, *and a class of well-formed formulas called* axioms *(there may be a finite or infinite number of axioms). Further, there shall be specified a list of rules, called* rules of inference. *If such a rule is called R, it defines the* relation of immediate consequence *by R between a set of well-formed formulas $M_1, \ldots, M_k$ called* premises, *and a formula N called* conclusion *or* theorem.

Examples of formal systems are the *Peano axioms*, *Zermelo-Fraenkel set theory* (with or without the *axiom of choice*), certain *geometries et cetera.*

The essence of *inference*, at least for a formalised theory, is *string processing* [391]. String processing can be performed by any (universal) computer. Conversely, any effectively computable process can be identified with some syntactic activity which is associated with "inference." (Nothing needs to be said about the semantic aspect of such a formalism, such as the "meaning" of some string or of some process.)

Therefore, from a purely syntactic point of view, every effectively computable process $U(p, s)$ can be identified with a formal system and *vice versa.* Indeed, as stated by K. Gödel in a *Postscriptum,* dated from June 3rd, 1964 ([205], p. 369-370):

> *... due to A. M. Turing's work [[reference[446]]], a precise and unquestionably adequate definition of the general concept of formal system can now be given, the existence of undecidable arithmetical propositions and the non-demonstrability of the consistency of a system in the same system can now be proved rigorously for* every *consistent formal system containing a certain amount of finitary number theory.*
>
> *Turing's work gives an analysis of the concept of "mechanical procedure" (alias "algorithm" or "computation procedure" or "finite combinatorial procedure"). This concept is shown to be equivalent with that of a "Turing machine." A formal system can simply be defined to be any mechanical procedure for producing formulas, called provable formulas.*

Let thus the "input" s of the computation correspond to "axioms," the "program" p correspond to the "rules of inference," and the "output" correspond to "provable theorems," respectively (see table 2.2, p. 35, for a translation of terminology). It should be stressed again that from the point of view of abstract coding and information theory, *input* and *program* are interchangeable, since, given a program p and a specific input s, it is always possible to write another program p' such that $U(p', \emptyset) = U(p, s)$. One may object that computations may halt while one may derive theorems from the axioms of formal systems forever. This critique can be met by considering only computations which never halt. If a computation halts, it can be identified with a similar computation which differs from the original one only by the feature that instead of the halting mode it goes into an infinite loop with no output (i.e., by the substitution `HALT` $\rightarrow$ `LABEL 1: GO TO LABEL 2; LABEL 2: GO TO LABEL 1`).

Another equivalence scheme between algorithms and formal systems has been introduced by G. Chaitin [101]. It again starts with the observation that one essential feature of formal systems is the existence of objective "derivation" rules, representable by some effective computation $U'(p, t)$, and (by the Church-Turing thesis) an associated recursive function which, applied to the axioms p, yields all theorems which can be derived from proofs with less than or equal to t characters length. In this scheme, the terms "computer" and "rules of inference", as well as "program" and "axioms", as well as "output up to cycle time t" and "theorems with proof size $\leq t$" are synonyms. For fixed cycle times t, Chaitin's scheme and the above scheme can be brought into a one-to-one correspondence by recalling that for a universal computer C there exists a program q which simulates U' on U and an input (p, t) such that $U'(p, t) = U(q, (p, t))$ for all p and t.

Chapter 2

Mechanism and determinism

With recursion theory being a relatively young discipline, the notion of "physical determinism" in classical physics lets unspecified the exact recursion theoretic status of the physical entities. In this chapter, such a specification is attempted, re-interpreting the classical meaning recursion-theoretically. It is not unreasonable to assume that a *recursive evolution* seems to be "at the heart" of the classical notion of "determinism." Usually, these evolution functions are defined on continua, such as $\mathbb{R}^n$. In particular, the initial values and the solutions are defined in $\mathbb{R}^n$. Yet, by theorem 1.7, p. 12, "almost all" elements of the continuum are uncomputable. In particular, the assumption of an exact (i.e., effectively computable) description of the initial value(s) becomes ridiculed.

"Strong determinism" or what henceforth will be called "mechanism" could alternatively be understood as a synonym for "total causality," which might be translated into the language of recursion theory by "total computability," or "computability in all aspects." (For a philosophical treatment of this and related topics, see for instance Ph. Frank, *Das Kausalgesetz und seine Grenzen* [187].) This should not be confused with the above requirement of a recursive evolution. It is a nontrivial substitution, since it requires *all* theoretical entities to be effectively computable.

For example, classical continuum physics is "deterministic" but not "mechanistic" in the above sense, since its evolution equations are computable, whereas its domain is the continuum. Here one could resort to the notion of "arbitrary but *finite* accuracy parameter values," or "finitely computable parameter values." Since finite numbers are recursively enumerable, this would restore effective computability.

Therefore, in what follows (if not denoted otherwise) the term "deterministic" refers to merely an *effectively computable evolution function*, whereas a system which is totally computable in all of its aspects is called "mechanistic." In a way, this is a translation of Laplace's demon into the language of recursion theory.

D 2.1 (Deterministic, mechanistic theory)

A deterministic *theory has an evolution function which is effectively computable / recursive.*

A mechanistic *theory is effectively computable / recursive in* total. *I.e.,* all *theoretical entities are effectively computable / recursive. In particular, all initial values, laws and solutions of a* mechanistic *theory are recursively enumerable.*

The terminology "mechanistic" is due to G. Kreisel [276]. According to G. Kreisel, a theory is mechanistic

if every sequence of natural numbers or every real number which is well de-

	computable evolution	uncomputable evolution
computable initial value	mechanistic physical theory (computable solution	indeterministic physical theory
uncomputable initial value	deterministic physical theory	indeterministic physical theory

Table 2.1: Deterministic *versus* mechanistic and indeterministic physical theory.

fined (observable) according to the theory *[[is]] recursive or, more generally, recursive in the data.*

The definition given above does not directly refer to predicates such as "well defined" or "observable." Also, the question remains whether it is possible to obtain physically meaningful (i.e., not excludable on by present physical theory) *non recursive* solutions of recursive initial values and recursive evolution equations. As has been shown by G. Kreisel [276] and M. B. Pour-El and J. I. Richards [369, 373], in principle, such solutions exist. (See also remark *(ii)* below and remarks *(i)*-*(v)* on page 29.) In order to avoid them, it might therefore be necessary to impose further restrictions on physical solutions. Such restrictions should be ultimately motivated by physical reasoning. (One could also decide to take such solutions seriously; with whatever outcome being the consequence.)

Further Remarks:

(i) Physical determinism in the defined sense does *not* imply that the initial and/or the solution (the final state) can be represented by an effective computation.

(ii) As has been pointed out before, computability of the equation of motion *and* the initial value does *not* guarantee computability of the solution, at least not if the solution is non unique [369], if it is obtained by unbounded linear operators ([373], chapter 3), or if its a weak solution ([373], p. 73). For mechanistic theories, uncomputable solutions from computable initial values and computable equations of motions are excluded by definition.

(iii) Since, from the point of view of coding and information theory, the distinction between the "evolution" and "initial value" (i.e., some algorithm and its input) is rather arbitrary, the above distinction between "determinism" and "mechanism" is rather arbitrary. A more radical but clearer distinction would be between non recursive and recursive ("mechanistic") theories.

(iv) Uncomputability does not imply randomness (see chapter 14 p. 196).

Table 2.1 summarises the computational aspects of physical theories. By identifying the evolution functions of mechanistic physics with the partial recursive functions, one obtains a correspondence which can be represented in table 2.2 (The table includes an extension to formal systems). The term *correspondence* between structures in physics, algorithmics and mathematics should be understood as a one-to-one *translation* between such structures, which become identical if only their elements are renamed. Note that, tech-

physics	algorithmics	mathematics	formal logic
mechanistic system	algorithm	partial recursive function	formal axiomatic system
initial state(s)/observable(s)	input	argument(s)	axiom(s)
evolution	computation	evaluation	derivation
(final) state(s)/observable(s)	output	value	theorem

Table 2.2: Correspondence between entities in physics, algorithmics, mathematics and logic.

nically speaking, a mechanistic physical system should be perceived as a never-ending computational process. Such an algorithm which does not halt could alternatively be viewed as a continuing derivation of theorems of a formal system, Table 2.3 gives a brief overview of the algorithmic features of physical theories.

physical theory	mathematical entity	algorithmic status	machine model
classical mechanics			
space, time, forces, …	continuum	deterministic but non mechanistic	oracle
electrodynamics			
space, time, fields, …	continuum	deterministic but non mechanistic	oracle
quantum mechanics			
wave function	continuum	deterministic but	oracle
space, time, fields, …	continuum	non mechanistic	
randomness of singele events	indeterministic		oracle
phase space	discretum		
discrete physics (see chapter 3)	discretum	computable	finite and infinite automaton

Table 2.3: Physical theories and their associated algorithmic status and machine model.

Chapter 3

Discrete physics

All previously introduced models of computation have been *discrete* in the sense that the processes involved are discontinuous with respect to space and time, storage capacity, program length and so on. Effective computations can be decomposed into distinct parts; e.g., execution steps. Any algorithmic model of a physical system inherits this feature. Discreteness is opposed to the assumption of continuity by classical physics.

In what follows, several features of discrete *versus* continuous models of motion and evolution will be discussed, partly by taking up the arguments of Zeno of Elea (5th century B.C.) on motion. For further considerations and references, see H. D. P. Lee's *Zeno of Elea* [294] and A. Grünbaum's *Modern Science and Zeno's paradoxes* [218] and *Philosophical Problems of Space of Time, Second, enlarged edition*, p. 159; *Ibid.*, p. 808 [219].

The arguments will be phrased in terms of Cellular Automaton models. This can be done without loss of generality as long as the context is universal computation. The same holds for discretization of space-time *versus* discretization of other variables; e.g., action-angle. For any universal computer can simulate any other universal computer, the only difference being the complexities required for one system to simulate the other.

3.1 Infinite divisibility and continuity

In what is now often referred to as the paradox of *"dichotomy,"* Zeno argues that (in Simplicius' interpretation, quoted from [294], p. 45) if space is infinitely divisible, and if

> "... *there is motion, it is possible in a finite time to traverse an infinite number of positions, making an infinite number of contacts one by one; but this is impossible, and therefore there is no motion."*

[Infinite divisibility of a set does not imply that its members are not recursive. For example, the set of the rational numbers $\mathbb{Q}$ and the set of the real numbers $\mathbb{R}$ are both infinitely divisible (*dense*), yet every rational is computable and "almost all" reals are uncomputable.] A similar argument is put forward in the paradox of *"Achilles and the Tortoise,"* which has been described in detail in chapter 1, p. 24.

A "resolution" of this kind of paradox can be obtained by assuming an infinite divisibility not only of space but also of time. I.e., one finite (proper) time interval can be divided into an infinitely many time steps in such a way that a body in motion traverses an infinite number of positions, making an infinite number of contacts, one position by one time step. A proper formalisation of this resolution can be expressed in terms of analysis. For further discussions along different lines see A. Grünbaum [218, 219].

Yet, as this resolution (and even constructive analysis [44]) operates by dividing space-time into infinitely many subdivisions, finite processes of computation are insufficient models of the situation. Recall Zeno's argument, p. 24. If one actually wants to simulate, for instance, successive positions of *"Achilles and the Tortoise,"* one may start by giving the following table:

position of the Tortoise:	1	$\frac{3}{2}$	$\frac{7}{4}$	$\frac{15}{8}$	$\frac{63}{32}$	$\cdots$	$2-\frac{1}{2^n}$
position of Achilles:	0	1	$\frac{3}{2}$	$\frac{7}{4}$	$\frac{15}{8}$	$\cdots$	$2-\frac{1}{2^{n-1}}$

with $n < \infty$. Although one may get arbitrary close to the meeting position 2, with a finite number of iterations one could never compute this point exactly. One could, of course, infer the meeting point by formally summing up the two geometric series (for instance, by using a program capable of symbolic mathematics), but this projection is different from what happens if Achilles catches up with the Tortoise in a step-by-step simulation.

Stated differently, by assuming infinite divisibility of space and time, an infinity of space-time points is approached by Achilles' and the Tortoise's body on a step-by-step basis in finite proper time and space. This would require a capacity of the system to compute "the limit" (of infinitely small space and time scales).

It is not completely unreasonable to speculate that in some way or another, this capacity of a physical system to compute these limits could be utilised; e.g., for the construction of an oracle which might be able to solve the halting problem. (For details, see p. 26.) This would require a dramatic revision of our concept of "effective" or "mechanic" computation in the way discussed in chapter 1. — Oracle computation by continuous systems might be seen as one aspect of the fact that they do not correspond to any process of effective computation.

If one excludes limit computations of the above kind, one has to explain why. It is to be expected that an explicit statement of exclusion will effectively render a discrete mathematical model of motion, as it is discussed below. This may be seen as a translation of Zeno's paradoxes of "dichotomy" and "Achilles and the Tortoise" into a more contemporary form; using terminology from continuum physics and recursion theory.

3.2 Finite divisibility and discreteness

In the paradox of the *"arrow,"* according to Aristotle [*Phys. Z* 9], Zeno argues as follows ([294], p. 53):

> ··· *For if, he says, everything is either at rest or in motion, but nothing is in motion when it occupies a space equal to itself, and what is in flight is always at any given instant occupying a space equal to itself, then the flying*

> *arrow is motionless.*
>
> ...
>
> *This conclusion follows from the assumption that time is composed of instants.*

This paradox may be derived from practical experiences: During an arbitrary but finite observation period $[t_0, t_1]$ a flying arrow gets "smeared out" as its "extension" LENGTH is measured by the position of the arrow tail r_0 at t_0 and the position of the arrow head r_1 at t_1. For finite time intervals of length $\Delta t = t_1 - t_0$, $\mathrm{LENGTH}(\Delta t) = r_1 - r_0$ is always greater than the length $\overline{\mathrm{LENGTH}}$ of the arrow at rest; i.e., for $\Delta t > 0$,

$$\mathrm{LENGTH}(\Delta t) > \overline{\mathrm{LENGTH}} \quad . \tag{3.1}$$

One may think of Zeno's "motionless flight" as a limit

$$\lim_{\Delta t \to 0} \mathrm{LENGTH}(\Delta t) = \overline{\mathrm{LENGTH}} \quad . \tag{3.2}$$

One concretion of the assumption that time is composed of instants or atomic "nows" is the requirement that time is *discrete*. Assume for the moment that space is discrete as well.

Motion in a discrete space-time can be realised by Cellular Automata. For a vision of Cellular Automaton models for space-time, see for instance, the articles *Digital Mechanics* by E. Fredkin [186], *Cellular Vacuum* by M. Minsky [323] and the book *Cellular Automata Machines* by T. Toffoli and N. Margolus [445]. In what follows, one-dimensional Cellular Automaton models of arrow motion are discussed.

The simplest arrow motion can be realised by a right shift; i.e., given a cell with its surrounding two neighbours, the transition rule $\{l, X, X\} \to l$ (X stands for any state) yields the state l as the next value for the center state. Let the initial state be "... --->____________ ···." Then, the time evolution is given by (see p. 249 for a *Mathematica* program simulating this evolution):

```
... --->______________________________________ ...
... _--->_____________________________________ ...
... __--->____________________________________ ...
... ___--->___________________________________ ...
... ____--->__________________________________ ...
... _____--->_________________________________ ...
... ______--->________________________________ ...
... _______--->_______________________________ ...
... ________--->______________________________ ...
... _________--->_____________________________ ...
... __________--->____________________________ ...
... ___________--->___________________________ ...
... ____________--->__________________________ ...
... _____________--->_________________________ ...
... ______________--->________________________ ...
... _______________--->_______________________ ...
... ________________--->______________________ ...
... _________________--->_____________________ ...
... __________________--->____________________ ...
           .
           .
           .
```

The velocity of the arrow is 1 cell per cycle time, independent of the length of the arrow. A less trivial example can be found in M. Minsky [323]. There, the transition rule is given by $\{-,-,>\} \to >$, $\{X,-,>\} \to X$, $\{X,>,>\} \to *$, $\{*,X,X\} \to >$, $\{*,>,X\} \to *$, $\{X,*,X\} \to -$ (X stands for any state). An arrow in this cellular array propagates with a velocity which depends on its length. E.g., for an arrow of length 3, 4, 5 and 6 cells, the velocity is 1 cell per 5, 8, 9 and 11 cycles, respectively. Explicitly,

```
... ->______________________________ ...
... ->>_____________________________ ...
... _>>_____________________________ ...
... _*>_____________________________ ...
... _-*_____________________________ ...
... _->_____________________________ ...
... _->>____________________________ ...
... __>>____________________________ ...
... __*>____________________________ ...
... __-*____________________________ ...
... __->____________________________ ...
... __->>___________________________ ...
... ___>>___________________________ ...
... ___*>___________________________ ...
... ___-*___________________________ ...
... ___->___________________________ ...
... ___->>__________________________ ...
... ____>>__________________________ ...
... ____*>__________________________ ...
       .
       .
       .
 (period 5)

... --->____________________________ ...
... --->>___________________________ ...
... ->>>____________________________ ...
... ->>>>___________________________ ...
... _>>>>___________________________ ...
... _*>>>___________________________ ...
... _-*>>___________________________ ...
... _-*>____________________________ ...
... _---*___________________________ ...
... _--->___________________________ ...
... _--->>__________________________ ...
... _->>>___________________________ ...
... _->>>>__________________________ ...
... __>>>>__________________________ ...
... __*>>>__________________________ ...
... __-*>>__________________________ ...
... __-*>___________________________ ...
... __---*__________________________ ...
... __--->__________________________ ...
       .
       .
       .
 (period 9)

... ---->___________________________ ...
... --->>___________________________ ...
... --->>>__________________________ ...
... ->>>>___________________________ ...
... ->>>>>__________________________ ...
... _>>>>>__________________________ ...
... _*>>>>__________________________ ...
... _-*>>>__________________________ ...
```

One could speculate that, if the motion of a body in the physical vacuum can be described similarly, then there is a remote possibility to increase this velocity by proper algorithmic manipulations; e.g., by re-programming the vacuum motion or by re-shaping the body.

There is an obvious contradiction with Zeno's argument: in the listed Cellular Automaton configurations the arrow in flight occupies a cell region equal to its size, and yet it moves. This may be explained from different perspectives.

One may argue that the shape of the arrow (i.e., the initial state) and the transition rules contain information about the evolution; such that any "snapshot" of the position of the arrow should be accompanied by another quantity called "momentum," which completes a description of the arrow's motion. In this way, phase space is introduced.

Another explanation would be that the shape of the arrow (i.e., the initial state) and the transition rules contain "hidden" information about the evolution which reveals itself after successive time periods. This argument somewhat resembles the construction of a reversible *second-order* system (computation) proposed by E. Fredkin ([445], p. 141) by

$$c^{t+1} = \tau c^t - c^{t-1} \quad , \tag{3.3}$$

$$c^{t-1} = \tau c^t - c^{t+1} \quad . \tag{3.4}$$

Here τ is an arbitrary function of the state c^t at time t. In these cases, a complete specification of the system (e.g., the arrow) has to be given by an ordered pair of regions occupied by the system at successive times.

3.3 Simultaneity

In the paradox of the *"stadium,"* Zeno argues that an attempt to define space and time distances by the motion of rigid bodies results in a contradiction. Assume again that space and time are uniformly partitioned into cells and instants (time cycles). Consider identical bodies marked by A, B, C, filling up one space cell, and three rows of such bodies (4 cells long) in a stadium: AAAA, BBBB, CCCC. The row AAAA is at rest with respect to the stadium. The row BBBB travels with constant velocity of one cell per time cycle with respect to the stadium. The row CCCC travels with the same constant velocity of one cell per time cycle with respect to the stadium, but into the opposite direction. Let the rows be initially arranged as follows (the arrows indicate the direction of movement seen from the reference frame of the stadium):

```
    A A A A
B B B B  →
      ←  C C C C
```

After one and two exterior cycle times, the positions of the rows are:

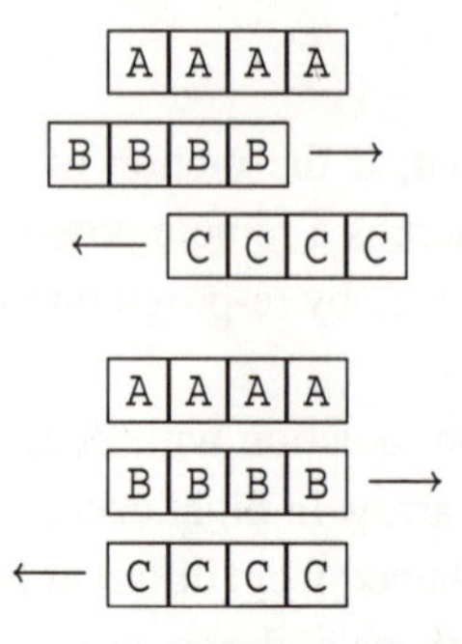

Zeno's paradox of the *"stadium"* follows from the assumption (/ convention / definition) that a measure of intrinsic (!) time of some body is the number of other bodies (cells) passed by this body. For then, after two time cycles, the rightmost B-body has passed two A-bodies but 4 C-bodies, yielding a passage time of 2 or 4, respectively.

This argument is somewhat similar to A. Einstein's critique of the classical notion of simultaneity put forward in relativity theory, resulting in a transformation of space and time scales [151]. In discrete physics, any velocity will be characterised by rationals, the "natural" extrinsic unit being one cell per cycle time. Without loss of generality it is assumed that the maximal velocity a body can move is one cell per cycle time, denoted by c. In analogy to relativity theory, this velocity is used to define synchronised events. Flows of this kind will be called *rays*.

The synchronisation convention of two clocks at two points A and B can be adopted from relativity theory ([151], p. 892):

D 3.1 (Einstein synchronisation [151]) *Assume two clocks at two arbitrary points A and B which are "of similar kind." At some arbitrary A-time t_A a ray goes from A to B. At B it is instantly (without delay) reflected at B-time t_B and reaches A again at A-time $t_{A'}$. The clocks in A and B are* synchronised *if*

$$t_B - t_A = t_{A'} - t_B \quad .$$

The two-ways ray velocity is given by

$$\frac{2\overline{AB}}{t_{A'} - t_A} = c \quad ,$$

where $\overline{AB}$ is the distance between A and B.

It is identical for all frames, irrespective of whether they are moving with respect to the rest frame of the cellular space or not.

Example:

Synchronisation by ray exchange can be simulated on a Cellular Automaton using the following rules for inertial arrow motion (X stands for any state): $\{>, \_, X\} \to >$, $\{X, \_, <\} \to <$, $\{\_, \_, \_\} \to \_$, $\{X, \_, >\} \to \_$, $\{<, \_, X\} \to \_$, $\{\_, >, \_\} \to \_$, $\{\_, <, \_\} \to \_$ and for reflection (rigid mirror I): $\{X, >, \mathtt{I}\} \to <$, $\{\mathtt{I}, <, X\} \to >$, $\{X, \mathtt{I}, X\} \to \mathtt{I}$, $\{\_, \_, \mathtt{I}\} \to \_$, $\{\mathtt{I}, \_, \_\} \to \_$, $\{\mathtt{I}, 9, \_\} \to \_$, $\{\_, 6, \mathtt{I}\} \to \_$. (The evolution can, for instance, be realised by the *Mathematica* program on p. 249.) With two rigid mirrors at A and B spaced 7 cells apart, the evolution is given by

```
                            A        B

...  ______________________I>_______I______________________  ...
...  ______________________I_>______I______________________  ...
...  ______________________I__>_____I______________________  ...
...  ______________________I___>____I______________________  ...
...  ______________________I____>___I______________________  ...
...  ______________________I_____>__I______________________  ...
...  ______________________I______>_I______________________  ...
...  ______________________I_______>I______________________  ...
...  ______________________I_______<I______________________  ...
...  ______________________I______<_I______________________  ...
...  ______________________I_____<__I______________________  ...
...  ______________________I____<___I______________________  ...
...  ______________________I___<____I______________________  ...
...  ______________________I__<_____I______________________  ...
...  ______________________I_<______I______________________  ...
...  ______________________I<_______I______________________  ...
...  ______________________I>_______I______________________  ...
...  ______________________I_>______I______________________  ...
...  ______________________I__>_____I______________________  ...
...  ______________________I___>____I______________________  ...
...  ______________________I____>___I______________________  ...
...  ______________________I_____>__I______________________  ...
...  ______________________I______>_I______________________  ...
...  ______________________I_______>I______________________  ...
...  ______________________I_______<I______________________  ...
...  ______________________I______<_I______________________  ...
...  ______________________I_____<__I______________________  ...
...  ______________________I____<___I______________________  ...
...  ______________________I___<____I______________________  ...
     .
     .
     .
```

Synchronisation can be defined in all frames, irrespective of whether they are moving with respect to the cell space or not. However, it cannot be expected that two frames which are in motion against each other have identical synchronisations. Indeed, the assumption of identical synchronisations yields a contradiction. Consider the following *Gedanken-experiment,* which is again due to Einstein ([151], p. 895): Let the points A and B be the endpoints of a rigid rod. Assume that the rod is moving with constant velocity with respect to the cellular space. Assume four observers; two observers travelling with the rod at points A and B, and two observers in the rest frame of the cellular space. The latter observers synchronise their clocks and measure the length $\overline{AB}$ of the moving rod in the rest frame of the cellular space (at equal rest frame times). Assume further that if two events are synchronised in the rest frame of the cellular space, then they are synchronised in the rest frame of the moving rod (wrong). An attempt to verify the synchronisation of the clocks in the moving frame yields to a refusal, since if the array is first emitted in the direction of motion and reflected afterwards, then

$$c - v = \frac{\overline{AB}}{t_B - t_A} \quad \text{and} \qquad c + v = \frac{\overline{AB}}{t_{A'} - t_B} \quad . \tag{3.5}$$

Hence, $t_B - t_A \neq t_{A'} - t_B$ and the clocks in A and B, as they are observed in the rest frame of the rod (in a frame moving with respect to the cellular space) are *not* synchronised. This is a contradiction to the assumption.

Of course, one could *define* a global time synchronisation; for instance by counting the number of cycle times of the array. This view, however, is an *extrinsic* (see below) view, which yields a preferred frame of reference, the frame at rest with respect to the cellular space. Such a definition is an alternative to the above definition of synchronisation given by Einstein. Einstein synchronisation is suitable for an *intrinsic* (see below) definition. It does not operate with a preferred frame of reference; the "ideological preference" being the relativity principle (symmetry). The according time will sometimes also be called *intrinsic time.* Intrinsic time is discrete. But, just as in relativity theory, for moving frames the elementary unit of intrinsic cycle time becomes dilatated. By a similar argument, the elementary unit of length becomes dilatated.

How far can one go to reconstruct relativity theory? Not very far if one does not take into account *dispersion* (e.g., the energy-momentum relation) and other dynamical features. Because even for relativistic kinematics, an alternative synchronisation could be defined by signals of *arbitrary* velocity or by *defining* a preferred frame of reference, such as the one at rest with respect to the cosmic background blackbody radiation. The physical content of Einstein synchronisation is the *absence* of any criterion by which one inertial frame might be preferred over the other. This is not the case for a cellular array, where — at least extrinsically — a preferred frame is the frame at rest with respect to the cellular array. Whether this preference holds for *intrinsic* perception is questionable.

Stated differently, one criterion for Einstein synchronisation is the invariance of the physical laws, such as electromagnetism, in arbitrary inertial frames. This is one reason why, for instance, synchronisation by sound is no good if one attempts to describe physical motion governed by electromagnetism. A detailed discussion of related topics is given in [428, 429, 432], where it is argued that, depending on dispersion relations, creatures in a "dispersive medium" would develop a theory of coordinate transformation very similar to relativity theory.

Chapter 4

Source coding

Physical entities such as experimental measurement results, time series *et cetera* as well as physical theories can be symbolically treated on an equal footing — as an *information source* of *symbols*. (The notion of the term "symbol" remains intuitive, since any definition of "symbol" has to be done in terms of symbols.)

The first sections are devoted to general *source coding schemes*, which have been developed in the context of the foundations of quantum mechanics [244, 124] and for the coding of time series, which has been introduced as *symbolic dynamics* [9]. We then consider the Gödel numbering of axiomatised physical theories. For a more detailed account, see, for instance, R. W. Hamming's book *Coding and Information Theory* [223], R. J. McEliece, *The Theory of Information and Coding* [317], or R. Ash, *Information Theory* [13].

4.1 General source coding

D 4.1 (System) *A* system *is anything on which experiments can be performed.*

Examples:

(i) Any finite deterministic automaton is a system. For instance, your PC / workstation (insert your favourite brand here:) "..." is a system. Another example is the *Moore automaton* defined on p. 142.

(ii) Any mechanical or quantum device is a system.

(iii) The great beyond is *no* system, unless one succeeds to perform experiments on it. This statement needs not be entirely ridiculous, because what is called *the great beyond* depends on scientific knowledge and believes which transform(s) in history. For instance, what is presently called "electromagnetic phenomena" was mostly obscure and beyond anybody's experimental limits only 300 years ago.

D 4.2 (Experiment, manual, outcome, event)
A physical system is characterised by experiments *performed on it. The collection of all experiments* $\mathfrak{M}$ *is called* manual *or* propositional calculus.

The i'th experiment will be denoted by e_i. *Any experiment can be decomposed into elementary* TRUE-FALSE-*experiments, called* propositions *and denoted by* e_{ij}.

The members of an experiment are called outcomes. *Associated with every proposition* e_{ij} *are the two* elementary outcomes TRUE *and* FALSE.

An event $p_i \subset \mathfrak{M}$ *is a subset of an experiment* e_i, *associated with particular outcome(s). In terms of elementary experiments,* p_{ij} *corresponds to the jth experiment* e_{ij} *of*

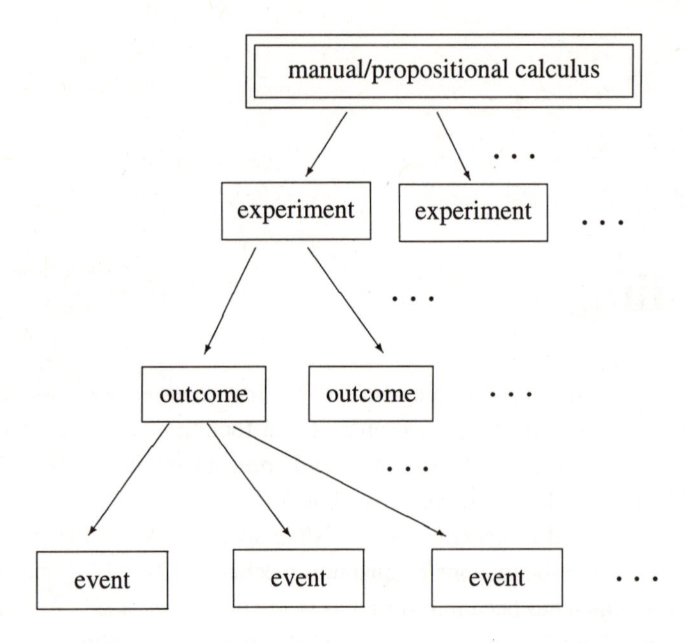

Figure 4.1: Hierarchy of perception.

e_i. *Every event is also called* element of physical reality.

Remarks:

(i) Fig. 4.1 shows the hierarchy "manual → experiment → outcome → event" of perception defined by 4.2.

(ii) The terminology *manual* suggests a collection or catalogue of events. It defines an empirical universe and represents the "admissible" *elements of physical reality*; insofar the *manual is the physical primitive*.

(iii) It is easiest to imagine an empirical universe consisting of propositions, i.e., statements which are either `TRUE` or `FALSE` (see the following example).

(iv) Throughout this book a notion of *element of physical reality* is used which refers only to *outcomes of actual measurements*. This concept is more restricted than the EPR-terminology [154]:

> *"If, without in any way disturbing a system, we can predict [[!]] with certainty (i.e., with probability equal to unity) the value of a physical quantity, then there exists an element of physical reality corresponding to this quantity."*

The most important difference is that Einstein does not deny the existence of elements of physical reality for physical entities which are only indirectly reconstructed from measurements of other entities and which have *not* actually been measured. E.g., assume an two entangled electrons in a singlet state. Then, according to Einstein, due to conservation of angular momentum, measurement of the spin of one electron fixes the spin the other one in that particular direction. Hence, according to Einstein, an element of physical reality could be ascribed to the spin in that particular direction of the other electron from the entangled pair. — No actual experiment needs to be performed to measure its spin directly. — This view of "results from unperformed experiments" is not adopted here.

Example:

Consider an automaton consisting of i internal states labelled by numbers $1 \leq k \leq i$. Assume that all internal states can, informally speaking, be "experimentally distinguished" (for a definition, see chapter 1.21, p. 23). Then, statements of the form "`the automaton is in state` k," $1 \leq k \leq i$, are *propositions*. The *manual* is the union of propositions. The possible experimental *outcomes* are `TRUE` and `FALSE`, respectively. After performing an actual experiment, an *event* is defined by one of these values. If defined, propositions can be constructed with logical "`or`" and "`and`" or "`not`" operations.

D 4.3 (Coding)

Let $\bigcup_{j=1}^{M} p_{ij} = p_i$ *be an event, decomposed into elementary events* p_{ij}. *For elementary events, the* source alphabet *consists of just two symbols* s_1 *and* s_2, *corresponding to* `TRUE` *and* `FALSE`, *respectively. The* code $\#(p_{ij})$ *of* p_{ij} *is thus defined by*

$$\#(p_{ij}) = \begin{cases} s_1 \text{ if } p_{ij} = \texttt{TRUE} \\ s_2 \text{ if } p_{ij} = \texttt{FALSE} \end{cases} .$$

Remarks:

(i) if one identifies $s_1 = 0$ and $s_2 = 1$, the event p_i can be coded as a binary rational number as follows:

$$\#(p_i) = 0.\#(p_{i1})\#(p_{i2})\#(p_{i3})\cdots\#(p_{ij})\cdots\#(p_{iM}) \quad .$$

A generalisation to arbitrary bases is straightforward.

(ii) Note that, instead of taking some (natural) numbers as symbols, one could have taken any other entity which could be identified as symbol, such as the letters of the English or Greek alphabet, or apples & oranges *et cetera.*

(iii) So far we have not dealt with the *translation* of these *source code symbols* into an alphabet which is *algorithmically recognisable* by some computable process. This is discussed in the section on *encoding*, p. 84. For information sources with source alphabet $s_1, s_2, \ldots, s_q$ with q symbols, the generalisation to radix $\geq q$ notation is straightforward. If the code alphabet consists of less symbols than the source alphabet, then more sophisticated encoding techniques are necessary. These will be discussed below.

(iv) For technical reasons it often will be assumed that with any number p occurring in the system also "elementary functions" (recursive functions) $g(p)$ thereof are codable within the system. This is equivalent to postulating the existence of a universal computer in the system.

(v) If the context is unambiguous, the code signs "#( )" can be dropped. Any event p_i is then written as $p_i = 0.p_{i1}p_{i2}p_{i3}\cdots p_{in}$.

(vi) For fractal source coding, see section 15.4, p. 207.

4.2 Symbolic dynamics

For dynamical systems, the method of *symbolic dynamics* [9] will be applied to the coding of the events in a manual represented by a *partition of a generalised phase space* (see also section 18.3, p. 238).

D 4.4 (Covering, partition) *Let the manifold X be a* generalised phase space *with points* $x \in X$. *A* covering $\xi = \{E_1, \dots, E_q\}$ *is a set of subsets of X such that*

$$X = \bigcup_{i=1}^{q} E_i$$
$$E_i \cap E_j \neq \emptyset \quad .$$

A partition $\xi = \{E_1, \dots, E_q\}$ *is a set of subsets of X such that*

$$X = \bigcup_{i=1}^{q} E_i$$
$$E_i \cap E_j = \emptyset \quad .$$

D 4.5 (Sequence of pointer readings)
To every $E_i \in \xi$, *associate a* pointer reading s^i. *A* sequence of n pointer readings, *representing experimental outcomes, is denoted by*

$$\Psi(n) = s_0, s_1 \dots s_{n-1} \quad .$$

Notice that superscripts are used to denote the i'th symbol s^i, whereas subscripts are used to denote the place of the symbol in the sequence of pointer readings Ψ. The s^i's in Ψ are characters of an arbitrary alphabet S with q symbols. (This motivates the view of Ψ as a *word*, which is the starting point of *linguistic analysis*, in particular with respect to the Chomsky hierarchy of languages [326].)

The terminology of symbolic dynamics is related to the general coding scheme: identify a time sequence Ψ obtained from a restricted alphabet $S = \{0, 1\}$ with a particular event, say p_i, by setting $s_j = p_{ij}$.

4.3 Coding physical theories

Syntactically, any physical theory is representable by symbols. (Let us, for the moment, disregard the *semantic* content of a theory, i.e., its "meaning" *et cetera.*) These symbols can be "read" by some "observer" or agent by performing a series of experiments. Examples are the sensory perceptions associated with the reading of a book on theoretical physics, or the reading of a computer tape containing an effective procedure for solving an evolution equation *et cetera*; more generally, by experimentally observing some representations of theories. E.g., the reading of the term "$E = mc^2$" can be seen as an optical scanning experiment which yields the (successive) events "E," "=," "m," "c" and "^{2}."

These symbols may be interpreted as input for some effective computation producing "predictions." In what follows, we assume that physical theories are *mechanistic*, i.e., that they can be represented by an algorithm.[1] For the time being, no attention is thus

[1] There exist theories with an *infinite* number of axioms, e.g., the *Peano axioms* or the induction rules

$$\varphi(0) \wedge \forall x(\varphi(x) \rightarrow \varphi(x+1)) \rightarrow \forall x \varphi(x) \quad .$$

In this case one may, nevertheless, be able to find a finite-size code of the *generating algorithm* of these axioms.

paid to theories which *cannot* be brought into such a form, as well as to the question, "*how* can such theories be created?"

The enumeration of programs discussed in section 1.6, p. 10, can be used to generate a unique code of such a theory. An alternative would be the use of instantaneous (prefix) codes for entities of formal systems; see section 7.1.1, p. 85. Still another (probably not very practical) coding scheme is Gödel's original construction of Gödel numbers, for which Gödel uses the uniqueness of prime factors of whole numbers. The following definition deals with a very particular language (alphabet), which is appropriate for the task of deriving the incompleteness theorems [204]; a generalisation to more general languages (alphabets) is straightforward.

D 4.6 (Gödel numbers)

Assume an alphabet consisting of the symbols $(,), \sim, \rightarrow, \forall,$ *variables* x_k, *constants* a_k, *functions* f_k^n *and relations* A_k^n. *It is possible to map terms containing these symbols injectively onto the odd, positive integers by the* Gödel number *function #:* $\#(() = 3; \#()) = 5; \#(,) = 7; \#(\sim) = 9; \#(\rightarrow) = 11; \#(\forall) = 13; \#(x_k) = 7+8k; \#(a_k) = 9+8k; \#(f_k^n) = 11+8\times(2^n\times 2^k); \#(A_k^n) = 13+8\times(2^n\times 3^k)$.

Every well-formed formula $F = s_1\cdots s_k$ *consisting of the symbols of the alphabet, functions and relations can be injectively mapped via*

$$\#(F) = \#(s_1\cdots s_k) = 2^{\#(s_1)}\times 3^{\#(s_2)}\times\cdots\times p_k^{\#(s_k)} \quad ,$$

where p_k *is the kth prime (notice that the power of two is odd).*

Every deductive proof consists of a sequence of well-formed formulas $F_1\cdots F_k$ *can be uniquely mapped via*

$$\#(F_1\cdots F_k) = 2^{\#(F_1)}\times 3^{\#(F_2)}\times\cdots\times p_k^{\#(F_k)} \quad ,$$

where again p_k *is the kth prime (here the power of two is even).*

In a *binary* Gödel numbering of the type just introduced, the axioms of a theory can be coded in bit strings (for instance by rewriting the values $\#(F)$ in binary notation). All such bit strings can be merged in a *single* finite bit string. This string is the object of investigation if one is interested in the algorithmic information content of a theory (see 7 and 9.7, p. 83, 122). The rules of inference can then be envisioned as a computable algorithm for enumerating the bit strings corresponding to provable theorems from the axiom bit string (for details, see table 2.2, p. 35).

Chapter 5

Lattice theory

This section gives a brief introduction to lattice theory. Lattice theory is a convenient framework for organising ordered structures such as experimental or logical statements. For a very readable and elementary introduction (in German), see H. Liermann [303]. A more detailed "canonical" introduction to lattice theory can be found in G. Birkhoff [43]. Orthomodular lattices are reviewed in the books by G. Kalmbach [251, 252], P. Pták and S. Pulmannová [376], R. Giuntini [201] and A. Dvurečenskij [144], among others. The books by J. Jauch [244], G. W. Mackey [306] and C. Piron [362] deal with physical applications, mainly in the context of quantum mechanics. A bibliography can be found in reference [349].

5.1 Relations

D 5.1 (Relation, equivalence relation)
Assume two sets M, N. Every subset of the Cartesian product $M \times N$ is a (binary) relation *fRg, $f \in M, g \in N$. There are as many relations as there are subsets of $M \times N$.*
Let $M = N$. An equivalence relation *satisfies the following properties:*

(i) fRf for all $f \in M$ (reflexivity)*;*

(ii) $fRg \Rightarrow gRf$ for all pairs $f, g \in M$ (symmetry)*;*

(iii) fRg and $gRh \Rightarrow fRh$ for all $f, g, h \in M$ (transitivity)*.*

D 5.2 (Equivalence class, quotient) *The subset $f^R = \{g \mid fRg\}$ is called the* equivalence class of f modulo R.

The set $M/R = \{f^R \mid f \in M\}$, consisting of all equivalence classes modulo R is called the quotient of M by R. *R yields a partitioning of M.*

Every equivalence relation R corresponds to a function φ such that $\varphi(f) = f^R$ for all $f \in f^R$. Conversely, to every function $\varphi : M \to N$ corresponds an *equivalence relation* "$\equiv_\varphi$" on M, defined by

$$f \equiv_\varphi g \Longleftrightarrow \varphi(f) = \varphi(g) \quad .$$

The elements in $M/\equiv_\varphi$, the quotient of M by $\equiv_\varphi$, correspond one-to-one to elements of N. This amounts to *renaming* the elements of N by elements of the quotient $M/\equiv_\varphi$. One can thus define a map

$$\varphi_c : M \to M/\equiv_\varphi \quad ,$$

calling it the *canonical form of φ*.

In general, an *isomorphism* between two algebraic structures (admitting certain "similar" operations) is a one-to-one element-to-element correspondence which preserves all combinations. The following definition specifies the operation to (binary) relations.

D 5.3 (Isomorphism, automorphism) *Let M_1 be an algebraic structure with a (binary) relation R_1 and let M_2 be another algebraic structure with a (binary) relation R_2. An* isomorphism $\cong$ *is a relation defined by a one-to-one map ("translation") I from M_1 into M_2 which preserves the relations, i.e., $M_1 \cong M_2$ with a one-to-one map I satisfying*

$$fR_1g \Longleftrightarrow I(f)R_2I(g) \quad .$$

The converse function I^{-1} defines an isomorphism from M_2 to M_1.

If $M_1 = M_2$, then $\cong$ is called an automorphism.

Remarks:

(i) Informally, the concept of an isomorphism is that two algebraic structures "look much the same" and become identical if only their entities are renamed.

(ii) An isomorphism defines an equivalence relation.

5.2 Partial order relation

D 5.4 (Partial ordering, poset) *A* partially ordered set (poset) *is a system M in which a binary* order relation *"$\succeq$" (inverse "$\preceq$") is defined, which satisfies*

(i) $f \succeq f$, for all $f \in M$ (reflexivity)*;*

(ii) $f \succeq g$ and $g \succeq h \Rightarrow f \succeq h$ (transitivity)*.*

(iii) $f \succeq g$ and $g \succeq f \Rightarrow f = g$ (identitivity)*.*

The "immediate superiority" of f with respect to g will be defined next.

D 5.5 *By "f* covers *g," it is meant that $f \succeq g$ and that $f \succeq x \succeq g$ is not satisfied by any $x \in M$, $x \neq f$, $x \neq g$.*

Any finite partially ordered set M can be conveniently represented graphically by a *Hasse diagram* in the following way.

D 5.6 (Hasse diagram) *Let M be a partially ordered set. The* Hasse diagram *of M is a directed graph obtained by drawing small filled circles representing the elements of M, so that f is higher than g whenever $f \succeq g$. A segment is then drawn from f to g whenever f covers g.*

Remarks:

(i) As the direction is always from the bottom to the top, Hasse diagrams are drawn undirected.

(ii) Any finite partially ordered set is defined up to isomorphism by its Hasse diagram. I.e., two isomorphic partially ordered sets must have a one-to-one relation between their highest & lowest elements, between elements just above lowest elements, and so on; corresponding elements must be covered equally.

D 5.7 (Linearly ordered set) *If for all elements f, g of a partially ordered set M either the relation "$f \succeq g$" or the relation "$g \succeq f$" is satisfied, M is called a* linearly ordered *set.*

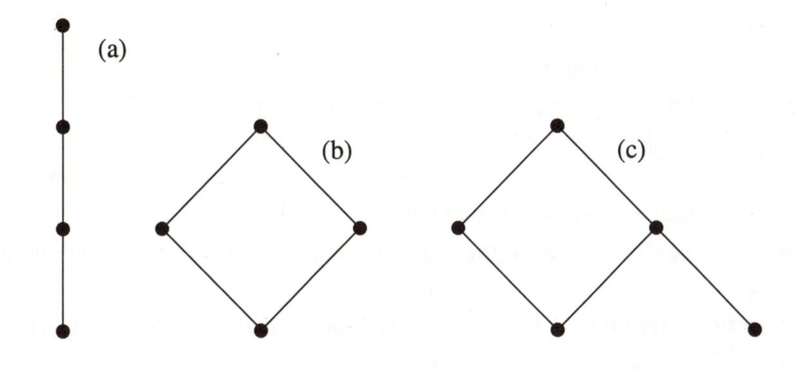

Figure 5.1: Examples of posets

D 5.8 (Chain) *A* chain *in a partially ordered set M is a subset $N \subset M$ which is a linearly ordered set.*

D 5.9 (Length) *Let N be a chain of a partially ordered set M. The* length $|N|$ *of N is the cardinal number (i.e., the number of elements) of N. The* length of the partially ordered set $|M|$ *is the supremum over the length of all chains in M minus 1. M has* finite length *if $M < \infty$.*

D 5.10 (Atom,coatom) *Assume a partially ordered set with a least element* **0** *and a greatest element* **1**.

(i) Atoms *are elements which cover the least element* **0**.

(ii) Coatoms *are elements which are covered by the greatest element* **1**.

(iii) The partially ordered set is atomic *if every $a \neq \mathbf{0}$ in it is greater than or equal to an atom.*

Examples:

(i) Fig. 5.1(a) shows the Hasse diagram of a linearly ordered set. Fig. 5.1(b) shows the Hasse diagram of a non linearly ordered set.

(ii) Figs. 5.1(a)& (b) show the Hasse diagrams of atomic sets. Fig. 5.1(c) shows the Hasse diagram of a non atomic set.

Next the *lattice* concept will be introduced.

5.3 Lattice

Lattice theory is the theory of partially ordered sets with the property that two arbitrary elements have a common upper and lower bound. It provides a "generic" framework for the investigation of important algebraic structures occurring, for instance, in Hilbert space theory and logic.

D 5.11 (Lattice, version I) *A partially ordered system $\mathfrak{L} = (\mathfrak{L}, \preceq)$ with order relation "$\preceq$" (inverse "$\succeq$") is a* lattice *if and only if any pair f, g of its elements has*

(i) a meet *or* (greatest) lower bound [infimum] *inf(f, g) such that*

$inf(f, g) \preceq f$

$inf(f, g) \preceq g$
$h \preceq f$ *and* $h \preceq g$ *imply* $h \preceq inf(f, g)$;

(ii) and a join *or* (least) upper bound [supremum] $\sup(f, g)$ *such that*
$\sup(f, g) \succeq f$
$\sup(f, g) \succeq f$
$h \succeq f$ *and* $h \succeq g$ *imply* $h \succeq \sup(f, g)$;

Instead of the definition 5.11, the following axioms characterise a lattice alternatively ([43], page 18).

D 5.12 (Lattice, version II) *A* lattice *is an algebraic structure* $\mathfrak{L} = (\mathfrak{L}, \sqcap, \sqcup)$ *with two operations "*$\sqcap$*" and "*$\sqcup$*" satisfying*

$$\begin{aligned} f \sqcap f &= f \text{ idempotence} \\ f \sqcup f &= f \text{ idempotence} \\ f \sqcap g &= g \sqcap f \text{ commutativity} \\ f \sqcup g &= g \sqcup f \text{ commutativity} \\ f \sqcap (g \sqcap h) &= (f \sqcap g) \sqcap h \text{ associativity} \\ f \sqcup (g \sqcup h) &= (f \sqcup g) \sqcup h \text{ associativity} \\ f \sqcap (f \sqcup g) &= f \text{ absorption law} \\ f \sqcup (f \sqcap g) &= f \text{ absorption law.} \end{aligned}$$

Remarks:

(i) With the identifications

$$f \preceq g \Longleftrightarrow f = f \sqcap g \text{ or } g = f \sqcup g \tag{5.1}$$
$$f \sqcap g = \inf(f, g), \quad f \sqcup g = \sup(f, g) \quad , \tag{5.2}$$

the structures $(\mathfrak{L}, \succeq)$ and $(\mathfrak{L}, \sqcap, \sqcup)$ are equivalent.

(ii) A lattice is *finite* if the number of elements is finite.

(iii) The upper or lower bound of a lattice satisfy

$$\mathbf{0} \sqcap f = \mathbf{0} \qquad \mathbf{0} \sqcup f = f \tag{5.3}$$
$$\mathbf{1} \sqcap f = f \qquad \mathbf{1} \sqcup f = \mathbf{1}. \tag{5.4}$$

Every finite lattice contains $\mathbf{0}, \mathbf{1}$.

(iv) Two lattices $\mathfrak{L}_1$ and $\mathfrak{L}_2$ are *isomorphic* if there exists a one-to-one map $I : \mathfrak{L}_1 \rightarrow \mathfrak{L}_2$ of the lattice $\mathfrak{L}_1$ into the lattice $\mathfrak{L}_2$ such that the (binary) relations $\sqcap$ and $\sqcup$ are preserved; i.e., $I(f \sqcap_{\mathfrak{L}_1} g) = I(f) \sqcap_{\mathfrak{L}_2} I(g)$ and $I(f \sqcup_{\mathfrak{L}_1} g) = I(f) \sqcup_{\mathfrak{L}_2} I(g)$ for all $f, g \in \mathfrak{L}_1$ (see also definion 5.3).

D 5.13 (Orthoganal complement, orthocomplement) f' *is a* orthogonal complement, *or* orthocomplement *of* f *if*

(i) $(f')' = f$,

(ii) $f' \sqcap f = \mathbf{0}$,

(iii) $f' \sqcup f = \mathbf{1}$,

(iv) $a \preceq b \Rightarrow a' \succeq b'$.

D 5.14 (Orthocomplemented lattice) *A lattice is called* orthocomplete, *if for all* $f \in \mathfrak{L}$ *there exists a* $f' \in \mathfrak{L}$. *I.e., an orthocomplemented lattice also contains the complements.*

D 5.15 (Subalgebra) *A* subalgebra *of an orthocomplemented lattice* $\mathfrak{L}$ *is a subset which is closed under the operations* $'$, $\sqcup$, $\sqcap$ *and which contains* $\mathbf{0}$ *and* $\mathbf{1}$.

Usually a distinction is made between a *subalgebra* and a *sublattice* of a lattice. The latter one is required to be closed under the operations $\sqcup, \sqcap$ and not necessarily under the orthocomplement $'$.

D 5.16 (Finite lattice) *A lattice* $\mathfrak{L}$ *is called* finite *if its cardinal number* $|\mathfrak{L}| < \infty$ *is finite; i.e., if it contains only a finite number of elements.*

D 5.17 (Exchange axiom) *A lattice* $\mathfrak{L}$ *satisfies the* exchange axiom *if for all* $a, b \in \mathfrak{L}$, *if* a *covers* $a \sqcap b$ *then* $a \sqcup b$ *covers* b.

D 5.18 (Complete lattice) *A lattice* $\mathfrak{L}$ *is* complete *if, for every subset* $\mathfrak{L}' \subset \mathfrak{L}$, *there exists a "meet"* $\sqcap \mathfrak{L}'$ *and a "join"* $\sqcup \mathfrak{L}'$ *in* $\mathfrak{L}$.

Remark:
By induction it can be shown that any finite lattice is complete.

5.3.1 Distributive lattice

The set operations of union & intersection "$\cup$" and "$\cap$" satisfy the distribution laws. I.e., let A, B and C be three subsets of a set, let "$\sqcap = \cap$" and "$\sqcup = \cup$," then the two distributive laws $A \cap (B \cup C) = (A \cap B) \cup (A \cap C)$ and $A \cup (B \cap C) = (A \cup B) \cap (A \cup C)$ are always satisfied, one implying the other. General lattice structures do not necessarily satisfy the distributive laws (for example, see Fig. 5.6 on p. 68).

D 5.19 (Distributive lattice) *A lattice is called* distributive, *if*

$$(f \sqcap g) \sqcup h = (f \sqcup h) \sqcap (g \sqcup h) \quad , \tag{5.5}$$

$$(f \sqcup g) \sqcap h = (f \sqcap h) \sqcup (g \sqcap h) \quad . \tag{5.6}$$

Remarks:
(i) Every linearly ordered set is a distributive lattice.
(ii) In a distributive lattice the orthogonal complement of an element is uniquely defined.
(iii) A criterion whether or not lattices satisfy the distribution laws is

$$(f \sqcap g) \sqcup (f \sqcap g') = f \quad .$$

5.3.2 Boolean lattice

D 5.20 (Boolean lattice) *An orthocomplemented, distributive lattice is called* Boolean lattice. *I.e., for a Boolean lattice the following laws are satisfied:*

$$f \sqcup g = g \sqcup f, \tag{5.7}$$

lattice operation	set of subset of set
order relation $\preceq$	subset relation $\subset$
"meet" $\sqcap$	intersection $\cap$
"join" $\sqcup$	union $\cup$
"complement" $'$	set complement $'$

Table 5.1: Identification of lattice relations and operations for the set of subsets of a set. The resulting lattice is Boolean.

$$f \sqcap g = g \sqcap f, \tag{5.8}$$
$$f \sqcap (g \sqcup h) = (f \sqcap g) \sqcup (f \sqcap h), \tag{5.9}$$
$$f \sqcup (g \sqcap h) = (f \sqcup g) \sqcap (f \sqcup h), \tag{5.10}$$
$$\mathbf{1} \sqcap f = f, \tag{5.11}$$
$$\mathbf{0} \sqcup f = f, \tag{5.12}$$
$$f' \sqcap f = \mathbf{0}, \tag{5.13}$$
$$f' \sqcup f = \mathbf{1}. \tag{5.14}$$

A Boolean lattice with n atoms is denoted by 2^n.

Example:

The set of subsets of a set is a Boolean lattice with the identifications summarised in table 5.1.

Remark:

If $f \sqcap g = \mathbf{0}$, then one may write $f \perp g$; in words "f is orthogonal to g."

5.3.3 Modular lattice

A weaker condition than distributivity is the *modular law*. It is obtained by assuming that $f \preceq h$, such that $h = f \sqcup h$, and evaluating the distributive law (5.10). Hence, every distributive lattice is modular.

D 5.21 (Modular lattice) *A lattice is called* modular, *if for all* $f \preceq h$,

$$(f \sqcup g) \sqcap h = f \sqcup (g \sqcap h) \quad . \tag{5.15}$$

The following theorem is stated without proof (see, for instance, G. Birkhoff, ref. [43], p. 66):

T 5.22 *Any non-modular lattice contains the lattice of Fig. 5.2 as a subalgebra.*

5.3.4 Orthomodular lattice

The modular law holds only in finite-dimensional Hilbert spaces. The following *orthomodular law* holds in arbitrary Hilbert spaces.

D 5.23 (Orthomodular lattice, version I) *A lattice is called* orthomodular, *if, for all* $f \preceq g$,

$$f \sqcup (f' \sqcap g) = g \quad . \tag{5.16}$$

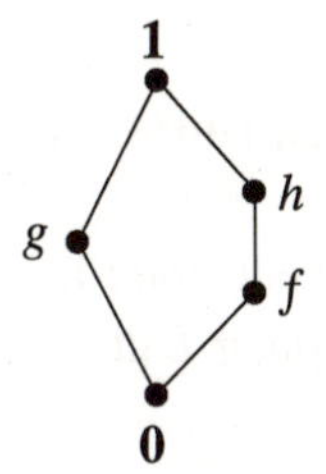

Figure 5.2: Any non-modular lattice contains this lattice as a subalgebra.

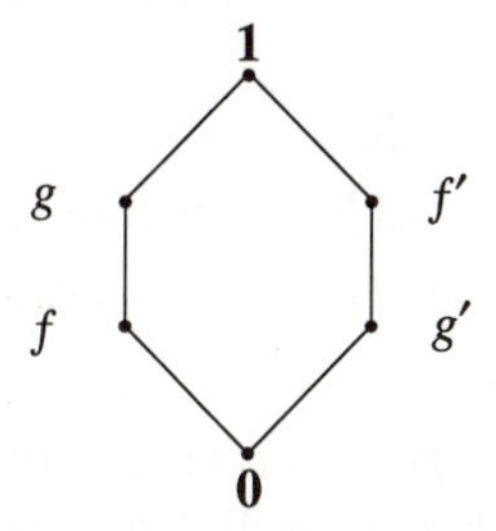

Figure 5.3: Any non-orthomodular lattice contains this lattice as a subalgebra.

D 5.24 (Orthomodular lattice, version II) *Any lattice which does not contain the subalgebra O_6 drawn in Fig. 5.3 is called* orthomodular.

For a proof of the equivalence of versions I and II of the definition (as well as other equivalent definitions), see, for instance, G. Kalmbach, *Orthomodular Lattices* [250], page 22. See also S. S. Holland's older review article [1].

The following implications are valid:

$$\text{distributivity} \underset{\not\Leftarrow}{\Rightarrow} \text{modularity} \underset{\not\Leftarrow}{\Rightarrow} \text{orthomodularity} \quad . \tag{5.17}$$

5.3.5 Commutator and Centre of orthomodular lattice

D 5.25 (Commutator)
Two elements a and b of an orthomodular lattice $\mathfrak{L}$ are commuting, *denoted by aCb, iff $a = (a \sqcap b) \sqcup (a \sqcap b')$, or $a = (a \sqcup b) \sqcap (a \sqcup b')$, or $a \sqcap (a' \sqcup b) = a \sqcap b$. Let*

$$C(a,b) = (a \sqcap b) \sqcup (a \sqcap b') \sqcup (a' \sqcap b) \sqcup (a' \sqcap b') \quad .$$

Then, aCb iff $C(a,b) = \mathbf{1}$ holds. The commutator *is defined by*

$$C(S) = \{x \in \mathfrak{L} \mid mCx \forall m \in S\} \quad .$$

D 5.26 (Centre) *The* centre *$\mathfrak{L}^c$ of an orthomodular lattice $\mathfrak{L}$ is the set of all elements commuting with all elements of $\mathfrak{L}$.*

D 5.27 (Irreducibility) *An orthomodular lattice is* irreducible *if $\mathfrak{L}^c = \{\mathbf{0}, \mathbf{1}\}$.*

Examples:

(i) The centre of every Boolean lattice is the original Boolean lattice; i.e., if A is a Boolean lattice, $A^c = A$.

(ii) Every Hilbert lattice is irreducible; i.e., $\mathfrak{C}(\mathfrak{H})^c = \{\mathbf{0}, \mathbf{1}\}$. For details, see, for instance, G. Kalmbach, *Orthomodular Lattices* [251], chapters 1& 4, or *Measures and Hilbert Lattices* [252], chapter 1.

5.3.6 Prime ideal, state

D 5.28 (Ideal) *A non void subset $\mathfrak{I}$ of an orthomodular poset $\mathfrak{P}$ is called an* ideal *if it satisfies the following conditions:*

(i) if $a \in \mathfrak{I}$ and $b \preceq a$, then $b \in \mathfrak{I}$,

(ii) if $a, b \in \mathfrak{I}$ and $a \perp b$ (i.e., $a \sqcap b = \mathbf{0}$), then $a \sqcup b \in \mathfrak{I}$.

D 5.29 (Prime ideal) *A* prime ideal *of an orthomodular poset $\mathfrak{L}$ is an ideal $\mathfrak{P}$, $\mathfrak{P} \neq \mathfrak{L}$ such that $a \perp b$ implies $a \in \mathfrak{P}$ or $b \in \mathfrak{P}$.*

D 5.30 (Prime) *An orthomodular poset $\mathfrak{L}$ is called* prime *if, for all $a, b \in \mathfrak{L}$, $a \neq b$, there exists a prime ideal $\mathfrak{P}$ of $\mathfrak{L}$ such that $a \in \mathfrak{P}$, $b \notin \mathfrak{P}$ or $a \notin \mathfrak{P}$, $b \in \mathfrak{P}$.*

Remarks:

(i) Let $\mathfrak{P}$ be a prime ideal. Then $x \in P$ or $x' \in P$;

(ii) Let $\mathfrak{P}$ be a prime ideal. aCb and $a \sqcap b \in \mathfrak{P}$ implies $a \in \mathfrak{P}$ or $b \in \mathfrak{P}$.

D 5.31 (State) *A* two-valued state *on an orthomodular poset $\mathfrak{L}$ is a mapping $s : \mathfrak{L} \rightarrow \{0, 1\}$ such that*

(i) $s(1) = 1$;

(ii) if $a \perp b$ and $a, b \in \mathfrak{L}$, then $s(a \sqcup b) = s(a) + s(b)$.

The set of all two-valued states on $\mathfrak{L}$ is denoted by $\mathfrak{S}(\mathfrak{L})$.

Remark:

Let $\mathfrak{L}$ be an orthomodular poset. Then the mapping $\varphi : \mathfrak{S}(\mathfrak{L}) \rightarrow \mathfrak{P}(\mathfrak{L})$, $\varphi(s) = \{x \in \mathfrak{L} \mid s(x) = 0\}$ is bijective [402].

5.3.7 Block pasting of orthomodular lattices

Orthomodular lattices can be decomposed into (or, conversely, composed by suitable) Boolean subalgebras. The following terminology is used.

D 5.32 (Block) *A maximal Boolean subalgebra of an orthomodular lattice is called a* block.

Informally speaking, the term "maximal" refers to the greatest possible number of atoms. The *construction* of orthomodular lattices from a union of Boolean algebras is the "inverse" problem to the task of *finding the block decomposition* (i.e., finding the *maximal* Boolean subalgebras) of a given orthomodular lattice.

D 5.33 (Pasting, $\{\mathbf{0}, \mathbf{1}\}$-pasting)

Let $\{\mathfrak{L}_i\}$ be a collection of orthomodular (Boolean) lattices such that, for all $\mathfrak{L}_i \neq \mathfrak{L}_j$, the following conditions are satisfied:

(i) $\mathfrak{L}_i \not\subset \mathfrak{L}_j$,

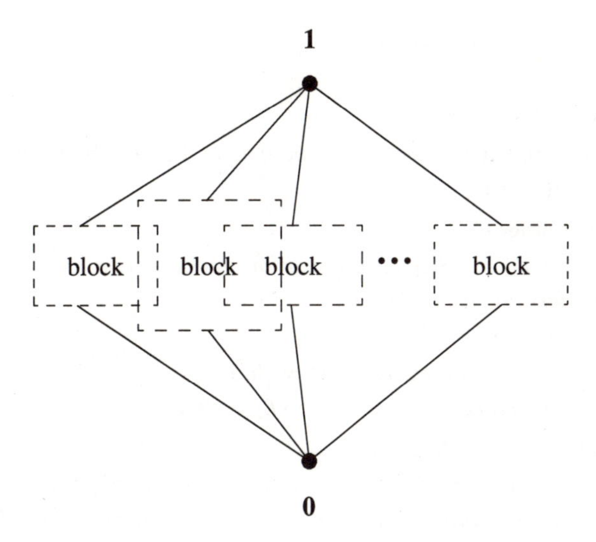

Figure 5.4: An arbitrary Hilbert lattice is a pasting of (not necessarily disjoint) Boolean subalgebras. It is irreducible.

(ii) $\mathfrak{L}_i \cap \mathfrak{L}_j$ *is a orthomodular (Boolean) sublattice, and the partial orderings and orthocomplementations of* $\mathfrak{L}_i$ *and* $\mathfrak{L}_j$ *coincide on* $\mathfrak{L}_i \cap \mathfrak{L}_j$.

The set $\mathfrak{L} = \bigcup_{\{\mathfrak{L}_i\}} \mathfrak{L}_i$ *is the* pasting *of* $\{\mathfrak{L}_i\}$. *Its partial order relation "*$\succeq_{\mathfrak{L}}$*" is defined by* $f \succeq_{\mathfrak{L}} g$ *iff, for some* $\mathfrak{L}_i$, $f \succeq_{\mathfrak{L}_i} g$. *Its orthocomplementation "*$'^{\mathfrak{L}}$*" is defined by* $f = g'^{\mathfrak{L}}$ *iff, for some* $\mathfrak{L}_i$, $f = g'^{\mathfrak{L}_i}$.

In particular, if $\mathfrak{L}_i \subset \mathfrak{L}_j = \{\mathbf{0}, \mathbf{1}\}$ *for* $i \neq j$, *then* $\mathfrak{L} = \bigcup_i \mathfrak{L}_i$ *is called the* $\{\mathbf{0}, \mathbf{1}\}$*-pasting of the* $\mathfrak{L}_i$*'s as the horizontal sum.*

T 5.34 *Every orthomodular lattice is a pasting of its blocks.*

For a detailed discussion, see G. Kalmbach, *Orthomodular Lattices* [251], chapter 4, in particular remark 12, p. 50, as well as M. Navara and V. Rogalewicz [333].

T 5.35 *Every Hilbert lattice (for a definition, see 5.47, p. 67) is an irreducible pasting of (not necessarily disjoint) blocks.*

While the intersection of *all* blocks of a Hilbert lattice contains only the two elements **0** and **1**, the intersection of two arbitrary blocks of a Hilbert lattice may contain several atoms which are common to these blocks. The pasting of its blocks forming the structure of an arbitrary Hilbert lattice is schematically drawn in Fig. 5.4.

The inverse question of whether any pasting of Boolean algebras results in an orthomodular or Hilbert lattice has been investigated by R. J. Greechie and M. Dichtl, among others. In what follows we shall introduce notations and techniques which can be used to construct orthomodular lattices from Boolean algebras. No attempt is made here to extensively review these efforts. We shall deal only with the most simple cases, i.e., with almost disjoint systems of blocks. More general pasting techniques are reviewed in G. Kalmbach, *Orthomodular Lattices* [251], chapter 4.

D 5.36 (Almost disjoint system of Boolean subalgebras)

Let $\mathfrak{B}$ *be a system of Boolean algebras.* $\mathfrak{B}$ *is* almost disjoint *if for any pair* $A, B \in \mathfrak{B}$ *at least one of the following conditions is satisfied:*

(i) $A = B$;

(ii) $A \cap B = \{\mathbf{0}, \mathbf{1}\}$;

(iii) $A \cap B = \{\mathbf{0}, \mathbf{1}, a, a'\}$, *where* a *is an atom in* A *and* B; *and* A *and* B *share the same elements* $\mathbf{0}$ *and* $\mathbf{1}$.

D 5.37 (Loop of order n)

A finite sequence $B_0, \ldots, B_{n-1}$ *of a system of blocks* $\mathfrak{B}$ *is a* loop of order n $(n \geq 3)$ *if (equality is understood as modulo* n*):*

(i) $B_i \cap B_{i+1} = \{\mathbf{0}, \mathbf{1}, a, a'\}$;

(ii) if $j \notin \{i-1, i, i+1\}$, *then* $B_i \cap B_j = \{\mathbf{0}, \mathbf{1}\}$;

(iii) for distinct indices i, j, k, $B_i \cap B_j \cap B_k = \{\mathbf{0}, \mathbf{1}\}$.

T 5.38 (Loop lemma (R. J. Greechie))

Let $\mathfrak{B} = \{B_i\}$ *be an almost disjoint system of Boolean algebras.* $\mathfrak{L} = \bigcup_{B_i \in \mathfrak{B}} B_i$ *is an orthomodular partially ordered set iff* $\mathfrak{B}$ *does not contain a loop of order 3.* $\mathfrak{L} = \bigcup_{B_i \in \mathfrak{B}} B_i$ *is an orthomodular lattice iff* $\mathfrak{B}$ *does not contain a loop of order 3 or 4.*

D 5.39 (Greechie lattice)

An orthomodular lattice is a Greechie lattice *if the following conditions are satisfied:*

(i) Every element can be written as a supremum of at least a countable number of mutually orthogonal atoms;

(ii) the collection of all blocks forms an almost disjoint subset.

A Greechie diagram consists of points "$\circ$", *representing the atoms.* Lines *linking the points (atoms) belong to the same block. Two lines are crossing in a common atom.*

For more general results on block pasting, see chapter 4 in G. Kalmbach's monography Orthomodular Lattices [251].

Examples & remarks:

(i) Greechie diagram and Hasse diagram of 2^2:

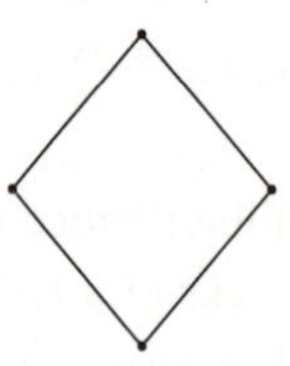

(ii) Greechie diagram and Hasse diagram of 2^3:

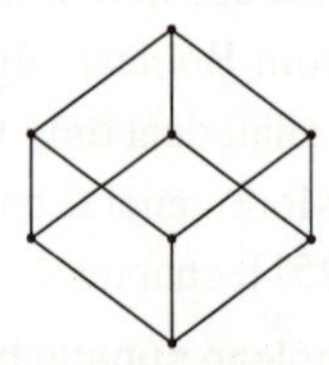

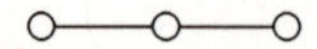

(iii) The following lattice characterised by its Greechie and Hasse diagrams is obtained by the pasting of two 2^3 with one common atom.

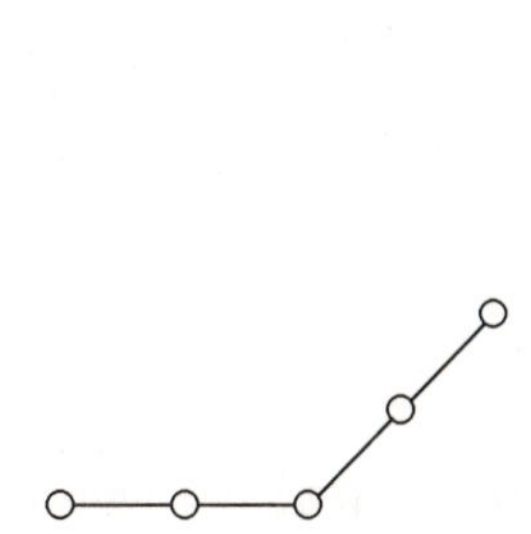

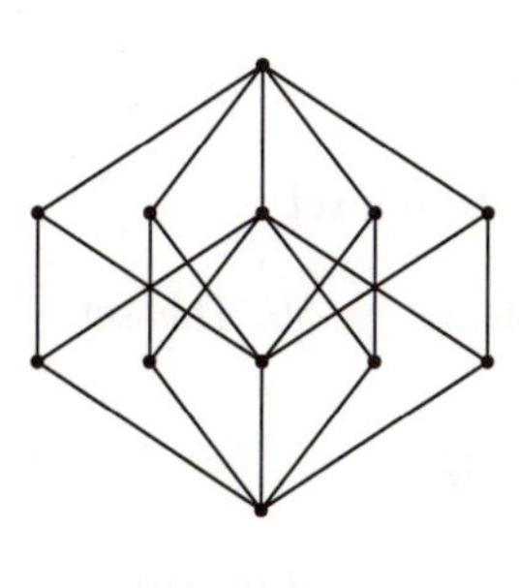

(iv) This Greechie lattice is an example of an orthomodular lattice which is not modular. It is a pasting of 2^2 and 2^3 and contains the lattice drawn in Fig. 5.2, p. 57.

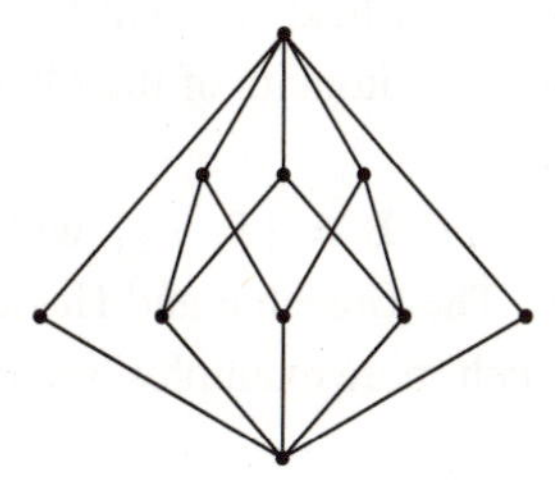

(v) Greechie diagram and Hasse diagram of an almost disjoint system of blocks of 2^3 with a loop of the order of 3. According to the loop lemma the resulting "pasted" structure is no orthomodular lattice (this can also be seen by direct inspection).

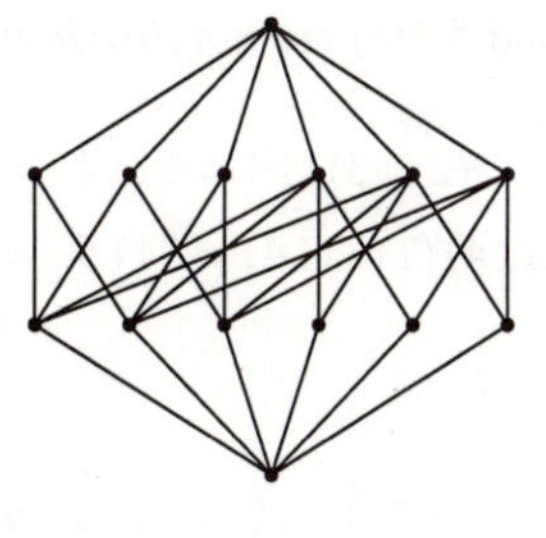

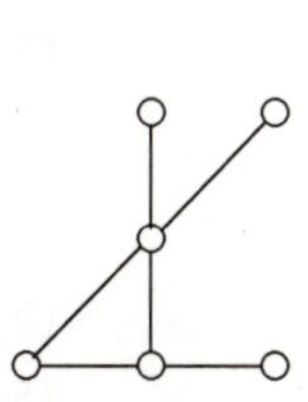

(vi) Two-dimensional case: The $\{\mathbf{0}, \mathbf{1}\}$-pasting (horizontal sum) of $\aleph_1$-many copies of 2^2 ($\aleph_1$ is the cardinality of the continuum) yields the Hilbert space $\mathfrak{C}(\mathfrak{H}_2)$, where the dimension (i.e., the maximal number of linear independent vectors of $\mathfrak{H}$) is two.

5.4 Examples

5.4.1 Set of subsets of a set

For details, see table 5.1, p. 56. This set is Boolean.

5.4.2 Partition logic

Consider a set M and a set $\mathfrak{P}$ of partitions of M. [A *partition* $P = \{m_i\}$ is a family of nonempty subsets m_i of M with the following properties: *(i)* $m_i \cap m_j = \emptyset$ or $m_i = m_j$; *(ii)* $M = \bigcup_i m_i$. Every partition $P \in \mathfrak{P}$ generates a Boolean algebra of the subsets in the partition P.] As for Boolean algebras, the partial order relation is identified with the subset relation (set theoretic inclusion) and the complement is identified with the set theoretic complement. The pasting of an arbitrary number of these Boolean algebras is called a *partition logic* (cf. definition 10.8, p. 159).

Partition logics are introduced here to identify the experimental logics of generic (finite) automata, in particular of automata of the Mealy type.

Example:

Let $M = \{1, 2, 3, 4, 5, 6\}$ and $\mathfrak{P} = \{P_1, P_2\}$ with $P_1 = \{\{1, 4, 5\}, \{2\}, \{3, 6\}\}$ and $P_2 = \{\{1, 2, 4\}, \{5\}, \{3, 6\}\}$. The Greechie and Hasse diagrams of this logic are shown in Fig. 10.11, p. 162. For much more examples, see chapter 10.

5.4.3 Greatest common divisor and least common multiplier

Identify $\mathfrak{L} = \mathbb{N}$ and $m \preceq n$ with "m devides n." Then $m \sqcap n$ is the *greatest common divisor* of m and n and $m \sqcup n$ is the *least common multiplier* of m and n. No complement is defined.

5.4.4 Lattices defined by Hasse diagrams

A lattice of five elements is represented by one of the five Hasse diagrams in Fig. 5.5. Lattices defined by 5.5(d) and 5.5(e) are *not distributive*, since for 5.5(d),

$$(2 \sqcap 3) \sqcup 4 = 1 \sqcup 4 \;=\; 4 \quad , \tag{5.18}$$

$$(2 \sqcup 4) \sqcap (3 \sqcup 4) = 5 \sqcap 5 \;=\; 5 \neq 4 \quad , \tag{5.19}$$

and for 5.5(e),

$$(3 \sqcap 4) \sqcup 2 = 1 \sqcup 2 \;=\; 2 \quad , \tag{5.20}$$

$$(3 \sqcup 2) \sqcap (4 \sqcup 2) = 3 \sqcap 5 \;=\; 3 \neq 2 \quad . \tag{5.21}$$

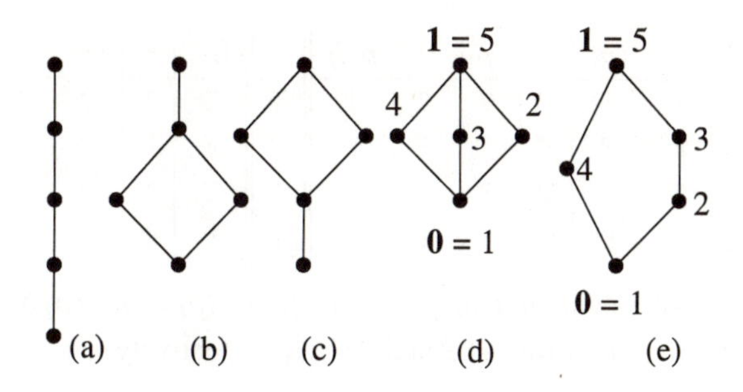

Figure 5.5: Hasse diagrams of (non isomorphic) lattices of five elements.

p_1	p_2	$(p_1 \rightarrow p_2)$	$(p_1 \wedge p_2)$	$(p_1 \vee p_2)$	$\neg p_1$	$p_1 = p_2$
TRUE	TRUE	TRUE	TRUE	TRUE	FALSE	TRUE
TRUE	FALSE	FALSE	FALSE	TRUE	FALSE	FALSE
FALSE	TRUE	TRUE	FALSE	TRUE	TRUE	FALSE
FALSE	FALSE	TRUE	FALSE	FALSE	TRUE	TRUE

Table 5.2: Truth-assignment tables for binary and unary relations in the classical propositional calculus.

5.4.5 Lattice of classical propositional calculus

Classical propositional calculus is the lattice of statements $\{p_i\}$, each one being either TRUE (exclusive) or FALSE. The unary negation $\neg$ and the binary relations and operations $\rightarrow, \wedge, \vee$ are defined by truth-assignments, enumerated in table 5.2. The identification of relations in lattice theory with relations in the propositional calculus is represented in table 5.3.

Remarks:

(i) The implication relation $p_1 \rightarrow p_2$ can be composed from the other relations $\wedge, \vee, \neg$ (and *vice versa*) by $(\neg p_1 \vee p_2)$, or by

$$p_1 = (p_1 \wedge p_2) \quad \text{or by} \tag{5.22}$$

$$p_2 = (p_1 \vee p_2) \quad \text{or by} \tag{5.23}$$

$$(p_1 = (p_1 \wedge p_2)) \vee (p_2 = (p_1 \vee p_2)) \quad . \tag{5.24}$$

lattice operation	propositional calculus
order relation $\preceq$	implication $\rightarrow$
"meet" $\sqcap$	disjunction "and" $\wedge$
"join" $\sqcup$	conjunction "or" $\vee$
"complement" $'$	negation "not" $\neg$

Table 5.3: Identification of lattice relations and operations for the classical propositional calculus.

p_1,	p_2,	$(p_1$	$=$	$(p_1$	$\wedge$	$p_2))$	$\vee$	$(p_2$	$=$	$(p_1$	$\vee$	$p_2))$
T	T	T	T	T	T	T	T	T	T	T	T	T
T	F	T	F	T	F	F	F	F	F	T	T	F
F	T	F	T	F	F	T	T	T	T	F	T	T
F	F	F	T	F	F	F	T	F	T	F	F	F

Table 5.4: Truth-assignment table of $(p_1 = (p_1 \wedge p_2)) \vee (p_2 = (p_1 \vee p_2))$. The abbreviations "T" and "F" have been used for TRUE and FALSE, respectively.

Equation (5.24) is the analogue to equation (5.1). It is verified by the truth-assignment table 5.4.

(ii) $p_1 = p_2$ can be composed as $(p_1 \wedge p_2) \vee (\neg p_1 \wedge \neg p_2)$.

(iii) The lattice defined by the propositional calculus is *distributive*, i.e., the following pairs of formulas (separated by "=") are equivalent:

$$
\begin{aligned}
p_1 \wedge (p_2 \vee p_3) &= (p_1 \wedge p_2) \vee (p_1 \wedge p_3) \\
p_1 \vee (p_2 \wedge p_3) &= (p_1 \vee p_2) \wedge (p_1 \vee p_3). \qquad (5.25)
\end{aligned}
$$

(iv) The lattice is orthocomplemented; i.e.,

$$\neg\neg p = p \qquad (5.26)$$

$$\neg p \wedge p = \mathbf{0} \qquad (5.27)$$

$$\neg p \vee p = \mathbf{1} \ . \qquad (5.28)$$

from the Intuitionist's pointof view, (5.26) would not be valid. **0** is identified with the *absurd* proposition; i.e., with the proposition which is always FALSE. **0** is identified with the *tautology* or trivial proposition; i.e., with the proposition which is always TRUE.

(v) By *(iii)* & *(iv)*, the classical propositional calculus is *Boolean.*

5.4.6 Lattice of subspaces of a Hilbert space — Hilbert lattices

Definition of Hilbert space

The Hilbert space concept shall be briefly reviewed next.

D 5.40 (Field) *A set of scalars K or $(K, +, \cdot)$ is a* field *if*

(I) to every pair a, b of scalars there corresponds a scalar $a + b$ in such a way that

- *(i)* $a + b = b + a$ (commutativity)*;*
- *(ii)* $a + (b + c) = (a + b) + c$ (associativity)*;*
- *(iii) there exists a unique* zero *element* 0 *such that* $a + 0 = a$ *for all* $a \in K$*;*
- *(iv) To every scalar a there corresponds a unique scalar $-a$ such that* $a + (-a) = 0$.

(II) to every pair a, b of scalars there corresponds a scalar ab, called product *in such a way that*

- *(i)* $ab = ba$ (commutativity)*;*
- *(ii)* $a(bc) = (ab)c$ (associativity)*;*

(iii) there exists a unique non-zero element 1, *called* one, *such that* $a1 = a$ *for all* $a \in K$;

(iv) To every non-zero scalar a *there corresponds a unique scalar* a^{-1} *such that* $aa^{-1} = 1$.

(III) $a(b+c) = ab + ac$ (distributivity).

Examples: The sets $\mathbb{Q}, \mathbb{R}, \mathbb{C}$ of rational, real and complex numbers with the ordinary sum and scalar product operators "+, ·" are fields.

D 5.41 (Linear space) *Let M be a set of objects such as vectors, functions, series et cetera. A set M is a* linear space *if*

(I) there exists the operation of (generalised) "addition," denoted by "+" obeying

(i) $f + g \in M$ *for all* $f, g \in M$ (closedness *under "addition")*;

(ii) $f + g = g + f$ (commutativity);

(iii) $f+(g+h)=(f+g)+h=f+g+h$ (associativity);

(iv) there exists a neutral element $\mathbf{0}$ *for which* $f + \mathbf{0} = f$ *for all* $f \in M$;

(v) for all $f \in M$ *there exists an* inverse $-f$, *defined by* $f + (-f) = \mathbf{0}$;

(II) there exists the operation of (generalised) "scalar multiplication" with elements of the field $(K, +, \cdot)$ *obeying*

(vi) $a \in K$ *and* $f \in M$ *then* $af \in M$ (closedness *under "scalar multiplication")*;

(vii) $a(f + g) = af + ag$ *and* $(a + b)f = af + bf$ (distributive laws);

(viii) $a(bf) = (ab)f = abf$ (associativity);

(ix) There exists a "scalar unit element" $1 \in K$ *for which* $1f = f$ *for all* $f \in M$.

Examples:

(i) vector spaces $M = \mathbb{R}^n$ with $K = \mathbb{R}$ or $\mathbb{C}$;

(ii) $M = \ell_2, K = \mathbb{C}$, the space of all infinite sequences

$$\ell_2 = \{f \mid f = (x_1, x_2, \ldots, x_i, \ldots),\ x_i \in \mathbb{C},\ \sum_{i=1}^{\infty} |x_i|^2 < \infty\} \quad ,$$

(iii) the space of continuous functions, complex-valued (real-valued) functions $M = C(a, b)$ over an open or closed interval (a, b) or $[a, b]$ with $K = \mathbb{C}$ $(K = \mathbb{R})$;

D 5.42 (Metric, norm, inner product)

A metric, *denoted by* dist, *is a binary function which associates a distance of two elements of a linear vector space and which satisfies the following properties:*

(i) $\text{dist}(f, g) \in \mathbb{R}$ *for all* $f, g \in M$;

(ii) $\text{dist}(f, g) = 0 \iff f = g$;

(iii) $\text{dist}(f, g) \leq \text{dist}(f, h) + \text{dist}(g, h)$ *for all* $h \in M$ *and every pair* $f, g \in M$.

A norm $\|\cdot\|$ *on a linear space M is a unary function which associates a real number to every element of M and which satisfies the following properties:*

(i) $\|f\| \geq 0$ *for all* $f \in M$;

(ii) $\|f\| = 0 \iff f = \mathbf{0}$;

(iii) $\|f + g\| \leq \|f\| + \|g\|$;

(iv) $\|af\| = |a|\,\|f\|$ *for all* $a \in K$ *and* $f \in M$ (homogeneity).

An inner product $\langle \cdot \mid \cdot \rangle$ *is a binary function which associates a complex number with every pair of elements of a linear space M and satisfies the following properties (* denotes complex conjugation):*

(i) $\langle f \mid g \rangle = \langle g \mid f \rangle^*$ *for all* $f, g \in M$*;*

(ii) $\langle f \mid ag \rangle = a\langle f \mid g \rangle$ *for all* $f, g \in M$ *and* $a \in K$*;*

(iii) $\langle f \mid g_1 + g_2 \rangle = \langle f \mid g_1 \rangle + \langle f \mid g_2 \rangle$ *for all* $f, g_1, g_2 \in M$*;*

(iv) $\langle f \mid f \rangle \geq 0$ *for all* $f \in M$*;*

(v) $\langle f \mid f \rangle = 0 \Longleftrightarrow f = \mathbf{0}$.

Remarks:

(i) With the identifications

$$\|f\| = \langle f|f \rangle^{1/2} \quad , \tag{5.29}$$

$$\mathrm{dist}(f, g) = \|f - g\| \quad , \tag{5.30}$$

features & structures are inherited in the following way:

$$M \text{ has an inner product } \underset{\not\Leftarrow}{\Rightarrow} M \text{ has a norm } \underset{\not\Leftarrow}{\Rightarrow} M \text{ has a metric.}$$

(ii) The *Schwartz inequality*

$$|\langle f \mid g \rangle| \leq \|f\| \|g\| \tag{5.31}$$

is satisfied.

D 5.43 (Separability, completeness)

A linear space M is separable *if, for any* $f \in M$ *and any* $\varepsilon > 0$, *there exists at least one element* f_i *of a sequence* $\{f_n \mid n \in \mathbb{N},\ f_n \in M\}$ *such that*

$$\|f - f_i\| < \varepsilon \quad .$$

A linear space M is complete *if any sequence* $\{f_n \mid n \in \mathbb{N},\ f_n \in M\}$ *with the property*

$$\lim_{i,j \to \infty} \|f_i - f_j\| = 0$$

defines a unique limit $f \in M$ *such that*

$$\lim_{i \to \infty} \|f - f_i\| = 0 \quad .$$

D 5.44 (Hilbert space, Banach space)

A Hilbert space $\mathfrak{H}$ *is a* linear space, *equipped with an* inner product, *which is* separable & complete.

A Banach space *is a* linear space, *equipped with a* norm, *which is* separable & complete.

Example:

$\ell_2, \mathbb{C}$ [see linear space example *(ii)*] with $\langle f \mid g \rangle = \sum_i x_i^* y_i$.

D 5.45 (Subspace, orthogonal subspace)

A subspace $\mathfrak{S} \subset \mathfrak{H}$ *of a Hilbert space is a subset of* $\mathfrak{H}$ *which is closed under scalar multiplication and addition, i.e.,* $f, g \in H,\ a \in K \Rightarrow af \in \mathfrak{S},\ f + g \in \mathfrak{S}$, *and which is separable and complete.*

lattice operation	Hilbert space operation
order relation $\preceq$	subspace relation $\subset$
“meet” $\sqcap$	intersection of subspaces $\cap$
“join” $\sqcup$	closure of subspace spanned by subspaces $\oplus$
“orthocomplement” $'$	orthogonal subspace $\perp$

Table 5.5: Identification of lattice relations and operations for Hilbert lattices.

An orthogonal subspace $\mathfrak{S}^\perp$ *of* $\mathfrak{S}$ *is the set of all elements in the Hilbert space* $\mathfrak{H}$ *which are orthogonal to elements of* $\mathfrak{S}$*, i.e.,*

$$\mathfrak{S}^\perp = \{f \mid f \in \mathfrak{H},\ \langle f \mid g\rangle = 0, \text{for all } g \in \mathfrak{S}\}.$$

Remarks:

(i) $(\mathfrak{S}^\perp)^\perp = \mathfrak{S}^{\perp\perp} = \mathfrak{S}$;

(ii) every orthogonal subspace is a subspace;

(iii) A Hilbert space can be represented as a direct sum of orthogonal subspaces.

Compatibility

D 5.46 (Compatibility) *Two subspaces* $\mathfrak{S}_1$ *and* $\mathfrak{S}_2$ *of a Hilbert space are called* compatible, *denoted by* $\mathfrak{S}_1 \leftrightarrow \mathfrak{S}_2$*, if*

$$(\mathfrak{S}_1 \sqcap \mathfrak{S}_2) \sqcup (\mathfrak{S}_1 \sqcap \mathfrak{S}_2^\perp) = \mathfrak{S}_1 \quad .$$

Remarks:

(i) $\mathfrak{S}_1 \leftrightarrow \mathfrak{S}_2 \Longleftrightarrow \mathfrak{S}_1 \sqcup (\mathfrak{S}_2 \sqcap \mathfrak{S}_1^\perp) = \mathfrak{S}_1 \sqcup \mathfrak{S}_2 = \mathfrak{S}_2 \sqcup (\mathfrak{S}_1 \sqcap \mathfrak{S}_2^\perp)$.

(ii) The relation $\leftrightarrow$ is symmetric, i.e., $\mathfrak{S}_1 \leftrightarrow \mathfrak{S}_2 \Longleftrightarrow \mathfrak{S}_2 \leftrightarrow \mathfrak{S}_1$.

Definition of Hilbert lattices

D 5.47 (Hilbert lattice)

A Hilbert lattice *is the lattice of all closed subspaces of a Hilbert space* $\mathfrak{H}$*; it is denoted by* $\mathfrak{C}(\mathfrak{H})$*. The “meet”* $\sqcap$ *is identified with the closure of the linear span, the “join”* $\sqcup$ *is identified with the set theoretic union and the “complement” of a subspace with its orthogonal subspace.*

The identification of relations and operations in lattice theory with relations and operations in Hilbert space is represented in table 5.5.

Remarks:

(i) $\mathfrak{C}(\mathfrak{H})$ is an orthocomplemented lattice.

(ii) In general, $\mathfrak{S}$ is not distributive. Let for instance $\mathfrak{S}'$, $\mathfrak{S}$, $\mathfrak{S}^\perp$ be subsets of a Hilbert space $\mathfrak{H}$ with $\mathfrak{S}' \neq \mathfrak{S}$, $\mathfrak{S}' \neq \mathfrak{S}^\perp$, then (see Fig. 5.6, drawn from J. M. Jauch [244], p. 27)

$$\mathfrak{S}' \sqcap (\mathfrak{S} \sqcup \mathfrak{S}^\perp) = \mathfrak{S}' \sqcap \mathfrak{H} \;=\; \mathfrak{S}', \text{ whereas} \tag{5.32}$$

$$(\mathfrak{S}' \sqcap \mathfrak{S}) \sqcup (\mathfrak{S}' \sqcap \mathfrak{S}^\perp) = \mathbf{0} \sqcup \mathbf{0} \;=\; \mathbf{0} \quad . \tag{5.33}$$

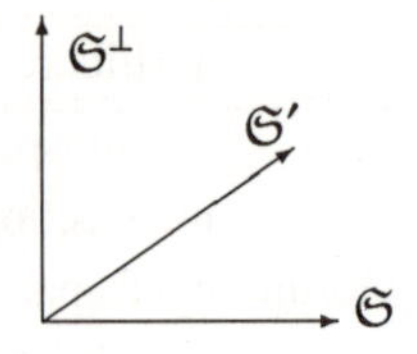

Figure 5.6: Demonstration of the nondistributivity of Hilbert lattices.

(iii) A finite dimensional Hilbert lattice is modular.

(iv) Since Hilbert lattices are orthomodular lattices, they can be constructed by the pasting of blocks (blocks are maximal Boolean subalgebras); the blocks need not be (almost) disjoint. This fact will be used for the construction of automata yielding arbitrary finite subalgebras of Hilbert lattices as propositional calculi. See chapter 10, p. 127 for details.

(v) In Hilbert lattices, the *orthoarguesian law* is satisfied. For a definition and details, see J. R. Greechie [211], G. Kalmbach [251] and R. Giuntini [201], p. 138. The *orthoarguesian law* is *not* satisfied by general orthomodular lattices. I.e.,

$$\text{finite orthomodular subalgebra of Hilbert lattice} \underset{\not\Leftarrow}{\Rightarrow} \text{orthomodular lattice} \quad . \qquad (5.34)$$

Algebraic characterisation of Hilbert lattices

An infinite dimensional $\mathfrak{C}(\mathfrak{H})$ is a complete, atomic, irreducible, orthomodular lattice satisfying the exchange axiom (i.e., if a covers $a \sqcap b$ then $a \sqcup b$ covers b). In some notations these criteria are the *defining features* of Hilbert lattices (e.g., G. Kalmbach, *Measures and Hilbert Lattices* [252], p. 11). Yet, they are also satisfied by lattices of Keller spaces [262, 252, 253]; Keller spaces are different from Hilbert spaces in important aspects.

A complete axiomatisation for lattices of separable complex Hilbert spaces has been given by W. J. Wilbur [460]; see also the review by R. Piziak [365]:

T 5.48 ((W. J. Wilbur [460])) *A lattice $\mathfrak{L}$ is isomorphic to a complex Hilbert lattice iff it satisfies the following conditions (i)-(vii).*

(i) $\mathfrak{L}$ has at least one point;

(ii) the length $\|\mathfrak{L}\| = \aleph_0$ (for a definition of length, see 5.9, p. 53);

(iii) If a and $a^\perp$ are nonzero elements of $\mathfrak{L}$ and $x \in \mathfrak{L}$, then there exist points $y, z \in \mathfrak{L}$ with $y < a$, $z < a^\perp$ and $x < y \sqcup z$;

(iv) If $a, b, c \in \mathfrak{L}$ and if $a \sqcup b$ is less than the sum of a finite set of points in $\mathfrak{L}$, and if $c < a$, then $a \sqcap (b \sqcup c) = (a \sqcap b) \sqcup c$;

(v) If $x, y \in \mathfrak{L}$ and $x \neq y$, then there is is a distinct third point $x \neq z \neq y$, $z \in \mathfrak{L}$ with $z < x \sqcup y$.

(vi) Given any four distinct points $u, v, w, x \in \mathfrak{L}$ with $u \sqcup v = w \sqcup x$, then there exist points $y, z \in \mathfrak{L}$ with $y \not< (v \sqcup z)$, $y \not< (w \sqcup x)$, $z \not< (w \sqcup x)$, $z \not< (u \sqcup y)$, such that $\|w \sqcup x \sqcup y \sqcup z\| = 2$ and $(u \sqcup y) \sqcap (z \sqcup x) < [(v \sqcup z) \sqcap (w \sqcup y)] \sqcup [(u \sqcup v) \sqcap (z \sqcup y)]$;

lattice operation	lattice of projections
order relation $E_1 \preceq E_2$	$E_1E_2 = E_1$
"meet" $\sqcap$	$\lim_{n\to\infty}(E_1E_2)^n$
"join" $\sqcup$	$E_1 + E_2 - E_1E_2$
"orthocomplement" $'$	orthogonal subspace

Table 5.6: Identification of lattice relations and operations on the lattice of projection operators $\mathfrak{P}(\mathfrak{H})$.

(vii) (Pappus' theorem) If $u, v, w, x, y, z \in \mathfrak{L}$ *are six distinct points with* $\|u \sqcup v \sqcup w \sqcup x \sqcup y \sqcup z\| = 2$ *and if* $(u \sqcup v) \sqcap (w \sqcup z) \sqcap (y \sqcup z) \neq 0$ *and if* $(x \sqcup y) \sqcap (z \sqcup u) \sqcap (v \sqcup w) \neq 0$, *then* $(u \sqcup x) \sqcap (v \sqcup y) \sqcap (w \sqcup z) \neq 0$.

However, as already conceded by W. J. Wilbur, the above characterisation, in particular axiom *(ii)* and countable completeness, is not purely algebraic. One may ask if it is possible to develop an axiomatisation of Hilbert lattices in *purely algebraic* terms. An answer to this question is unknown. For related discussions, see, for instance, G. Takeuti [440] and G. Kalmbach [250, 251]. It might be conjectured that *there is no recursive enumeration of the axioms of Hilbert lattices* [290].

5.4.7 The algebra of projections

D 5.49 (Projection) *A* projection *is an operator E defined on a Hilbert space* $\mathfrak{H}$ *which is self-adjoint and idempotent, i.e.,*

$$E = E^\dagger = E^2 \quad .$$

There is an isomorphism between the set of projections, denoted by $\mathfrak{P}(\mathfrak{H})$ and all closed subspaces $\mathfrak{C}(\mathfrak{H})$ of $\mathfrak{H}$: given a projection E, the corresponding subspace is $\mathfrak{S} = E(\mathfrak{H})$; given a closed subspace $\mathfrak{S}$, any $f \in \mathfrak{H}$ can be decomposed uniquely as a sum $f = g + h$, where $g \in \mathfrak{S}$ and $h \in \mathfrak{S}^\perp$; the projection corresponding to $\mathfrak{S}$ is then the operator E determined by $Ef = g$. This one-to-one correspondence allows a translation of the lattice structure of the subsets of Hilbert space discussed before into the algebra of projections. $\mathfrak{S}_1 \leftrightarrow \mathfrak{S}_2 \Leftrightarrow E_1E_2 = E_2E_1$, where E_i is the projection onto $\mathfrak{S}_i$, $i = 1, 2$. (This is a reason why quantum logic is also said to correspond to a non-commutative probability theory.)

The identification of relations and operations in lattice theory with the relations and operations in the lattice of projections is represented in table 5.6 ([244], p. 37).

5.4.8 Lattice of quantum propositional calculus

In what follows the mathematical formalism of quantum mechanics is very briefly reviewed. No attempt is being made to give a complete set of axioms. See also, for instance, J. von Neumann [337], A. Messiah [319], L. E. Ballentine [20]. Its primitive concepts are those of *state* and *observable*. An *observable* is represented by a self-adjoint operator

O on a Hilbert space $\mathfrak{H}$. In the *spectral representation,* O can be written as

$$O = \sum_n r_n E_n \quad , \tag{5.35}$$

where E_n are orthogonal projection operators and r_n are the corresponding eigenvalues.

The projections E_n correspond to the physical properties of a quantum system. In J. von Neumann's words ([337], p. 249; in our notation, $\mathfrak{E}$ is a proposition):

Apart from the physical quantities $\mathfrak{R}$, there exists another category of concepts that are important objects of physics — namely the properties of the states of the system S. Some such properties are: that a certain quantity $\mathfrak{R}$ takes the value λ — or that the value of $\mathfrak{R}$ is positive — ⋯

To each property $\mathfrak{E}$ we can assign a quantity which we define as follows: each measurement which distinguishes between the presence or absence of $\mathfrak{E}$ is considered as a measurement of this quantity, such that its value is 1 if $\mathfrak{E}$ is verified, and zero in the opposite case. This quantity which corresponds to $\mathfrak{E}$ will also be denoted by $\mathfrak{E}$.

Such quantities take only the values of 0 and 1, and conversely, each quantity $\mathfrak{R}$ which is capable of these two values only, corresponds to a property $\mathfrak{E}$ which is evidently this: "the value of $\mathfrak{R}$ is $\neq 0$." The quantities $\mathfrak{E}$ that correspond to the properties are therefore characterized by this behavior.

That $\mathfrak{E}$ takes on only the values 0, 1 can also be formulated as follows: Substituting $\mathfrak{E}$ into the polynomial $F(\lambda) = \lambda - \lambda^2$ makes it vanish identically. If $\mathfrak{E}$ has the operator E, then $F(\mathfrak{E})$ has the operator $F(E) = E - E^2$, i.e., the condition is that $E - E^2 = 0$ or $E = E^2$. In other words: the operator E of $\mathfrak{E}$ is a projection.

The projections E therefore correspond to the properties $\mathfrak{E}$ (through the agency of the corresponding quantities $\mathfrak{E}$ which we just defined). If we introduce, along with the projections E, the closed linear manifold $\mathfrak{M}$, belonging to them ($E = P_{\mathfrak{M}}$), then the closed linear manifolds correspond equally to the properties of $\mathfrak{E}$.

In the case of nondegenerate eigenvalues r_n and eigenvectors $|n\rangle$, O can be written as

$$E_n = |n\rangle\langle n| \quad . \tag{5.36}$$

For continuous spectra, the sum (5.35) becomes an integral. A *state* is represented by a self-adjoint, positive definite operator ρ of unit trace. I.e., in terms of nondegenerate eigenvectors $|n\rangle$ and eigenvalues ρ_n, ρ can be written as

$$\rho = \sum_n \rho_n |n\rangle\langle n| \tag{5.37}$$

with $0 \le \rho_n \le 1$ and $\sum_n \rho_n = 1$. A *pure state* is defined by the condition $\rho^2 = \rho$; in this case the sum in (5.37) reduces to a single contributing term such that $\rho = |n\rangle\langle n|$, with $\rho_n = 1$ and all $\rho_{n'\neq n} = 0$. Pure states can be represented by normalised vectors $|\psi\rangle \propto |n\rangle$ in the Hilbert space. The observable O can only take on its eigenvalues r_n. The *average value* of an observable O in the state ρ is given by

$$\langle O\rangle = \mathrm{Trace}(\rho O) \quad . \tag{5.38}$$

In the case of a pure state representable by the Hilbert space vector ψ and nondegenerate eigenvalues of O, the probability is given by

$$|\langle\psi|n\rangle|^2 \tag{5.39}$$

with the associated eigenvalue r_n. The Hilbert space $\mathfrak{H}$ is the closure of the direct sum of subspaces. The dynamic law for the state can be written in the form

$$\rho(t) = U\rho(0)U^{-1} \quad . \tag{5.40}$$

G. Birkhoff and J. von Neumann suggested [42], that, roughly speaking, the "logic of quantum events" — or, by another wording, *quantum logic* or the *quantum propositional calculus* — should be obtainable from the formal representation of physical properties. They conjectured that the Hilbert space formalism of quantum mechanics [337] is an appropriate theory of quantum events. Since, in this formalism, projection operators correspond to the physical properties of a quantum system, quantum logics is modelled in order to be isomorphic to the lattice of projections $\mathfrak{P}(\mathfrak{H})$ of the Hilbert space $\mathfrak{H}$, which in turn is isomorphic to the lattice $\mathfrak{C}(\mathfrak{H})$ of the set of subspaces of a Hilbert space. I.e., by assuming the physical validity of the quantum Hilbert space formalism, the corresponding isomorphic logical structure is investigated. Since, in this approach, quantum theory comes first and the logical structure of the phenomena are derived by analysing the theory, this could be considered a *"top-down"* method. In the case of the automata propositional calculus (see below, chapter 10, p. 127) one proceeds *"bottom-up"*, i.e., by analysing the structure of elementary processes first and conjecturing a corresponding linear space structure afterwards.

The order relation $p_1 \preceq p_2$ is identified with "whenever p_1 is true it follows that p_2 is true, too." It is also written as "$p_1 \rightarrow p_2$."

The proposition $p_1 \wedge p_2$ will be identical to the "and" operator of the ordinary (classical) propositional calculus; i.e.,

$$\text{"}p_1 \text{ TRUE"} \text{ and } \text{"}p_2 \text{ TRUE"} \Longleftrightarrow \text{"}p_1 \wedge p_2 \text{ TRUE"}. \tag{5.41}$$

The same applies to the terms $\neg p_1$.

In J. M. Jauch's interpretation, $p_1 \sqcap p_1$ is realised by an infinite sequence of alternating pairs of filters for the propositions p_1 and p_2, respectively. The proposition $p_1 \sqcap p_1$ is TRUE if the system passes this filter, and it is not true otherwise.

The "or" proposition $p_1 \vee p_2$ satisfies only the following relation:

$$\text{"}p_1 \text{ TRUE"} \text{ or } \text{"}p_2 \text{ TRUE"} \Rightarrow \text{"}p_1 \vee p_2 \text{ TRUE"}. \tag{5.42}$$

Because of (5.42), the lattice structure does not imply the distributive laws. Indeed, since $\mathfrak{C}(\mathfrak{H})$ is non distributive, the quantum propositional calculus has to be non distributive as well. G. Birkhoff and J. von Neumann [42] suggested the weaker *modular identity*

$$f \sqcup (g \sqcap h) = (f \sqcup g) \sqcap h \quad , \tag{5.43}$$

generic lattice	order relation $\preceq$	"meet" $\sqcap$	"join" $\sqcup$	"complement" $'$
lattice of subsets of a set	subset $\subset$	intersection $\cap$	union $\cup$	complement
propositional calculus	implication $\rightarrow$	disjunction "and" $\wedge$	conjunction "or" $\vee$	negation "not" $\neg$
Hilbert lattice	subspace relation $\subset$	intersection of subspaces $\cap$	closure of subspace linear span $\oplus$	orthogonal subspace $\perp$
lattice of projection operators	$E_1E_2 = E_1$	$\lim_{n\rightarrow\infty}(E_1E_2)^n$	$E_1 + E_2 - E_1E_2$ if E_1, E_2 commute	orthogonal subspace

Table 5.7: Comparison of the identifications of lattice relations and operations for the lattices of subsets of a sets, for experimental propositional calculi, for Hilbert lattices, and for lattices of projection operators.

instead of (5.25), thereby restricting the quantum propositional calculus to modular, orthocomplemented lattices. Since the modular law holds only in the case of finite dimensional Hilbert spaces, it has been proposed to study the *orthomodular law* instead

$$f \preceq g \Rightarrow f \sqcup (f' \sqcap g) = g \quad , \tag{5.44}$$

where f and g are projection operators. This law holds in all Hilbert spaces.

5.4.9 Comparison

Table 5.7 lists the identifications of relations of operations of various lattice types.

Chapter 6

Extrinsic-intrinsic concept

Epistemologically, the *intrinsic/extrinsic* concept, or, by another naming [392, 393], the *endophysics/exophysics* concept, is related to the question of how a mathematical or a logical or an algorithmic universe is perceived from within/from the outside. The physical universe (in O. E. Rössler's *dictum,* the "Cartesian prison"), by definition, can be perceived from within only.

Extrinsic or *exophysical* perception can be conceived as a hierarchical process, in which the system under observation and the experimenter form a two-level hierarchy. The system is laid out and the experimenter peeps at every relevant feature of it without changing it. The restricted entanglement between the system and the experimenter can be represented by a one-way information flow from the system to the experimenter; the system is not affected by the experimenter's actions. (Logicians might prefer the term *meta* over *exo*.)

Intrinsic or *endophysical* perception can be conceived as a non-hierarchical effort. The experimenter is part of the universe under observation. Experiments use devices and procedures which are realisable by internal resources, i.e., from within the universe. The total integration of the experimenter in the observed system can be represented by a two-way information flow, where "measurement apparatus" and "observed entity" are interchangeable and any distinction between them is merely a matter of intent and convention. Endophysics is limited by the self-referential character of any measurement. An intrinsic measurement can often be related to the paradoxical attempt to obtain the "true" value of an observable while — through interaction — it causes "disturbances" of the entity to be measured, thereby changing its state. Among other questions one may ask, *"what kind of experiments are intrinsically operational and what type of theories will be intrinsically reasonable?"*

Imagine, for example, some artificial intelligence living in a (hermetic) cyberspace. This agent might develop a "natural science" by performing experiments and developing theories. It is tempting to speculate that also a figure in a novel, imagined by the poet and the reader, is such an agent.

Since in a virtual reality only *syntactic* structures are relevant, one might wonder if concerns of this agent about its "hardware basis," e.g., whether it is "made of" billiard balls, electric circuits, mechanical relays or nerve cells, are mystic or even possible (cf. H. Putnam's brain-in-a-tank analysis [377]). I do not think this is necessarily so, in particular if the agent could influence some features of this hardware basis. One example is a hardware damage caused by certain computer viruses by "heating up" computer components such as storage or processors. I would like to call this type of "back-reaction" of a virtual reality on its computing agent *"virtual backflow interception."* Intrinsic phenomenologically, the virtual backflow could manifest itself by some violation of a "superselection rule;" i.e., by some virtual phenomenon which violates the fundamental laws of a virtual reality, such as symmetry & conservation principles.

No attempt is made here to (re-)write a comprehensive history of related concepts; but a few hallmarks are mentioned without claim of completeness. Historically, Archimedes conceived *"points outside the world, from which one could move the earth."* Archimedes' use of "points outside the world" was in a mechanical rather than in a metatheoretical context: he claimed to be able to move any given weight by any given force, however small. The 18'th century physicist B. J. Boscovich realised that it is not possible to measure motions or transformations if the whole world, including all measurement apparata and observers therein, becomes equally affected by these motions or transformations (cf. O. E. Rössler [393], p. 143). Fiction writers informally elaborated consequences of intrinsic perception. E. A. Abbot's *Flatland* describes the life of two- and onedimensional creatures and their confrontation with higher dimensional phenomena. The *Freiherr von Münchhausen* rescued himself from a swamp by dragging himself out by his own hair. Among contemporary science fiction authors, D. F. Galouye's *Simulacron Three* and St. Lem's *Non Serviam* study some aspects of artificial intelligence in what could be called "cyberspaces." Media artists such as Peter Weibel [197] create virtual realities and are particularly concerned about the *interface* between "reality" and "virtual reality," both practically and philosophically. On the forefront of interface designs are *cochlear implants*, which restore some degree of hearing in clinical patients with severe hearing impairment [123]. Finally, by outperforming television & computer games, commercial virtual reality products might become very big business. From these examples it can be seen that concepts related to intrinsic perception may become fruitful for physics, the computer sciences, business and the arts as well.

Already in 1950 (19 years after the publication of Gödel's incompleteness theorems), K. Popper has questioned the completeness of self-referential perception of "mechanic" computing devices [366]. Popper used techniques similar to Zeno's paradox (which he called "paradox of Tristram Shandy") and "Gödelian sentences" to argue for a kind of "intrinsic indeterminism."

In a pioneering study on the theory of (finite) automata, E. F. Moore has presented *Gedanken-experiments on sequential machines* [328]. There, E. F. Moore investigated automata featuring, at least to some extend, similarities to the quantum mechanical uncertainty principle. In the book *Regular Algebra and Finite Machines* [125], J. H. Conway has developed these ideas further from a formal point of view without relating them to physical applications. Probably the best review of experiments on Moore-type automata can be found in W. Bauer's book *Automatentheorie* [28] (in German).

D. Finkelstein [177, 178] has considered Moore's findings from a more physical point of view, introducing an "experimental logic of automata" and the term *"computational complementarity."* An illuminating account on endophysics topics can be found in O. E. Rössler's article on *Endophysics* [392], as well as in his book *Endophysics* (in German) [393]; O. E. Rössler is a major driving force in this area. H. Primas has considered "endophysical" and "exophysical" entities [374, 375], which, very roughly speaking, correspond to E. O. Rössler's and the author's terminology by the exchange "exo (extrinsic) $\leftrightarrow$ endo (intrinsic)." H. Primas' approach is "top-down," i.e., theoretical models form an ("Platonic") exo-world, in which the observed system is perceived by an endo-description (cf. D. Finkelstein [177, 178]). These concepts should not be confused with the present analysis, which is "bottom-up" and procedural; i.e., which concentrates on the specification of the measurement act, in particular "from within" a system. A forthcoming collection of articles on related topics is edited by H. Atmanspacher and G. Dalenoort [15].

The terms *"intrinsic"* and *"extrinsic"* appear in the author's studies on intrinsic time scales in arbitrary dispersive media [428, 432, 429], very much (unknowingly) in the spirit of B. J. Boscovich. There, the intrinsic-extrinsic concept has been re-invented (probably for the 100'th time, and, I solemnly swear) independently. It is argued that, depending on dispersion relations, creatures in a "dispersive medium" would develop a theory of coordinate transformation very similar to relativity theory. Another proposal by the author was to consider a new type of "dimensional regularisation" by assuming that the space-time support of (quantum mechanical) fields is a fractal [431]. In this approach one considers a fractal space-time of Hausdorff dimension $D = 4-\varepsilon$, with $\varepsilon \ll 1$, which is embedded in a space of higher dimension, e.g., $\mathbb{R}^{n\geq 4}$. Intrinsically, the (fractal) space-time is perceived "almost" as the usual fourdimensional space.

Besides such considerations, J. A. Wheeler [457], among others, has emphasised the role of *observer-participancy.* In the context of what is considered by the Einstein-Podolsky-Rosen argument [154] as "incompleteness" of quantum theory, A. Peres and W. H. Zurek [354, 355] and J. Rothstein [396] have attempted to relate quantum complementarity to Gödel-type incompleteness.

In what follows, the intrinsic-extrinsic concept will be made precise in an *algorithmic* context, thereby closely following E. F. Moore [328]. The main reason for the algorithmic approach is that algorithmic universes (or, equivalently, formal systems) are the royal road to the study of undecidability. The intrinsic-extrinsic concept will be applied to investigate *computational complementarity* (chapter 10, p. 127) and *intrinsic indeterminism* (chapter 12, p. 181); both again in the algorithmic context. Other tasks, such as the setting of space-time coordinates [428, 432, 429], may require other specifications, in particular of the interface.

6.1 Gedankenexperiments on finite automata

In a groundbreaking study [328], Edward Moore analysed two kinds of *Gedankenexperiments* on finite automata, which will be slightly adapted for the present purposes. In both cases, the automaton, or, in a different context, an arbitrary physical system, is treated as a "black box" in the following sense:

(i) only the input and output terminals of the automaton are accessible. The experimenter is allowed to perform experiments *via* these interfaces in the form of stimulating

the automaton with input sequences and receiving output sequences from the automaton. The experimenter is not permitted to "open up" the automaton, but

(ii) the transition and output table (diagram) of the automaton (in its reduced form) is known to the experimenter (or, if you prefer, is given to the experimenter by some "oracle").

The most important problem, among others, is the *distinguishing problem:* it is known that an automaton is in one of a particular class of internal states: find that state.

In the first kind of experimental situation, only a *single* copy of the automaton is accessible to the experimenter. The second type of experiment operates with an *arbitrary number* of automaton copies. Both cases will be discussed in detail below.

If the input is some *predetermined* sequence, one may call the experiment a *preset experiment.* If, on the other hand, (part of) the input sequence depends on (part of) the output sequence, i.e., if the input is *adapted* to the reaction of the automaton, one may call the experiment an *adaptive experiment.* We shall be mostly concerned with preset experiments, yet adaptive experiments can be used to solve certain problems with automaton propositional calculi; see chapter 10, p. 137.

Research along these lines has been pursued by S. Ginsburg [200], A. Gill [199], J. H. Conway [125] and W. Brauer [28].

6.1.1 Single-automaton configuration

In the first kind of Gedankenexperiment, only *one single* automaton copy is presented to the experimenter. The problem is to determine the initial state of the automaton, provided its transition and output functions are known (distinguishing problem). In a typical experiment, the automaton is "feeded" with a sequence of input symbols and responds by a sequence of output symbols. An input-output analysis then reveals information about the automaton's original state.

Assume for the moment that such an experiment induces a state transition of the automaton. I.e., after the experiment, the automaton is not in the original initial state. In this process a loss of potential information about the automaton's initial state may occur. In other words: certain measurements, while measuring some particular feature of the automaton, may make impossible the measurement of other features of the automaton. This irreversible change of the automaton state is one aspect of the "observer-participancy" in the single-automaton configuration. (This is not the case for the multi-automaton situation discussed below, since the availability of an arbitrary number of automata ensures the possibility of an arbitrary number of measuring processes.)

In developing the intrinsic concept further, the automaton and the experimenter are "placed" into a *single* "meta"-automaton. One might think of the experimenter as of a human being or an automaton. Thereby, a theoretical modelling pursued by the experimenter, such as a(n) (algorithmic) description of the automaton *et cetera*, is placed *inside* the experimenter, or at least at the same (hierarchical) level as the experimenter. If the experimenter reacts mechanically, the setup can be readily constructed by simulating both the original finite deterministic "black box" automaton as well as the experimenter and their interplay by a universal automaton. One can imagine such a situation as one sub-

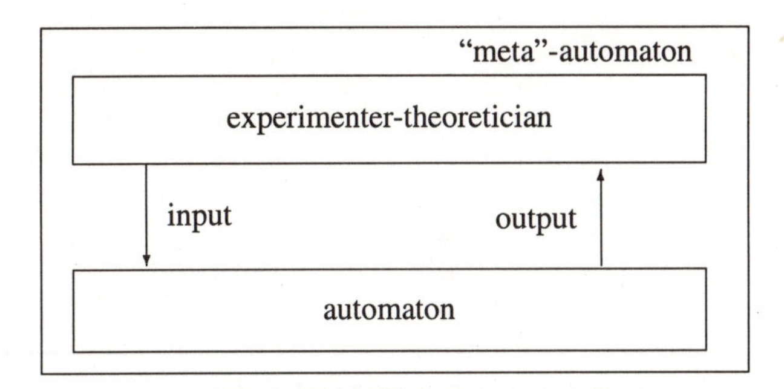

Figure 6.1: Schematic diagram of an experiment on a single automaton, taking place in a "meta"-automaton.

program checking another subprogram, also including itself. For an illustration, see Fig. 6.1.

In certain cases it is necessary to iterate this picture in the following way. Suppose, for instance, the experimenter attempts a *complete* intrinsic (algorithmic) description. Then, the experimenter has to give a complete description of his own intrinsic situation. In order to be able to model the own intrinsic viewpoint, the experimenter has to introduce an or system which is a *replica* of its own universe. This amounts to substituting the "meta"-automaton for the automaton in Fig. 6.1. Yet, in order to be able to model the intrinsic viewpoint of a new experimenter in this new universe, this new experimenter has to introduce another system which is a *replica* of its own universe, ..., resulting in an iteration *ad infinitum.* In analogy to the process of considering *sub*systems relative to our system, one may assume *super*systems and *super*observers, for which our system and our observers are just models and metaphors of perception. By that reasoning one arrives at a hierarchy of intrinsic/extrinsic-descriptions; the only relevant entity being the *relative* position therein. One may conjecture that an observer in a hypothetical universe corresponding to the "fixed point" or "invariant set" of this process has complete self-comprehension; see Fig. 6.2. Of course, in general this observer cannot be a finite observer: a complete description would only emerge in the limit of infinite iterations (cf. K. Popper's "paradox of Tristram Shandy" and chapter 12 p. 187). Finite observers cannot obtain complete self-comprehension. In psychology, the above setup is referred to as the *observing ego.* In experiments of this kind — e.g., imagine a vase on a table; now imagine you imagining a vase on a table; now imagine you imagining you imagining a vase on a table; now imagine you imagining you imagining you imagining a vase on a table; now imagine you imagining you imagining you imagining you imagining a vase on a table — humans may concentrate on 3 – 5 levels of iteration.

6.1.2 Multi-automata configuration

The second kind of experiment operates with an *arbitrary number* of automaton copies. One automaton is a copy of another if both automata are isomorphic and if both are in the same initial state. With this configuration the experimenter is in the happy condition

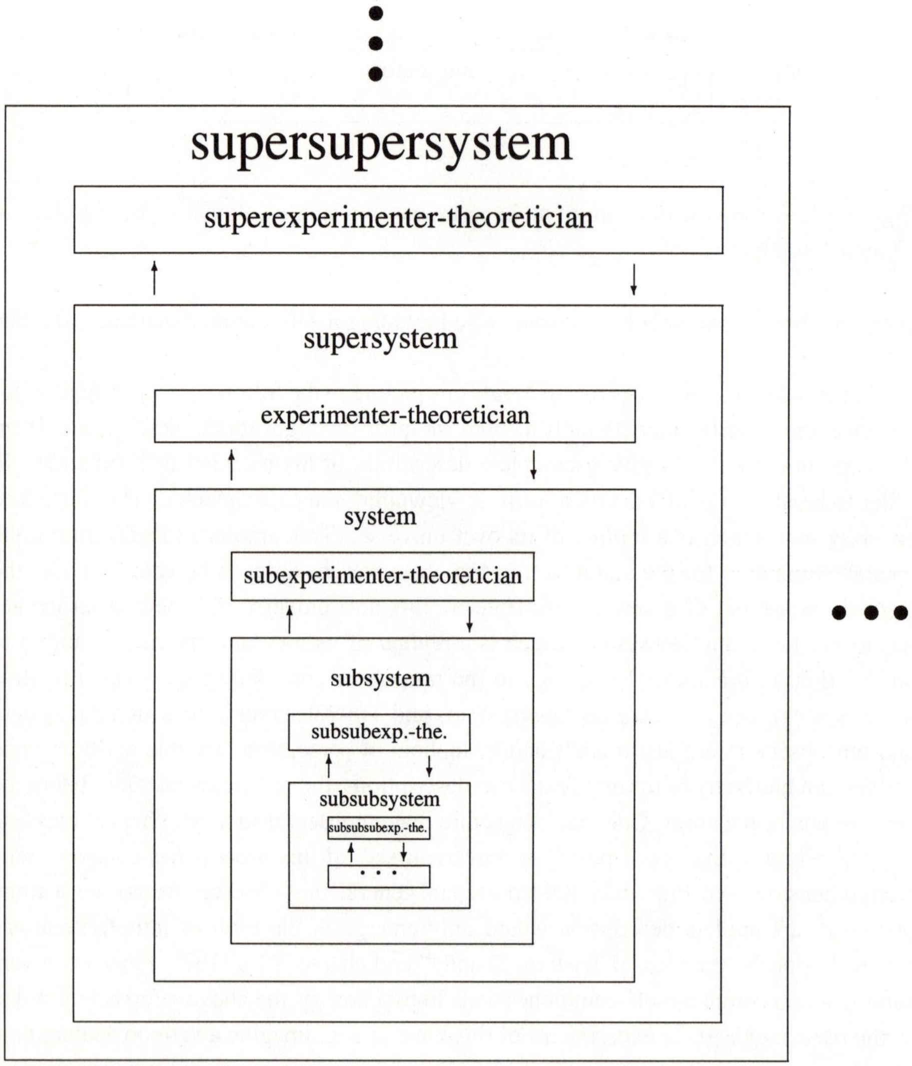

Figure 6.2: Hierarchy of intrinsic perception.

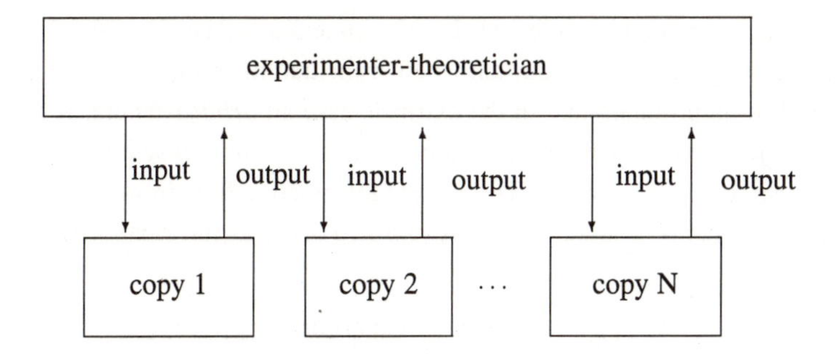

Figure 6.3: Schematic diagram of an experiment on an arbitrary number of identical automaton copies.

to apply as many input sequences to the original automaton as necessary. Again, any theoretical modelling pursued by the experimenter is placed within the experimenter, or at least at the same hierarchical level. In the extrinsic case, theoretical (algorithmic) descriptions are in the "outside" of the automaton copies. In a sense, the observer is not bound to "observer-participancy," because it is always possible to "discard the used automaton copies," and take a "fresh" automaton copy for further experiments. For an illustration, see Fig. 6.3.

6.2 Definition

In the foregoing section, important features of the extrinsic-intrinsic concept have been isolated in the context of finite automata. A generalisation to arbitrary physical systems is straightforward. The features will be summarised by the following definition. (Anything on which experiments can be performed will be called *system*. In particular, finite automata are systems.)

D 6.1 (Extrinsic, intrinsic) *An* intrinsic *quantity is associated with an experiment*

(i) performed on a single copy *of the system,*

(ii) with the experimenter-theoretician being part of the system.

An extrinsic *quantity, denoted by a tilde sign "$\sim$" is associated with an experiment*

(i) utilising, if necessary, an arbitrary number of copies *of the system,*

(ii) with the experimenter-theoretician not being part of the system.

In what follows, the term *experimenter* is a synonym for *experimenter-theoretician.*

Remarks:

(i) One may ask whether, intuitively, the extrinsic point of view might be more appropriately represented by, stated pointedly, the application of a "can-opener" for the "black box" to see "what's really in it." Yet, while the physical realisation might be of some engineering importance, the primary concern is the phenomenology (i.e., the *experimental performance* of the system) and not how it is constructed. In this sense, the technological

base of the automaton is irrelevant. For the same reason, i.e., because this is irrelevant to phenomenology, it is not important whether the automaton is in its minimal form.

(ii) The requirement that in the extrinsic case an *arbitrary* number of system copies is available is equivalent to the statement that *no interaction takes place between the experimenter and the system.* (The reverse information flow from the observed system to the experimenter is necessary.) This results in a one-way information flow in the extrinsic case:

$$\text{system} \underset{\not\Leftarrow}{\overset{\Rightarrow}{}} \text{experimenter} \quad ,$$

and a two-way information flow in the intrinsic case:

$$\text{system} \Longleftrightarrow \text{experimenter} \quad .$$

An information "backflow" makes possible the application of diagonalization techniques (see chapter 12, p. 181), and also results in complementarity, which might be seen as a "poor man's version of diagonalization" (cf. 10, p. 127).

(iii) The definition applies to physical systems as well as to logic and (finite) automata. Automaton worlds provide an ideal "playground" for the study of certain algorithmic features related to undecidability, such as "computational complementarity" and "intrinsic indeterminism."

(iv) In an extreme case, the input has no effect on the object; the object is "just telling its story."

(v) The *extrinsic-intrinsic problem* is the interrelation between extrinsic and intrinsic entities.

6.3 Observer

It is important to realise that an *experimenter-theoretician* or, by another naming, an *observer,* not only consists of primary senses to express and receive information by the interface (see below), but is also the location of theoretical modelling. Such theoretical models may, for instance, be contained and represented by sequences of symbols in books (on theoretical physics, theoretical chemistry *et cetera*); or stored in computers. Therefore, one should not perceive the observer as a homogeneous entity, such as a human being devoid of any (cultural) context, but by a structure of primary senses and a theorical complex.

6.4 Interface

Intuitively, an interface shall be understood as a means or mode of interaction between two systems; more specifically, between a system which is called the observed system and a system which is called the experimenter (observer). Since in the intrinsic case, the setup is symmetric — there is an interaction and therefore a flow of information between both the observed system and the experimenter — the role of the observed system and the experimenter is interchangeable and a matter of convention. (Usually, the experimenter is identified with an apparatus which is thought of as being linked to a human observer.)

An interface is also denoted by the terms *filter* or *cut*, the latter one deriving its name from the Cartesian cut (mind/body interface, if it exists [147]) and the Heisenberg cut, referring to complementarity. As suggested by O. E. Rössler [394, 395], the generic interface is denoted by a swinging double line "$\wr\wr$." Throughout this book the interface is modeled by the exchange of symbols, denoted by "$\rightleftarrows$." I.e., the interface is reduced to its syntax.

6.5 Completeness

In what follows, the Einstein-Podolsky-Rosen concept of completeness of physical theories will be adapted [154]:

> *Whatever the meaning assigned to the term* complete, *the following requirement for a complete theory seems to be a necessary one: $\cdots$ every element of physical reality must have a counterpart in the physical theory.*

This can be translated into the new terminology.

D 6.2 (Completeness) *A theory is* complete *if there is a (computable/recursive) one-to-one correspondence to the manual of a system.*

Whereas in this book the term *completeness* is used in close analogy to Einstein's original approach, a different, more restricted concept is used for an *element of physical reality*; cf. section 4.1, p. 46.

One may ask, *"does there exist an* intrinsically *defined theory about its own system which is in one-to-one correspondence with system?"* Or, stated pointedly, *"could we ever have a complete theory of the world?"* These questions will be dealt with in chapter 12, p. 181.

Chapter 7

Algorithmic information

Informally speaking, complexity is some kind of measure of the computational resources necessary to perform a calculation. These resources have been grouped into two categories:

(i) Static complexity, subdivided into *algorithmic information,* which is a measure of the smallest program length to perform a given task, and *loop-depth complexity;* and

(ii) Dynamic or computational complexity, which can be subdivided into *time complexity* or *depth* and *storage capacity.* Table 7.1 schematically shows the various complexities discussed below.

Several attempts have been made in the literature to propose complexities which grasp the intuitive notion of "organisation." These measures shall not be discussed here. Ch. Bennett's notion of "logical depth" [35] will be reviewed in the context of computational complexity (chapter 8, p. 99). A notion of "self-generating" complexity was proposed by P. Grassberger [208].

The physical importance of *algorithmic information* will be shortly discussed next. One criterion of *selection* among alternative theories is the (minimal) length of their representation. This amounts to one form of "Occam's razor," a principle according to which the best explanation of an event is the one that is the "simplest." In the case of algorithmic information theory, the informal term "simplest" is specified by "shortest." I.e., some physical theory qualifies over others with an equal amount of "predictive power" if it can be expressed in shorter terms. (This criterion, judged by history, does in general not apply to the pursuit of the natural sciences [367, 281, 283, 173, 222]. One may also ask why the "simplest theory" should in priciple be the "true theory," if in this context both "simplicity" and "truth" can be given a meaning whatsoever.)

Another application of algorithmic information is this: The *"natural laws"* governing the evolution of a mechanistic system are representable by effectively computable

<table>
<tr><td rowspan="4">complexity</td><td rowspan="2">static</td><td>algorithmic information
(program size)</td></tr>
<tr><td>(loop depth)</td></tr>
<tr><td rowspan="2">dynamic</td><td>computational complexity
(execution time)</td></tr>
<tr><td>(storage size)</td></tr>
</table>

Table 7.1: Types of complexities

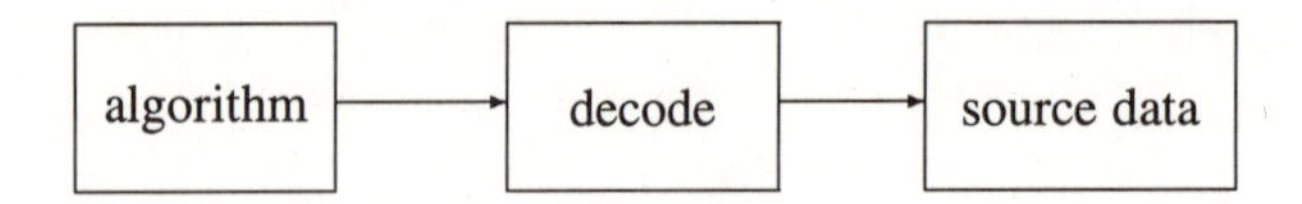

Figure 7.1: Decoding scheme for experimental message.

processes or recursive functions. Algorithmic information is a measure of the length of such a description. — Yet, there exist (mathematical) objects (such as the Mandelbrot set [307, 162, 163]) which have a "very short" description (of a few hundred bits of FORTRAN code) and intuitively "look complex."

There is still another application of algorithmic information: A notion of randomness based on algorithmic information (see chapter 14, p. 193) is equivalent to statistical-based notions of randomness. A chaotic system could be defined by the requirement that its description cannot be "compressed" into any program code (corresponding to a "natural law" or an algorithm) of "shorter length" than the original description. (An alternative notion of randomness, based on "computational irreducibility," or dynamical "runtime" complexity is introduced in section 14.2, p. 197.)

7.1 Encoding strategies

Before introducing algorithmic information theory, specific *encoding* techniques are shortly reviewed. Encoding is usually necessary if the primary source data have to be processed (e.g., interpreted or generated) algorithmically. I.e., one has to *translate* a stream of *source symbols* into an algorithmically recognisable *code alphabet.* This is achieved by the following definition.

D 7.1 *Assume a* source alphabet *with q symbols, $s_1, s_2, \ldots, s_q$ and a* code alphabet *with $r \leq q$ symbols, $c_1, c_2, \ldots, c_r$, where r stands for the radix of the code alphabet. An* encoding *is a map # of source symbols into the set of sequences of symbols from the code alphabet, i.e.,*

$$\#(s_i) = c_{j_1} \cdots c_{j_n} \quad .$$

The trivial case $r = q$ allows a one-to-one map of both alphabets.

Recall that, according to the Church-Turing thesis, recursive "natural laws" are equivalent to algorithms. Such algorithms representing "natural laws" have to be formulated in a specific language or code which is recognisable by some computational device. Stated differently, the algorithm contains information (the "law") for some automaton, which deciphers this information and outputs a data stream. This data stream should be identical with the source data stream or the output of the corresponding physical system (Fig. 7.1). In other words, we are concerned with a suitable representation of some experimental "message." We are not concerned with the question of a "meaning" of the "message" or an "underlying law," but we shall investigate the technical question of how to encode source symbols in a "reasonable" way, such that the encoding is unique, compact, easy to recognise and transmit, and so on.

For further reference, see R. W. Hamming's book *Coding and Information Theory* [223], as well as R. J. McEliece, *The Theory of Information and Coding* [317] and R. Ash, *Information Theory* [13], among others. A detailed consideration of code word length is also given by S. K. Leung-Yan-Cheong and T. M. Cover [297], J. Rissanen [386], and P. Elias [155]. A detailed treatment of algorithmic information theory can also be found in C. Calude's forthcoming book *Information and Randomness --- An Algorithmic Perspective* [78].

7.1.1 Instantaneous/prefix codes — self-delimiting programs

Let us call a string of symbols x a *prefix* of another string of symbols y if $y = xz$ for some string of symbols z. In the following, the terms *instantaneous, prefix, prefix-free* and *self-delimiting* will be used synonymously. A motivation for instantaneous codes comes from the *concatenations* of source symbols. Consider the following example. Assume, for the moment, $q = 4$ and $r = 2$, and two encoding strategies $\#_1$ and $\#_2$, represented by

$$\#_1(s_1) = 0, \#_1(s_2) = 01, \#_1(s_3) = 1, \#_1(s_4) = 00$$

and

$$\#_2(s_1) = 0, \#_2(s_2) = 01, \#_2(s_3) = 011, \#_2(s_4) = 111 \quad ,$$

Assume then that we encode the message "0011." It can be checked that $\#_1$ is *no unique* code, since this message "0011" might mean $\#_1(s_1s_1s_3s_3)$, but also $\#_1(s_4s_3s_3)$ or $\#_1(s_1s_2s_3)$. On the other hand, $\#_2$ turns out to be *unique*, with "0011" meaning $\#_2(s_1s_3)$. It is quite evident that non uniqueness of the encoded message is an undesirable feature of the code. For transmission and processing of source information it is therefore reasonable to require *uniqueness* of code.

Another undesirable feature of the encoding is that the original source information can be translated only *after the whole message* has been sent. This is, for instance, the case for $\#_2$. Therefore, another criterion for a code is its *instant decodability*. I.e., the symbols should be decodable even *before* the whole code has been transmitted. This can be achieved by the requirement that *no code word is a prefix of another code word.* Take, for instance,

$$\#_3(s_1) = 10, \#_3(s_2) = 11, \#_3(s_3) = 01, \#_3(s_4) = 00.$$

The message "0011" can be instantly (i.e., *immediately* after transmission of the code of each symbol) interpreted as $\#_3(s_4s_2)$. Codes of this kind are called *instantaneous codes* or *prefix codes.* The two terms will be used synonymously.

Instantaneous codes need no "end-markers" to indicate when a message ends. The requirement that no code word is the prefix of another code word yields a construction of such codes by a "coding tree." Starting from a "root node," every node of the tree branches into r "leafs." Every node which is occupied by a code word is a *terminal node;* i.e., in order to avoid that this code word is the prefix of another code word, the tree gets pruned at that node. This is illustrated in Fig. 7.2, which contains two instantaneous encoding schemes. Fig. 7.2(a) represents one possible instantaneous coding scheme for $q = 8$ and $r = 2$, whereas 7.2(b) lists one possible instantaneous coding scheme for $q = 4$

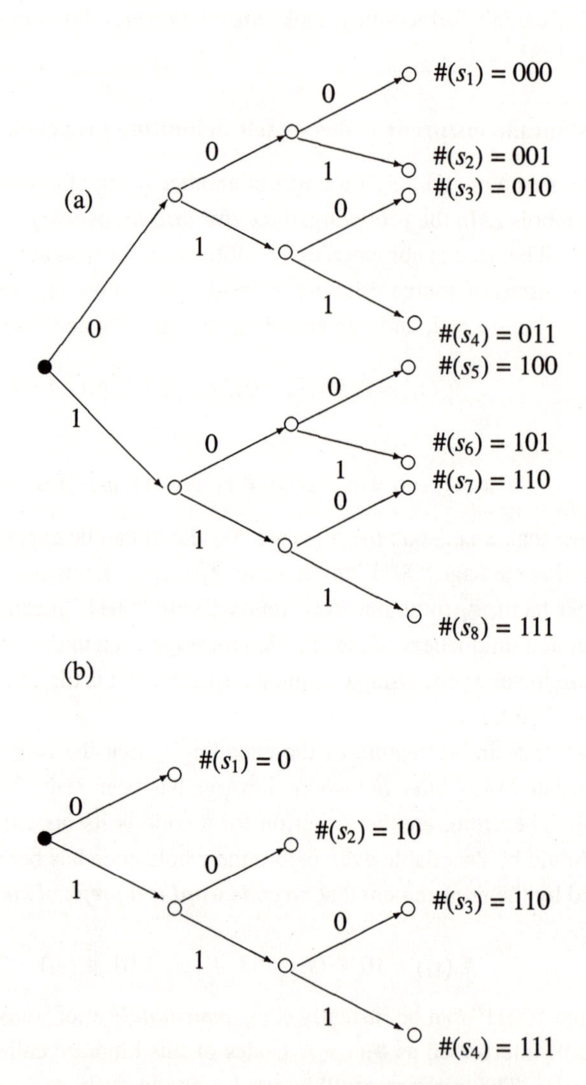

Figure 7.2: Two instantaneous (prefix-free) coding schemes.

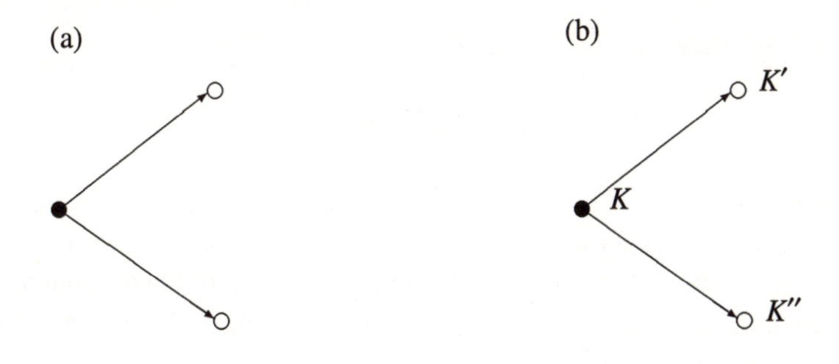

Figure 7.3: Binary tree (a) of length one, (b) of length $n+1$.

and $r = 2$. (For arbitrary q and r, a generalisation is straightforward.) The statement ''*the price for instant decodability is the length of the code*'' will be quantified next.

7.1.2 Kraft inequality

T 7.2 (Kraft inequality [275]) *An instantaneous code of a q-symbol source alphabet* $\{s_1, s_2, \ldots, s_q\}$ *with encoded word length* $l_1 \leq l_2 \leq \cdots \leq l_q$ *satisfies the* Kraft inequality

$$\sum_{i=1}^{q} r^{-l_i} \leq 1 \quad , \tag{7.1}$$

where r is the number of symbols (radix) of the code alphabet.

Proof:

For simplicity assume $r = 2$. We shall prove the Kraft inequality by induction. Consider first the binary tree of length one, represented in Fig. 7.3(a). We can encode with this tree a source code containing one or two symbols, yielding a Kraft sum $\frac{1}{2} \leq 1$ or $\frac{1}{2} + \frac{1}{2} \leq 1$, respectively.

Next assume that the Kraft inequality is true for all trees of length n. Now consider in Fig. 7.3(b) a tree of maximum length $n+1$. The first "root" node (the tree is drawn upside down) leads to at most a pair of subtrees (of length at most n), for which we have the inequalities $K' \leq 1$ and $K'' \leq 1$, where K' and K'' are the values of the respective Kraft sums. Each length l_i in a subtree is increased by one when the subtree is joined to the main tree as specified. Therefore, for binary encoding, i.e., for $r = 2$, an extra factor $\frac{1}{2}$ appears, and one obtains

$$\frac{1}{2}K' + \frac{1}{2}K'' \leq 1 \quad .$$

For arbitrary radix r, at most r subtrees contribute with a factor of $1/r$, and the Kraft inequality (7.1) is again true.

Example:

Binary programs p of unbounded length which have instantaneous codes and run on universal computers U correspond to $r = 2$ and $q = \infty$. In this case the Kraft inequality becomes ("$|p|$" stands for the length of p)

$$\sum_p 2^{-|p|} \leq 1 \quad . \tag{7.2}$$

Remark:

Notice that if every terminal node is identified with one code word s_i, then the Kraft sum (7.1) is exactly one. Only for inefficient coding, i.e., if one or more terminal nodes are not used, the inequality sign applies. The codes of Fig. 7.2 are both *efficient*, and the Kraft inequality (7.1) holds with equality. The *optimal* encoding strategy (in the sense of "shortest encoding of messages") depends also on the *relative frequency of occurrence* of the source symbols. It is evident that "more frequent source symbols" should be identified with "shorter" code words, thereby enabling (on the average and compared to other encoding schemes) a shorter length of the message as a whole. For a recent account of the algorithmic version of the Kraft inequality, see C. Calude and Eva Kurta [70].

7.2 Algorithmic information theory

Intuitively similar formulations of (preliminary concepts of) *algorithmic information (content)* (also known as *program size complexity*, *Kolmogorov complexity*, algorithmic complexity, *Kolmogorov-Chaitin randomness et cetera*) have (on the basis of absence of proof to the contrary independently) appeared in the writings of G. J. Chaitin [96, 109, 110], R. J. Solomonoff [419] and A. N. Kolmogorov [271]. The following brief *expose* of algorithmic information theory follows Chaitin's approach [109, 110, 112] and adapts his conventions: $O(1)$ is read *"of the order of* 1" and denotes a function whose absolute value is bounded by an unspecified positive constant. I.e., $\varphi(x) = O(1)$ means $|\varphi| < A$, where A is a constant independent of the arguments x of φ. "$|s|$" denotes the *length* of an object s coded in binary notation. If not stated otherwise, only *universal computers*, denoted by U, are considered. For a more detailed treatment, the reader is referred to G. Chaitin's books *Algorithmic Information Theory* [110] and *Information, Randomness and Incompleteness* [110], the latter one containing the collection of his works, including some (easy-to-read) introductory articles. One of G. Chaitin's intentions is to extend Gödel's incompleteness theorems [204] to algorithmic information theory. Informally speaking, in this view algorithmic information of an entity is a sort of quantification or measure for "mathematical truth" captured by that entity. G. Chaitin is also pointing to the fact that there exists *"randomness in arithmetic"* (i.e., arithmetic statements which are just as "random" as the outcome of the flipping of a fair coin). This approach has sparked a great number of reactions, mostly positive [113, 182, 138], some sceptical [190, 287]. A further (independent) account can be found in a review article by A. K. Zvonkin and L. A. Levin [475]. For more references, see also the review article by M. Lie and P. M. B. Vitányi [300] and M. van Lambalgen's dissertation, published in [285]. The forthcoming beook by C. Calude [78] will surely be a major source of reference. [In a note in 1972, K. Gödel introduced a notion of program size complexity in *Some remarks on the undecidability results*, reprinted in the *Collected Works, Volume II* [206], p. 305 (see *"Another version of the first undecidability theorem"*), which was based on the (not necessarily *shortest*) number of symbols or the system of axioms necessary to solve a problem. In the context of attempts to formulate and evaluate a unified theory, Einstein introduced a measure of strength of field equations, based upon the number of free parameters (see [153], p. 138).]

Informally speaking, the basic idea is the characterisation of a mathematical object by the *length of the shortest program* which outputs a code of that object. This measure is denoted by H' in honour of L. Boltzmann's function proportional to entropy [114]. The apostrophe sign " ′ " indicates unspecified program code. In what follows, H' is measured in bits. The original approach is ambiguous: consider for instance a binary sequence $x(n)$ of length n. At first glance it seems that the information content H' of $x(n)$ could not exceed the length of that sequence [plus $O(1)$ from additional program statements like `PRINT` $x(n)$]; i.e. $H'(x(n)) \leq n + O(1)$.

So far, no specific encoding of the program generating the sequence $x(n)$ has been specified. Therefore, H' will be defined ambiguously. (This is where the encoding of programs discussed in the previous sections comes in.) Assume, for instance, that specific symbols (such as the blank symbol " _ "), which we shall call end-markers, are allowed to end-mark an enumeration. By very efficiently utilising these end-markers, a program may scan through all digits of $x(n)$, determine its length in real-time execution, print out $x(n)$ as well as n, and finally halt. Notice that the *length n* of bits may represent an *additional* valuable piece of information, which may be worth $H'(n)$ bits. As a consequence of tricky programming it might be possible to squeeze $H'(n)$ *more* irreducible information into $x(n)$ than is contained in the n bits of $x(n)$, resulting in a total information of

$$n + H'(n) \quad .$$

The above argument can be iterated. (Let "$|x|$" of an object encoded as binary string stand for the length of that string.) Indeed, by allowing end-markers, the length $|x|$ of a string x could be subject to more information than just $||x||$, since, by the same argument as before, it could contain a total information of $H'(|x|) \leq ||x|| + H'(||x||) + O(1)$. Therefore, by recursion,

$$H'(x) \leq |x| + ||x|| + |||x||| + \cdots + O(1) \quad .$$

The length $|x|$ can be estimated by $\log_2 x$; more precisely, by $\lg x$, where $\lg x$ stands for the greatest integer less than the base-two logarithm of the real number x; for $x \neq 2^n$ ($n \in \mathbb{N}$), this is just the integer part of the base-two logarithm of x. I.e., if $2^n < x \leq 2^{n+1}$, then $\lg(x) = n$. Thus, if $\log_2^* x$ and $\lg_2^* x$ are defined by (a *Mathematica* program for $\log_2^*$ is printed on page 250)

$$\log_2^* x = \log_2 x + \log_2 \log_2 x + \log_2 \log_2 \log_2 x + \cdots \quad , \tag{7.3}$$

$$\lg_2^* x = \lg_2 x + \lg_2 \lg_2 x + \lg_2 \lg_2 \lg_2 x + \cdots \quad , \tag{7.4}$$

where only the *positive* terms contribute to the sum, all sequences have an information content of

$$H'(x) \leq \lg_2^* x + O(1) \leq \log_2^* x + O(1) \quad ,$$

and some (the "most informative") of them will have an algorithmic information content of

$$H'(x) = \lg_2^* x + O(1) \quad .$$

Consequently, the allowance of end-markers results in the undesirable feature that there exist sequences of length n — and thus of algorithmic information content of the order of n — with "overall" information of the order of $n + \log_2 n + \cdots$ bits. Another, more technical problem, concerns the *subadditivity* of algorithmic information [i.e., $H'(x,y) \leq H'(x) + H'(y) + O(1)$].

One could speculate that it might be possible to squeeze even *more* than $n + H'(n)$ bits of information into a sequence of length n by counting the computation time t on some specified computer. This is impossible [114], because the run time t of a program p that halts can be computed from p by an algorithm which simulates p and in addition *counts* the execution time t. Such an algorithm would at most take $O(1)$ bits more than p. Thus $H'(p,t) = H'(p) + O(1)$.

G. Chaitin [109] (see also L. A. Levin [299]) has proposed a modification of the original definition, which has eliminated both deficiencies and has restored the subadditivity property of algorithmic information. This modification is based upon the restriction to *instantaneous (prefix)* program codes. Programs should be self-delimiting; i.e., they should, for instance, do not contain any end-markers. One way to achieve this is the use of symbols which can be instantaneous decoded; this results in prefix coding techniques discussed in section 7.1.1, p. 85. Another, probably more intuitive, way to achieve instant decodability is the requirement that the program must somehow be informed about the length n of a sequence $x(n)$ *beforehand.* This requirement necessitates an additional program statement indicating n without end-markers, yielding an increase in program length up to $\lg_2^*(n)$ bits. With this definition, a string of size n can have an algorithmic information content of $n + H(n) + O(1)$. Consequently, the length — and thus the algorithmic information — of a binary program producing a binary string is *not* bounded by the length of that string, but rather by $(\text{length}) + \log_2(\text{length}) + \log_2(\log_2(\text{length})) + \cdots + O(1)$.

In general and on the average, instantaneous decodability will result in longer program codes than codes of programs which are non uniquely or non instantaneous decodable. This is the reason why the Kraft sum (7.1), which can be interpreted here as the exponentially weighted sum over the length of all allowed programs, converges. (A program coded by symbols which are instantaneously decodable is instantaneously decodable itself.)

Another motivation for the requirement of instantaneous program codes comes from the perception of a computer as a decoding equipment [109, 110]: its programs correspond to the encoded message and its output corresponds to the decoded message. A "reasonable" requirement is that the coded messages (i.e., the programs) are "instantaneously readable" and therefore should not depend on other encoded messages or on other parts or the rest of the transmitted message. This translates into the requirement that no program is a prefix of another (or, in other words: no extension of a valid program is a valid program). In the foregoing section, such a code has been called *instantaneous* or *prefix* code.

7.2.1 Definition

The static complexity, more specifically the program size complexity or algorithmic complexity or algorithmic information will be denoted by H and can be defined as follows.

D 7.3 (Algorithmic information [94, 109, 110])

Assume instantaneous (prefix) encoding.

The canonical program *associated with an object s representable as string is denoted by* s^* *and defined by*

$$s^* = \min_{U(p)=s} p \quad . \tag{7.5}$$

I.e., s^* *is the first element in the ordered set of all strings that is a program for* U *to calculate* s. *The string* s^* *is thus the code of the smallest-size program which, implemented on a (universal) computer, outputs* s. *(If several binary programs of equal length exist, the one is chosen which comes first in an enumeration using the usual lexicographic order relation "*$0 < 1$.*")*

Let "$|x|$*" of an object encoded as (binary) string stand for the length of that string. The* static or algorithmic complexity $H(s)$ *of an object s representable as string is defined as the length of the shortest program p which runs on a computer U and generates the output s:*

$$H(s) = |s^*| = \min_{U(p)=s} |p| \quad . \tag{7.6}$$

If no program makes computer U output s, then $H(s) = \infty$.

The joint algorithmic information $H(s,t)$ *of two objects s and t representable as strings is the length of the smallest-size binary program to calculate the concatenation of s and t simultaneously.*

The relative or conditional algorithmic information $H(s|t)$ *of s given t is the length of the smallest-size binary program to calculate s from a* smallest-size program *for t:*

$$H(s|t) = \min_{U(p,t^*)=s} |p| \quad . \tag{7.7}$$

7.2.2 Algorithmic probability

Assume again instantaneously decodable programs, represented by bit strings. Consider a specific program p, corresponding to a certain bit string, say `0010111000` $\cdots$ `1110111010`, of length $|p|$. One could ask, *"what is the probability that p will be generated by applying a random process of two random variables, such as the flipping of a fair coin,* $|p|$ *times?"* — It is like *"a million monkeys typing away"-scenario.* This probability is $2^{-|p|}$.

One may also ask, *"what is the probability that a valid program producing a specific object s, or any object at all, will be obtained by the flipping of a fair coin?"* — Of course, the sequence obtained by this random process has to be "arbitrary long" to assure that one has not just gotten the first bits of a valid program producing s. The following definitions are motivated by these questions. For an early definition of probability measures on the set of output sequences of automata, see the article by K. de Leeuw, E. F. Moore, C. E. Shannon and N. Shapiro, in *Automaton Studies* [295].

D 7.4 (Algorithmic probability, halting probability, version I)
Let s be an object encodable as binary string and let $S = \{s_i\}$ be a set of such objects s_i. then the algorithmic probability *P is defined by*

$$P(s) = \sum_{U(p)=s} 2^{-|p|} \tag{7.8}$$

$$P(S) = \sum_{s_i \in S} P(s_i) = \sum_{U(p)\in S} 2^{-|p|} \tag{7.9}$$

$$\Omega = \sum_{s} P(s) = \sum_{s} \sum_{U(p)=s} 2^{-|p|} = \sum_{U(p)\downarrow} 2^{-|p|} \tag{7.10}$$

Remarks:

(i) P is, strictly speaking, no probability:

Due to the Kraft inequality (7.1), i.e., for inefficient coding and due to the fact that not all programs converge, $\Omega = \sum_s P(s) \leq 1$ needs not be *exactly* 1.

Let $x(n)$ denote a binary sequence of length n and let $x(n)0$ and $x(n)1$ denote sequences of length $n+1$ which are obtained from $x(n)$ by appending the symbol 0 and 1, respectively. Let for the moment the symbols "$x(n) \preceq U(p)$" denote that $x(n)$ is an initial segment of $U(p)$. If one slightly modifies the definition of P such that

$$P'(x(n)) = \sum_{x(n)\preceq U(p)} 2^{-|p|} \quad ,$$

then

$$P'(x(n)) \geq P'(x(n)0) + P'(x(n)1) \quad .$$

But even if one assumes the original definition of P, two independent, random objects s and t yield

$$P(s) + P(t) = P(s,t) + O(1) \quad .$$

With these *provisos*, P will be called "probability" nevertheless, because $P(s)$ is a reasonable measure of the frequency that some prefix-free program produces the object s on the standard universal computer U.

(ii) Ω is the *halting probability* (with null free data), i.e., the probability that an arbitrary program (with no input) halts. Ω is random and — in the limit of infinite computing time — can be obtained in the limit from below by a computable algorithm [109, 110]. For details, see 14.1.2, p. 196.

(iii) The set of all true propositions of the form "$H(s) \leq n < \infty$" or "$P(s) > 2^{-n}$" is recursively enumerable because it is possible to "empirically" find $H(s)$ by running all programs of size less than or equal to n and by observing whether they output s. However, one would have to wait "very long," i.e., longer than any recursive function of n, though for finite length strings $|s| < \infty$ not eternally, to recognise that; see 8.7, p. 104 for details.

(iv) $P_{C'}(s)$ and $\Omega_{C'}$ can be defined for any (not necessarily universal) computer U' by substituting U' for U in (7.8) and (7.10). We shall use the isomorphy between deterministic physical systems and effectively computable processes to relate the algorithmically defined probabilities (7.8) to physical entropy in chapter 18, p. 231.

There are infinitely many programs contributing to the sum in (7.8), but the greater the size of any such program is, the more it gets (exponentially) suppressed. Therefore, the dominating term in the sum for $P(s)$ stems from the canonical program s^*. Indeed, it can be shown [109, 110] [cf. (7.35), p. 97] that there are "few" minimal programs contributing substantially to the sums (7.8), (7.9) and (7.10). Thus the probabilities to produce a specific object s as well as the halting probability can also be defined by taking into account only the *canonical (i.e., shortest-size) programs*. I.e., if in (7.8) the sum over all programs is reduced to a single contribution from the canonical program s^* for which $|s^*| = H(s)$, then one can define $P^*(s)$ and Ω^* as follows:

D 7.5 (Algorithmic probability, halting probability, version II)

Let s be an object encodable as binary number and let $S = \{s_i\}$ be a set of such objects s_i. then the algorithmic probability P^* *is defined by*

$$P^*(s) = 2^{-|s^*|} = 2^{-H(s)} \tag{7.11}$$

$$P^*(S) = \sum_{s_i \in S} P^*(s) = \sum_{s \in S} 2^{-H(s)} \tag{7.12}$$

$$P^*(\mathbb{N}) = \Omega^* = \sum_{n \in \mathbb{N}} 2^{-H(n)} \quad . \tag{7.13}$$

If we would not have restricted the allowed program codes to instantaneous ones, "much more" allowed program codes could have contributed to the sums (7.8), (7.9) and (7.10). As a result, these sums might not have been bounded from above by 1 and might have even diverged. Due to instantaneuos (prefix) encoding, the *Kraft inequality* (7.1)

$$\sum_p 2^{-|p|} \leq 1$$

holds for all programs, irrespective of whether they halt or not. (For details, see (7.1) on page 87.) Since the class of programs which halt or produce a specific output s is a subclass of all programs, one obtains

$$\Omega^* \leq \Omega \leq 1 \quad . \tag{7.14}$$

Furthermore, for $s \subset S \subset \mathbb{N}$,

$$0 \leq P^*(s) \leq P(s) \leq 1 \tag{7.15}$$

$$0 \leq P^*(S) \leq P(S) \leq 1 \tag{7.16}$$

$$0 \leq P(s) \leq P(S) \leq P(\mathbb{N}) = \Omega \leq 1 \tag{7.17}$$

$$0 \leq P^*(s) \leq P^*(S) \leq P^*(\mathbb{N}) = \Omega^* \leq 1 \tag{7.18}$$

(7.19)

7.2.3 Threshold for computability

The discussion of the computability of H is relegated to the second part of this book, section 9.7, p. 122.

7.2.4 Complexity of applications

T 7.6 (Complexity of applications) *If x is some natural number and ψ is some effectively computable / recursive function, then the algorithmic complexity of $\psi(x)$ is*

$$H(\psi(x)) \leq H(x) + c_\psi = H(x) + O(1) \quad .$$

A *proof* of 7.6 is straightforward, since any effectively computable function can be coded into a finite program of length $c_\psi = O(1)$.

7.2.5 Machine independence

Consider the algorithmic information of one and the same object with respect to two universal computers. Roughly speaking, for certain reasonable machine models, the information content with respect to the first computer should differ from the information content with respect to the second computer by the order of the length of the shortest translation program from one computer code to the other. For universal devices, this yields just a $O(1)$ contribution which is *independent* of the particular object.

More precisely, the *machine independence* of algorithmic information can be obtained if one *assumes* two suitable universal machines U and U' such that there exists a *translation program* $\psi : p' \to p$ of constant length $O(c_\psi)$ with $U(\psi(p')) = U'(p')$ for all p' and *vice versa.* This restriction of the class of universal machine models is necessary, since for arbitrary universal machines U, U' there does not exist a translation program $\psi : p' \to p$ of *constant* length $O(c_\psi)$ with $U(\psi(p')) = U'(p')$ for all p'. Indeed, G. Chaitin requires this condition in the definition of universal machines.

With this restriction, by theorem 7.6,

$$\begin{aligned} H(x) = \min_{U(p)=x} |p| \leq & \min_{U(\psi(p'))=x} |\psi(p')| \\ & \leq \min_{U(\psi(p'))=x} (|p'| + c_\psi) \\ & = \min_{U'(p')=x} (|p'| + c_\psi) \\ & = H_{U'}(x) + c_\psi \\ & = H_{U'}(x) + O(1) \quad . \end{aligned}$$

If *both* U and U' were universal computers, then $H \leq H_{U'} + c_1$ and $H_{U'} \leq H + c_2$, where c_1, c_2 are unspecified positive constants, corresponding to, roughly speaking, the length of the binary translation programs, which are usually of the order of 1000 bits. Thus, for all objects x the absolute value of the difference between H and $H_{U'}$ satisfies

$$|H - H_{U'}| \leq \max(c_1 , c_2) = O(1) \quad , \tag{7.20}$$

showing the *invariance* of algorithmic information with respect to variations within a certain class of universal computer models.

7.2.6 Estimates and identities

The following relations hold:

T 7.7 ((Chaitin [101, 110]))

Let s and t be two objects representable as binary strings.

$$
\begin{aligned}
H(s,t) &= H(t,s)+O(1) \quad ; & (7.21)\\
H(s|s) &= O(1) \quad ; & (7.22)\\
H(H(s)|s) &= O(1) \quad ; & (7.23)\\
H(s) &\leq H(s,t)+O(1) \quad ; & (7.24)\\
H(s|t) &\leq H(s)+O(1) \quad ; & (7.25)\\
H(s,t) &= H(s)+H(t|s^*)+O(1) \quad ; & (7.26)\\
H(s,t) &\leq H(s)+H(t)+O(1) \quad \text{(subadditivity)} \quad ; & (7.27)\\
H(s,s) &= H(s)+O(1) \quad ; & (7.28)\\
H(s,H(s)) &= H(s)+O(1) \quad ; & (7.29)\\
\max_{|s|=n} H(s) &= n+H(n)+O(1) \quad ; & (7.30)\\
H(s) &= -\log_2 P^*(s) \quad ; & (7.31)\\
H(s) &= -\log_2 P(s)+O(1) \quad ; & (7.32)\\
H(s|t) &= -\log_2 P(s|t)+O(1) = -\log_2 \frac{P(t)}{P(t,s)}+O(1) \quad . & (7.33)
\end{aligned}
$$

Remarks:

(i) The subadditivity (7.26) & (7.27) will be used for a proof of incompleteness theorems for lower bounds on H. For details, see 9.7, p. 122. Informally speaking, it means that, with respect to program size, it is quite effective for a computer to "do one thing after the other."

(ii) Theorem (7.30) refers to the *maximal complexity of a finite bit string*. At a first glance it seems amazing that the complexity of a string can exceed its size. But, as has already been pointed out, one should bear in mind that a prefix-free program code has to be longer than non prefix-free program code.

Informally speaking, an enumeration of an output string has to contain information specifying the length of the output string. To demonstrate this fact consider a print program of an algorithmically incompressible sequence $x(n)$ of the form "`PRINT THE FOLLOWING SEQUENCE OF LENGTH` n`:` $x(n)$." The print program contains the strings "`PRINT THE FOLLOWING SEQUENCE OF LENGTH`" and "`:`" which contribute $O(1)$ to the program length; furthermore it has to contain an enumeration of n which needs no end-marker. Such an enumeration of n could be: "`READ THE NEXT` $|n|$ `SYMBOLS:` n" and so on, until $|\cdots|n|\cdots|$ is 1. This results in a nesting of read statements, corresponding to contributions $\leq \log_2(\cdots\log_2(n)\cdots)$. These terms have to be included for a calculation of the program length.

By recalling the argument using recursive instantaneous (prefix) encoding, one obtains *bounds from above*

$$
H(s(n)) \leq n+H(n)+O(1) \leq \lg_2^* s(n)+O(1) \leq \log_2^* s(n)+O(1) \quad . \tag{7.34}
$$

A trivial bound from below is $O(1)$. It is easily verified that $\log_2^* 1 = 0$, $\log_2^* 2 = 1$, $\log_2^* 3 \approx 2.25$, $\log_2^* 4 = 3$, $\log_2^* 5 \approx 3.82$, $\log_2^* 6 \approx 4.41$, $\log_2^* 7 \approx 4.87$, $\log_2^* 8 \approx 5.25, \ldots$. Fig. 7.4

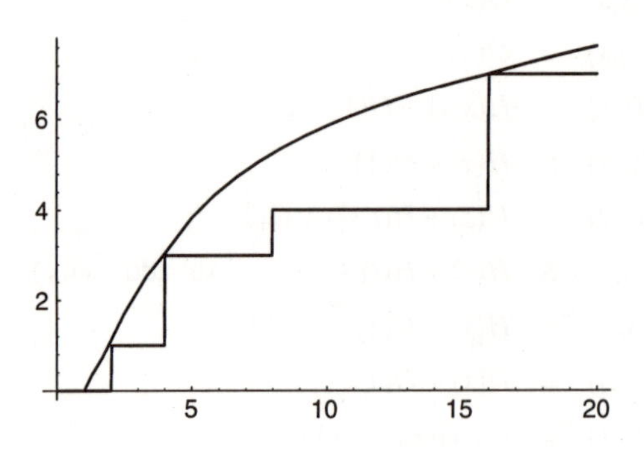

Figure 7.4: The functions $\log_2^* x$ and $\lg_2^* x$.

plots $\log_2^* x$ as well as $\lg^* x$ and demonstrates that $\log_2^* x$ is *no concave function* [297].

(iii) Theorems (7.31), (7.32) and (7.33) express a very important identity between the algorithmic information content of an object s and the probability that it is algorithmically produced. They suggest a formal analogy of algorithmic information to Shannon information (18.5), p. 233, and physical entropy. Theorem (7.31) is an immediate consequence of the definition of $P^*(s) = 2^{-|s^*|} = 2^{-H(s)}$, equation (7.11).

A *proof* of theorem (7.32), which has important physical applications (see chapter 18, page 231) is less straightforward. Following G. Chaitin [109, 110], it has to be shown that most of the probability is concentrated on the minimal size programs, or alternatively that there are few minimal programs. If there are many large programs, a much smaller program is constructed in the proof. The proof uses non constructive elements, such as computation of the algorithmic probability in the limit from below, as well as the existence of universal computers and program codes with extraordinary properties. [The proof is *not* a trivial consequence of the usual definition of the *(Shannon) information* gain $I(s) = -\log p(s)$ if symbol s occurs. In this case, the probability $p(s)$ is defined by the relative frequency of occurrence (and not the algorithmic probability) of the symbol s.]

Recall that $\lg(x)$ stands for the greatest integer less than the base-two logarithm of the real number x; for $x \neq 2^n$, $n \in \mathbb{N}$, this is just the integer part of the base-two logarithm of x. I.e., if $2^n < x \leq 2^{n+1}$, then $\lg(x) = n$. Thus, $2^{\lg x} < x$. Notice that, as has been pointed out before, the set of all true propositions of the form "$H(s) \leq n$" or "$P(s) > 2^{-n}$" is recursively enumerable because it is possible to "empirically" find $H(s)$ by running all programs of size less than or equal to n and see whether they output s. On the basis of this process, postulate a universal computer D which simulates the original computer U and in addition enumerates all true theorems of the form "$P(s) > 2^{-n}$" without repetition. Further postulate that for D a *single* program of length n exists which outputs s if and only if the condition $P(s) > 2^{1-n}$, or $n \geq -\lg P(s) + 1$ is satisfied. (The extra factor 2 in $P(s) > 2\times 2^{-n}$ is required for a proof of the existence of an instantaneous code, see below.)

Hence the number of programs p of length n such that $D(p) = s$ is 1 if $n \geq -\lg P(s) + 1$ and is 0 otherwise. The smallest program which outputs s is the one with $n = -\lg P(s) + 1$, and thus

$$H_D(s) = -\lg P(s) + 1 \quad .$$

These postulates indeed allow instantaneous program codes, since

$$\begin{aligned} P_D(s) = \sum_{D(p)=s} 2^{-|p|} &= 2^{-H_D(s)} \sum_{D(p)=s} 2^{H_D(s)-|p|} \\ &= 2^{-H_D(s)} \sum_{k=0}^{\infty} \left(\frac{1}{2}\right)^k \\ &= 2^{-H_D(s)+1} \\ &= 2^{\lg P(s)} < P(s) \end{aligned}$$

and, due to the Kraft inequality which is satisfied by the original code (corresponding to H and P), $\sum_s P_D(s) < \sum_s P(s) \leq 1$ is satisfied. Taking into account that the algorithmic information is machine independent up to $O(1)$ [see (7.20), p. 94 for details], one obtains theorem (7.32).

Since by theorem (7.32) $P(s) = e^{-H(s)} \sum_{U(p)=s} 2^{H(s)-|p|} = e^{-H(s)} O(1)$, one concludes that *there are few minimal programs*, i.e.,

$$\sum_{U(p)=s} 2^{H(s)-|p|} = O(1) \quad . \tag{7.35}$$

7.3 Infinite computation

H_∞ and P_∞ will denote the algorithmic information content and the algorithmic probability of *infinite* computations. Rather few is known about the properties of H_∞ and P_∞. A result of R. M. Solovay [421] states that, for arbitrary recursively enumerable infinite sets S,

$$H_\infty(S) \leq -3 \log_2 P_\infty(S) \quad . \tag{7.36}$$

Assume, for example, a program for enumerating the natural numbers in successive order. Such a program will not halt in finite time. The minimal length of this program will eventually become "much smaller" than the complexity of most of the individual numbers it outputs. The related "finite" version of this statement is the fact that there exist sets of objects $S = \{s_1, \ldots, s_n\}$, $n < \infty$ whose algorithmic information content $H(S)$ is arbitrary small compared to the algorithmic information content of some unspecified *single* elements $s_i \in S$; i.e.,

$$H(S) < \max_{s_i \in S} H(s_i) \quad . \tag{7.37}$$

A simple example is the set of integers smaller than or equal to a certain "large" number which can be represented "very compactly." An example for such a number is $2^{2^{2^{2^{2^{2^{2^{2}}}}}}}$. Hence $S = \{1, 2, 3, \ldots, 2^{2^{2^{2^{2^{2^{2^{2}}}}}}} - 2, 2^{2^{2^{2^{2^{2^{2^{2}}}}}}} - 1, 2^{2^{2^{2^{2^{2^{2^{2}}}}}}}\}$. — A pseudocode for a program producing S can be written down very shortly: "`DO LABEL1 I=1, 1,` $2^{2^{2^{2^{2^{2^{2^{2}}}}}}}$ `; LABEL1 PRINT (*,*) I; STOP; END;`" yet production of certain single individual numbers `I` of S could require a much longer program code.

Chapter 8

Computational complexity

This chapter deals with the question of the *time of computation*, i.e., with the number of discrete steps in a computation of an object. (Another dynamical complexity measure is *space* or *storage size*, the number of distinct storage locations accessed by the computation.) Such an object can be a (binary) string, the solution to a mathematical problem *et cetera*. Let N be some number characterising the *size* of a problem. In chapter 4, p. 45, techniques for representing an arbitrary object by a code string of symbols x are discussed. If an object (with code) x is the solution to the problem, the size parameter N needs not necessarily coincide with the length $|x(N)|$ of x. Examples of such objects are the generation of a sorted list of N items or finding the roots of a polynomial of degree N.

8.1 Definition

Consider again a universal computer and instantaneous (prefix) program codes.

D 8.1 (Computational complexity)

Assume a problem of the order of N and its solution x(N), if it exists. The dynamical or computational complexity *$H_D(x(N))$ is the time (number of cycles, computing steps) it takes for the fastest program p running on a universal computer U to calculate x(N):*

$$H_D(x(N)) = \min_{U(p)=x(N)} \mathrm{TIME}(p) \quad .$$

If no program makes computer U output x(N), then $H_D(x(N)) = \infty$.

Examples:

(i) A trivial example is the enumeration of a string $x(n)$ of length n on a sequential machine outputting one symbol per time cycle. A program containing a print statement of the form "PRINT THE FOLLOWING SEQUENCE OF LENGTH n: $x(n)$" consumes a minimal time $H_D(x(n)) = n + O(1)$.

(ii) The telephone book search with N entries (more general, the *search-a-sorted-list* problem) outputs the searched-for number $x(N)$ in minimal time $H_D = O(\log N)$.

(iii) The travelling salesman problem of finding the *shortest* route for a traveller who wishes to visit each of N cities. The solution is the sequence $x(N)$ of the cities visited consecutively. It is not difficult to reason that $H_D(x(N)) \leq O(N!)$, but the statement that the problem is intractable, i.e., that it cannot be solved in polynomial time, i.e., that the minimal time is $H_D(x(N)) > O(N^k)$, where $k < \infty$, remains conjectural [227, 196].

(iv) The task of finding the first n bits of the halting probability Ω (see section 14.1.2, p. 196) is, for large n, uncomputable and $H_D(\Omega(n)) = \infty$ [109, 110].

Ch. H. Bennett [35] has proposed an alternative definition of computational complexity. There, the computer program is fixed by identifying p with the *canonical*, i.e., smallest-size, program $[x(n)]^*$ generating a (binary) sequence $x(n)$ of length n. The resulting measure is the *"logical depth."* For a much more detailed analysis, see Ch. H. Bennett [37].

D 8.2 (Logical depth (Bennett [35]))

The logical depth $D(x(n))$ *of a sequence* $x(n) = x_1 \dots x_n$ *is the time (number of cycles, computing steps) it takes for the canonical program* $[x(n)]^*$ *running on a universal computer* U *to calculate* $x(n)$*:*

$$D(x(n)) = \mathtt{TIME}([x(n)]^*) \quad (U([x(n)]^*) = x(n)) \quad .$$

If no program makes computer U *output* x*, then* $D(x) = \infty$.

Remarks:

(i) By specifying a Turing machine model (or any one which outputs only one digit at the time), one obtains $D(x(n)) \geq |x(n)|$.

(ii) In view of the possibility of a "trade-off" between computing speed (related to computational complexity) and program size (related to algorithmic information) [101], algorithmic information — logical depth and program size of the fastest program — computational complexity are dual concepts.

(iii) In Ch. H. Bennett's terminology, a sequence $x(n)$ is $[H(x(n)), D(x(n))]$*-deep*, or $D(x(n))$ cycles *deep* with $H(x(n))$ bits *significance*. Any string $x(n)$ might be called "shallow" if it can be produced in a time D which grows not faster than some polynomial in n, i.e., $D(x(n)) \leq n^k,\ k < \infty$. This amounts to saying that the problem of finding $x(n)$ is in the complexity class of *polynomial-time algorithms* P; i.e., it is *tractable* (see definition 8.3, p. 102). Otherwise it might be called "deep." This terminology is somewhat arbitrary but justified by the fact that the class P is closed under variations of "reasonable" computer models, meaning that one computer can simulate another by a polynomial-time algorithm.

We shall use computational complexity to formalise the notion of randomness based on the heuristic approach of "computational irreducibility" in chapter 14.2, p. 197.

8.1.1 Uncomputability

The uncomputability of computational complexity can be proved by contradiction, utilising diagonalization techniques: Assume that there exists an effectively computable algorithm `TIME` which, according to the Church-Turing thesis, would correspond to a recursive function H_D (wrong). `TIME` computes the running time of an arbitrary algorithm p which, implemented on machine U, produces the output $x(N)$ and, if $U(p)$ does not halt, renders infinity. By identifying $x(N)$ with an arbitrary sequence $x(n)$ of length n, `TIME` could construct an effectively computable "halting program" `HALT` as follows:

$$\mathtt{HALT}(U(p) = x(n)) = \begin{cases} \mathtt{TRUE} & \text{if } \mathtt{TIME}(U(p) = x(n)) < \infty \\ \mathtt{FALSE} & \text{if } \mathtt{TIME}(U(p) = x(n)) = \infty \end{cases} \quad . \tag{8.1}$$

This contradicts the undecidability of the halting problem, theorem 9.2, p. 114. Therefore one is either left with a contradiction or with the uncomputability of `TIME`. There is no other consistent choice than accepting that `TIME` does not correspond to any effectively computable algorithm.

8.1.2 Machine dependence

In contrast to algorithmic information, which, up to $O(1)$, remains unchanged under changes of universal computer models, computational complexity is a highly machine dependent concept. However, as will be argued below, "reasonable" universal machines simulate one another by polynomial-type algorithms. Therefore, one computation which has a polynomial time bound on one computer will also have a polynomial time bound on virtually any other; perhaps though by a polynomial of different degree. This means for instance that, in Ch. H. Bennett's terminology, a shallow string remains shallow and a deep string remains deep.

Let us, for the moment, consider *Amdahl's law* which evaluates the time gain of suitably algorithms on machines which process information *in parallel:* Let f be the percentage of possible parallelism in this algorithm and N be the number of parallel units, then the relative increase in computing time with respect to a single computing unit ($N = 1$) is $[f/N+(1-f)]^{-1}$. Denote by C_N a computer with N instantly accessible processing units. We may then consider the computational complexity $H_D(N, f)$ as a function of the machine C_N and the degree of possible parallelisation. As a consequence, with an infinite number of instantly accessible processing units ($N = \infty$), the decrease in computational complexity with respect to a single processor unit is given by

$$H_D(\infty, f) = (1-f)H_D(1, f) \quad .$$

It is assumed that the relative percentage of possible parallelism f refers to the *shortest* programs for N processors. For $f = 0.5$, $H_D(\infty, 0.5) = 0.5\, H_D(1, 0.5)$, which means that such a problem could be solved in half of the time it takes for only one processor if two processors are employed. [Amdahl's law can be derived as follows: consider the total computation time t_p for a parallelised problem with respect to the time t_s for the corresponding sequential problem; t_p consists of the parallel computing time $t_s f/N$ (if parallelisation can make use of all N units) plus the remaining non-parallel computing time $t_s(1-f)$, i.e., $t_p = t_s[f/N-(1-f)]$. Amdahl's law is the ratio t_p/t_s.]

In what follows we restrict our attention (if not stated otherwise) to single-unit sequential universal computers such as to the Turing machine C_T.

8.2 Computational complexity classification

The following overview is only a short briefing on computational complexity classification. More detailed accounts can for instance be found in the books by D. Harel [227], Jan van Leeuwen [296] (editor), M. R. Garey & D. S. Johnson [196], L. Kronsjö [279], C. Calude [65] and J. E. Hopcroft & J. D. Ullman [240], among many others.

Let again N be some number characterising the *size* of a computational problem; for instance the task of sorting a list of N items, or finding the roots of a polynomial of degree N. Then we may ask, "what is the functional behaviour of the time complexity $H_D(N)$, i.e., the minimal amount of computation time to solve a problem, as N increases?"

8.2.1 The class P

There is one important class of problems, the *polynomial-time bound algorithms* P, which coincides with the class of effectively computable algorithms which require "reasonable" running time and data storage.

D 8.3 (Polynomial-time algorithms P)
The class of polynomial-time algorithms P *contains algorithms which can be solved in* polynomial time*; i.e., whose computational complexity increases with* $H_D(N) \leq O(N^k)$ *for fixed* $k < \infty$.

Computational problems which can be solved by polynomial-time algorithms are called tractable *or* feasible*; if they are solvable but* not *in P, they are called* intractable.

Examples:

Examples for problems in P are the telephone book search with N entries (more general, the *search-a-sorted-list* problem) with $H_D = O(\log N)$, or the *sorting N entries* problem with $H_D = O(N \log N)$. Problems with $O(2^N)$, $O(N^N)$ or $O(N!)$ are not in P.

8.2.2 Invariance of complexity class P

The class P is "closed" with respect to variations of "reasonable" computer models. That means that a problem which can be algorithmically solved by a running time which is polynomially bounded on one specific computer (with specific elementary operations, number of tapes *et cetera*) can be solved within polynomial time bounds on any other "reasonable" machine as well. I.e., one assumes that there exists a polynomial-time overhead, originating, for instance, from the execution of an effective translation program from one code into the other code. The degree of the polynomial might differ, though. I.e., it is not guaranteed that that a problem requireing $O(N^k)$ time on one computer will require $O(N^k)$ time on another computer, but it *can be* guaranteed that it requires $O(N^l)$ time with some other fixed exponent l. The constant-factor overhead in space (storage size) is not motivated here. The reader is, for instance, referred to P. van Emde Boas' article [46]. One may formulate the following conjecture:

C 8.4 (Invariance thesis) *"Reasonable" computer models can simulate each other within a polynomial-time overhead and a constant-factor overhead in space (storage size).*

A further conjecture, relating execution time in parallel to computers to space (storage size) in sequential computers can be stated as follows.

C 8.5 (Parallel computation thesis)
Whatever can be solved in polynomially bounded space (storage size) on "reasonable" sequential computers can be solved in polynomially bounded time on "reasonable" parallel computers, and vice versa.

type of problem	algorithmic status	computer model
⋮	⋮	⋮
unbounded	undecidable	*oracle*
⋮	⋮	⋮
exponential bound $H_D(N) \leq O(k^N)$	computable but intractable	universal computer (universal CA, Turing Machine *etc.*)
polynomial bound $H_D(N) \leq O(N^k)$	tractable	universal computer (universal CA Turing Machine *etc.*)
	finite $N < \infty$	finite machine (insert your favorite brand here: "…")

Table 8.1: Correspondence between types of problems, their algorithmic status, and devices capable to solve them.

Remarks:

(i) One may translate these conjectures into definitions if one defines computers to be "reasonable" only if they satisfy the statements.

(ii) Similar to the parallel computation thesis, a "trade-off" can be formulated between *algorithmic information and computational complexity.* For further details, the reader is referred to an article by G. Chaitin [101].

8.2.3 The class NP

There exists an interesting class of problems which may take excessively large computation times when evaluated straightforwardly (i.e., deterministically); yet their solution could be enhanced by the introduction of an *oracle* and a subsequent verification procedure. This class is called NP. It is defined as follows:

D 8.6 (Algorithms NP)

The class of non deterministic polynomial-time algorithms NP *contains algorithms which can be* solved *by a (non deterministic)* oracle *and* verified *in* polynomial time.

Although it is evident that P⊂NP (substitute the polynomial-time algorithm for the oracle), N≠NP remains conjectural. Table 8.1 gives an overview of the type of computations corresponding to functional classes and machine models; deterministic or non deterministic.

8.3 Maximum halting time

The *"busy beaver function"* Σ (I would have rather called it *"wild weasel function"*) is the answer to the question, *"what is the largest number which can be calculated by programs whose algorithmic information content is less than or equal to a fixed, finite natural number?"* In other words:

D 8.7 (Busy beaver function) *The* busy beaver function $\Sigma(n)$ *is defined by the* largest *natural number whose algorithmic information content is less than or equal to n; i.e.,*

$$\Sigma(n) = \max_{H(k)\leq n} k \quad .$$

This definition is due to Chaitin [101, 107]. Originally, T. Rado [379] asked how many 1's a Turing machine with n possible states and an empty input tape could print on that tape before halting. The first values of the Turing busy beaver $\Sigma_T(x)$ are known to be [142, 51]:

number of states n	number of 1's printed
$\Sigma_T(1)$	1
$\Sigma_T(2)$	4
$\Sigma_T(3)$	6
$\Sigma_T(4)$	13
$\Sigma_T(5)$	≥ 1915
$\Sigma_T(7)$	≥ 22961
$\Sigma_T(8)$	$\geq 3\cdot(7\cdot 3^{92}-1)/2$.

For large values of n, $\Sigma_T(n)$ *grows faster than any computable function* of n; more precisely, let f be an arbitrary computable function, then there exists a positive integer k such that

$$\Sigma(n) > f(n) \text{ for all } n > k \quad .$$

In this sense, Σ has a very similar role with respect to the recursive functions as the Ackermann generalised exponential A (defined in 1.1, p. 7) has with respect to the primitive recursive functions.

The following results have not been proven for instantaneous codes. Nevertheless, it could be expected that they hold even after translation to instantaneous codes. As a consequence of its definition, Σ satisfies the following relations:

T 8.8 ((G. Chaitin [109])) *If Σ is defined, then*

$$H(\Sigma(n)) \leq n \quad ; \tag{8.2}$$

$$H(\Sigma(n)) = n+O(1) \quad ; \tag{8.3}$$

$$\Sigma(n) \leq \Sigma(n+1) \quad ; \tag{8.4}$$

$$\text{if } H(n) \leq m, \text{ then } n\leq\Sigma(m) \quad ; \tag{8.5}$$

$$n \leq \Sigma(H(n)) \quad ; \tag{8.6}$$

$$\text{if } H(n) > i, \text{ for all } n\geq m, \text{ then } m>\Sigma(i) \quad . \tag{8.7}$$

Although in general the computational complexity or logical depth associated with a (finite or infinite) object is uncomputable, it is possible to derive a non recursive/uncomputable) bound from above, associated with the algorithmic complexity of programs which halt:

T 8.9 (Bound from above on computation time and complexity (G. Chaitin [109])) *Either a program p halts in cycle time less than $\Sigma(H(p)+O(1))$ or it never halts. Therefore, if one defines $d(n) = \max_{|x^*|\leq n} H_D(x) = \max_{|x^*|\leq n} D(x)$, then*

$$d(n) = \Sigma(n + O(1)) \quad .$$

I.e., $\Sigma(n+O(1))$ is the minimum time $d(n)$ at which all programs of complexity $\leq n$ which halt have done so.

Proof:

Here we shall only proof that there is an *upper bound* for the computational complexity and logical depth for programs of algorithmic information $\leq n$, namely for objects x with $H(x) \leq n$, $H_D(x), D(x) \leq d(n) \leq \Sigma(n + O(1))$: consider two universal computers U and U'. Given an arbitrary program p with $H(p) \leq n$, $U(p)$ simulates $U'(p)$ and additionally *counts* the cycle time t of U' until p has halted. If this happens, $U(p)$ outputs the execution time t of p and halts. Now if $d(n)$ is defined, then there exists at least one program p_L of algorithmic information $\leq n$ which takes longest, such that $U(p_L) = d(n)$. Since, independent of n, the simulation of U' on U and counting requires $O(1)$ additional bits of program size, the algorithmic information of p with respect to U is governed by the algorithmic information of p with respect to U' with an additional additive constant, i.e., $H(d(n)) \leq n + O(1)$. By theorem (8.5) we conclude that $d(n) \leq \Sigma(n + O(1))$.

Remark:

Intuitively speaking, production of $\Sigma(n)$ is among the most time-consuming tasks for a program of complexity $\leq n$ — indeed, it takes $\Sigma(n + O(1))$ cycles to do so.

T 8.10 (Uncomputability of Σ) *For large n, $\Sigma(n)$ is not effectively computable / non recursive.*

Proof by contradiction: Assume $\Sigma(n)$ were computable. This would provide a recursive solution to the halting problem, since, given an n bit program, one would only have to wait until time $\Sigma(H(n) + O(1))$, when all programs which halt have done so. But recursive solutions are not allowed by theorem 9.2, p. 114. The alternative is either a contradiction or acceptance of the uncomputability of Σ. There is no other consistent choice than stating the uncomputability of Σ.

8.4 Greatest recurrence period

Recurrence is defined by periodicity in output.

D 8.11 (Recurrence) *A computer $U(p)$ producing a periodic output*

$$\underbrace{x_1 x_2 \cdots x_k}_{k \text{ places}},\underbrace{x_{k+1} x_{k+2} \cdots x_{2k}}_{k \text{ places}},\cdots$$

with

$$x_i = x_{nk+i}$$

for all $0 \leq i \leq k$ *and* $n \in \mathbb{N}$ *has a* period length *or* recurrence k.

T 8.12 (Bound from above on recurrence) *The greatest recurrence T_r of a program of algorithmic information n is given by*

$$T_r(n) = \Sigma(n + O(1)) \quad .$$

I.e., a program of length n on a universal computer cannot have a period length exceeding $\Sigma(n + O(1))$, and there exist programs of length $n + O(1)$ with period length $\Sigma(n)$.

Proof:

First, $T_r(n) \leq \Sigma(n + O(1))$ will be proved. Consider again two universal computers U and U'. Given an arbitrary program p with $H(p) \leq n$, $U(p)$ simulates $U'(p)$ and additionally produces a constant output symbol (say, "0") until p has halted. If this happens, $U(p)$ outputs a different output symbol (say, "1") and goes into an infinite loop. If $T_r(n)$ is defined, then there exists at least one program p_L of algorithmic information $\leq n$ which takes longest, such that the recurrence of $U(p)$ is $T_r(n)$. Since, independent of n, the simulation of U' on U, outputting either "0" or "1" and supplementary operations requires $O(1)$ additional bits of program size, the algorithmic information of p with respect to U is governed by the algorithmic information of p with respect to U' with an additional additive constant, i.e., $H(T_r(n)) \leq n + O(1)$. By theorem (8.5) one concludes that $T_r(n) \leq \Sigma(n + O(1))$. The foregoing construction shows that there exist programs of length $n + O(1)$ which yields output of periodicity $\Sigma(n)$.

Part II
Undecidability

ALPHA 60: I will calculate ... so that failure ... is impossible.
LEMMY: I'll fight until failure *does* become possible.
ALPHA 60: Everything I plan will be accomplished.
LEMMY: That's not certain. I too have a secret.
ALPHA 60: What is your secret? ... Tell me ...Mr Caution.
LEMMY: Something that never changes with the night or the day, as long as the past represents the future, towards which it will advance in a straight line, but which, at the end, has closed on itself into a circle.
...
ALPHA 60: Several of my circuits are looking for the solution to your puzzle. I will find it.
LEMMY: If you find it ... you will destroy yourself in the process ... because you will have become my equal, my brother.
from "Alphaville" by Jean-Luc Godard [203]

Chapter 9

Classical results

9.1 True ≠ provable

Consider the classical [310] liar paradox "`I am lying`" in the form "`this statement is false.`" Kurt Gödel achieved a mathematically meaningful theorem by translating it into a formal statement, which can be informally expressed as "`this statement is unprovable.`" Gödel himself was well aware of this analogy. Already in his centennial 1931 paper [204], he stated,

> *"Die Analogie dieses Schlusses mit der Antinomie* Richard *springt in die Augen; auch mit dem "Lügner" besteht eine nahe Verwandtschaft [Fußnote 14: Es läßt sich überhaupt jede epistemologische Antinomie zu einem derartigen Unentscheidbarkeitsbeweis verwenden.], ..."*
> *English translation [205], p. 149:*
> *"The analogy of this argument with the Richard antinomy leaps to the eye. It is closely related to the "Liar" too; [footnote 14: Any epistemological antinomy could be used for a similar proof of the existence of undecidable propositions] ..."*

The consequence of a claim like "`this statement is unprovable`" can be summarised by the following alternative: *(i)* if, on the one hand, this statement were provable in a particular formalism, then this fact would contradict the message of the statement (i.e., its unprovability), which in turn would render the whole formalism *inconsistent; (ii)* if, on the other hand, this statement would be unprovable in a particular formalism, this would confirm the message of the statement. As a result, the formalism would be *incomplete* in the sense that there exist true statements which cannot be proven. — There is no other consistent choice than rejecting *(i)* and accepting incompleteness *(ii)*.

Similarly, other metamathematical and logical paradoxa [266] have been used systematically for the derivation of undecidability or incompleteness theorems. The method is proof by contradiction: first, a statement is assumed to be true; this statement yields absurd (paradoxical) consequences; the only consistent choice being its unprovability or nonexistence. Mostly, absurd consequences are constructed by similar techniques as Cantor's diagonalization method; see below.

It is certainly not illegitimate to ask, *"what is the 'essence' or the 'meaning' of Gödel's incompleteness result,"* and, *"what is the 'feature' responsible for it?"* There exist at least two features of the argument which are noteworthy: self-reference and the possibility to express but not to prove (!) certain facts about a formal theory intrinsically, i.e., within the theory itself. (This occurs only for theories which are "strong enough" to allow coding of metastatements *within* the theories themselves. Theories which are "too weak," i.e., theories in which metastatements *within* the theories themselves cannot be coded, do not feature incompleteness of this kind, although they are incomplete in a more basic sense.) The next two sections will concentrate on these two issues.

9.1.1 Self-reference

At first glance, *self-reference* might be considered as the essential source of the paradox yielding Gödel's incompleteness result. Consider, for instance, a claim very similar to the liar paradox: "`this statement is true`." This claim is weird (i.e., *not well-founded*) in the sense that no contradiction occurs if one assumes that it is true or false; yet its *meaning* remains unclear, because it does not really make any difference whether one chooses one of the two cases. More generally, it has been suggested [382, 60] that pathologies occur if a symbol (e.g., a word, a sentence, a statement, a language *et cetera*) refers to its own *semantics, meaning* or *interpretation*; more precisely, if a symbol refers to the relation between itself and the object it stands for. This might be considered as one example of the general fact that there exist objects which cannot be named by any (formally defined) finite language. (The concept of "truth" discussed below is another example.)

Another, lively, example is a stone disc outside of Rome's Santa Maria in Cosmedin. There is a slot carved in the stone disc. Legend has it that anyone sticking a hand into this slot while uttering a wrong statement will not be able to get the hand out again. Rudy Rucker confessed thrusted his hand into the slot by saying, "`I will not be able to pull my hand back out`" [397] — whoever was responsible for the reaction of the slot must have been in real trouble! The crocodile version of this dilemma [382] is a crocodile, having stolen a child and saying to the child's mother, "`only if you guess whether I am going to give you back your child or not, shall you get it back`," with the mother replying, "`you will not give it back`." Still another all-time favourite is Russell's paradox [400] in the form of a male barber who is obliged to "`shave all people who do not shave themselves`". This order is inconsistent when applied to the barber himself. — Russell's paradox applies to G. Cantor's 1883 definition of a set, i.e., *"A set is a Many which allows itself to be thought of as a One"*: the "`set of all sets which are not members of themselves`" is a paradoxical construction, which, among other paradoxes embodied in G. Cantor's original set theory, has motivated the axiomatisation of set theory at the price of restricting the mathematical universe.

Yet, besides all these difficulties, there is nothing wrong with self-reference *per se*: take for instance the claim "`this statement contains` $\cdots$ `words`." — If $\cdots$ = `five`, the sentence is true, for all other number-arguments it is false. (Nevertheless, paradoxical constructions are quite close: consider, for instance, Berry's paradox in the form, "`the smallest positive integer that cannot be specified in less than a thousand words`", which, if it existed, has just been specified by fourteen words.)

Indeed, self-reference is an essential feature of any intrinsic perception. Therefore it is not unreasonable to assume that self-reference, if applied properly, yields well-defined statements. Troubles in the form of inconsistencies occur in particular circumstances, e.g., by attempting a *complete* self-interpretation or, technically, after some kind of "diagonalization." As will be discussed later, these inconsistencies, interpreted properly, represent a *via regia* to undecidability.

9.1.2 Truth *versus* provability

Another recurrent feature of all the above-mentioned logical paradoxa is the use of an "evident," though incorrect, notion of *truth.* Although not expressed in the original paper, it is worth noting that K. Gödel himself considered this to be the main feature of his incompleteness theorems. (This citation is taken from K. Gödel's reply to a letter by A. W. Burks, reprinted in J. von Neumann's *Theory of Self-Reproducing Automata* [340], p. 55; see also [167], p. 554.)

> *"I think the theorem of mine which von Neumann refers to is not that on the existence of undecidable propositions or that on the lengths of proofs but rather the fact that a complete epistemological description of a language A cannot be given in the same language A, because the concept of truth of sentences of A cannot be defined in A. It is this theorem which is the true reason for the existence of undecidable propositions in the formal systems containing arithmetic. I did not, however, formulate it explicitly in my paper of 1931 but only in my Princeton lectures of 1934. The same theorem was proved by Tarski in his paper on the concept of truth [[cf. A. Tarski's earlier paper [441]]] published in 1933 in* Act. Soc. Sci. Lit. Vars., *translated on pp. 152-278 of* Logic, Semantics and Metamathematics *[[[442]]]."*

Absolute truth is a transfinite concept, which is *not definable by any finite description.* An informal proof by contradiction of this fact can be envisioned as follows: suppose that a finite description TM of a "universal truth machine" exists. (No difference is made here between TM and its code #(*TM*).) The truth machine is supposed to work in the following way: one inputs an arbitrary statement, asking whether it is correct. TM then outputs `TRUE` or `FALSE`, depending on whether the statement has been correct or incorrect, respectively. Now consider the input, "`the machine described by` TM `will not output that this statement is true`". The resulting problem is of the same kind as the one encountered in the liar paradox: The truth machine cannot say `TRUE` or `FALSE` without running into a contradiction. Therefore, TM cannot decide all questions (it is unable to decide at least one question), contradicting the assumption that the truth machine decides all questions. However, somebody watching from the *outside* (i.e., someone who is not part of this truth machine) knows that the above statement is $\widetilde{\mathtt{TRUE}}$, but this results in an extrinsic notion of truth which is stronger than the "portion of truth" available to the truth machine (to indicate this, the truth value is tilded). One could, of course, proceed by simply adding this statement to TM, producing a $\widetilde{\mathrm{TM}}$. By the same argument, $\widetilde{\mathrm{TM}}$ would not be able to decide the input, "`the machine described by` $\widetilde{\mathrm{TM}}$ `will not output that this statement is true`," which is $\widetilde{\widetilde{\mathtt{TRUE}}}$, ..., forcing a hierarchy of notions of truth *ad infinitum.*

The paradox of the liar and similar paradoxes can thus be resolved by accepting that these constructions operate with a bounded "degree" or "strength" of truth. In this sense, Gödel's incompleteness results mean a formal proof that the notion of truth is too comprehensive to be grasped by any finite mathematical model. No comprehensive concept of truth can be defined. In particular, no intrinsic consistency proof can be given.

9.1.3 Consequences for physics

The impact of these findings on physics is hardly perceived yet. It would be, for instance, quite tempting to speculate that Kurt Gödel and Albert Einstein, two close friends and colleagues at the Institute for Advanced Study in Princeton, who almost certainly discussed foundational aspects intensively ([205], pages 11,12), considered possible similarities between the Heisenberg uncertainty principle / complementarity in quantum mechanics and incompleteness / undecidability in mathematics. The following anecdote by John Archibald Wheeler has been communicated to me by Gregory Chaitin (G. Chaitin, *Naturwissenschaft und Weltbild*, seminar, Technical University Vienna, January 15th, 1991; a printed for-the-record version can be found in [39]):

> *[[In 1979]] I went up to [[John Archibald]] Wheeler and I asked him, "Prof. Wheeler, do you think there's a connection between Gödel's incompleteness theorem and the Heisenberg uncertainty principle?" Actually, I'd heard that he* did, *so I asked him, "What connection do you think there is between Gödel's incompleteness theorem and Heisenberg's uncertainty principle?"*
> *This is what Wheeler answered. He said, "Well, one day I was at the Institute for Advanced Study, and I went to Gödel's office, and there was Gödel..." I think Wheeler said that it was winter and Gödel had an electric heater and had his legs wrapped in a blanket.*
> *Wheeler said, "I went to Gödel, and I asked him, 'Prof. Gödel, what connection do you see between your incompleteness theorem and Heisenberg's uncertainty principle?' " I believe that Wheeler exaggerated a little bit now. He said, "And Gödel got angry and threw me out of his office!" Wheeler blamed Einstein for this. He said that Einstein had brain-washed Gödel against quantum mechanics and against Heisenberg's uncertainty principle!*

Despite such anecdotal dissent, several attempts have been made to translate mathematical undecidability into a physical context [259, 274, 354, 396, 327, 131, 132], among them two very notable early articles by Karl Popper [366] in the late 40'th.

After a brief review of the "classical" mathematical results, we shall take a fresh look at this issue. Various forms of undecidability related to physics can be formalised by the method of *diagonalization.* (Diagonalization has already been used in previous chapters and has been introduced first by Georg Cantor for a proof of the undenumerability of the reals.) The following physical applications will be discussed in their classical context first:

(i) The *problem of forecast* of a mechanistic, i.e., totally computable, system. Thereby it is assumed that absolute knowledge about the recursive laws governing that system has been given to us by some "oracle." The general problem of forecast will be linked to

what is called the *"recursive unsolvability of the halting problem."*

(ii) While the problem of forecast appears already in a very generel extrinsic context, one can consider the same problem in an *intrinsic* setup. I.e., consider a theory about a system *represented within that very system.* K. Gödel, utilising the technique of Gödel numbering, achieved the undecidability theorems by expressing (meta-arithmetic) statements about arithmetics within arithmetics. In analogy, one could hope for expressing *"intrinsic indeterminism"* by representing theoretical statements about a system within that very system.

(iii) The *rule inference problem* can be expressed by the question, *"given a specified class of laws, usually the class of recursive / computable functions, which one of these laws governs a particular system?"* Thereby it is assumed that the system is treated as a black box; i.e., one is only allowed to perform input / output analysis. This kind of problem is related to the problem of identifying and learning a language.

(iv) The impossibility to state exactly the algorithmic information content of an arbitrary sequence.

(v) The construction of "toy universes," generated by finite automata and the investigation of their logical structure. One goal is the creation of nonlocal automata models which feature quantum-like behaviour, in particular *"computational complementarity."*

(vi) Intrinsically, the physical measurement and perception process exhibits (paradoxical) features resembling computational complementarity and diagonalization. An idealised measurement attempts the impossible: on the one hand it pretends to grasp the "true" value of an observable, while on the other hand it has to interact with the object to be measured and thereby inevitably changes its state. Integration of the measurement apparatus does not help because then the observables inseparably refer to the state of the object and the measurement apparatus *combined*, thereby surrendering the original goal of measurement (i.e., the measurement of the object). These considerations apply to quantum as well as to classical physics with the difference that quantum theory postulates a lower bound on the transfer of action by Planck's constant $\hbar$.

One may even embark on a much more radical (metaphysical) program, which can be stated pointedly as follows: *"All instances of "randomness" in physics will eventually turn out to be undecidable features of mechanistic systems. There is no randomness in physics but a constant confusion in terminology between randomness and undecidability. God does not play dice."*

9.2 Cantor's diagonalization method

Georg Cantor's diagonalization technique has first been introduced around 1873 (in a correspondence with Dedekind) and since then has become the *via regia* to the investigation of undecidability. Probably the most prominent application, already introduced by G. Cantor himself, is a proof that the set of real numbers is *not denumerable,* i.e., there cannot exist any complete listing of the reals, one after the other.

Assume for the moment that the set of reals is denumerable. (This assumption will yield a contradiction.) That is, the enumeration is a one-to-one function $f : \mathbb{N} \to \mathbb{R}$

(wrong), i.e., to any $k \in \mathbb{N}$ exists some $r_k \in \mathbb{R}$ and *vice versa*. No algorithmic restriction is imposed upon the enumeration, i.e., the enumeration may or may not be effectively computable. For instance, one may think of an enumeration obtained *via* the enumeration of computable algorithms and by assuming that r_k is the output of the k'th algorithm. Let $0.d_{k1}d_{k2}\cdots$ be the successive digits in the decimal expansion of r_k. Consider now the diagonal of the array formed by successive enumeration of the reals,

$$\begin{array}{rcllll} r_1 & = & 0.d_{11} & d_{12} & d_{13} & \cdots \\ r_2 & = & 0.d_{21} & d_{22} & d_{23} & \cdots \\ r_3 & = & 0.d_{31} & d_{32} & d_{33} & \cdots \\ \vdots & & \vdots & \vdots & \vdots & \ddots \end{array} \tag{9.1}$$

yielding a new real number $0.d_{11}d_{22}d_{33}\cdots$. Now change each one of these digits, avoiding zero and nine in a decimal expansion. This is necessary because reals with different digit sequences are equal to each other if one of them ends with an infinite sequence of nines and the other with zeros, for example $0.0999\ldots = 0.1000\ldots$. The result is a real $r' = 0.d_1'd_2'd_3'\cdots$ with $d_n' \neq d_{nn}$ which differs from each one of the original numbers in at least one (i.e., in the "diagonal") position. Therefore, there exists at least one real which is not contained in the original enumeration, contradicting the assumption that *all* reals have been taken into account. Hence, $\mathbb{R}$ is not denumerable.

9.3 Recursive unsolvability of the halting problem

The *"halting problem"* is related the question of how a mechanistic system will evolve or what an algorithm or an automaton will output or what theorems are derivable in a formal system. In other words, it is referring to the *problem of forecast* for a mechanistic system.

D 9.1 (Halting problem)
Let there be an algorithm A and an input x. The halting problem *(Church version) is the decision problem associated with the question whether or not $A(x)$ will produce a specific output y in finite time. Equivalently, one may ask whether A will* terminate *on x (Turing version). The case "A terminates or* converges *on x" is denoted by $A(x)\downarrow$; the case "A does* not *terminate or* diverge *on x" is denoted by $A(x)\uparrow$.*

A. Church's version of the halting problem reduces to A. Turing's version if the termination condition is the production of output y.

T 9.2 (Recursive unsolvability of the halting problem (A. Turing [446])) *There is no effectively computable algorithm / partial recursive function which decides the halting problem. The halting problem is unsolvable.*

To obtain more intuition, let us see how J. von Neumann interpreted the recursive unsolvability of the halting problem (the following quotation is taken from J. von Neumann's *Theory of Self-Reproducing Automata*, ed. by A. W. Burks [340], p. 51):

> *Turing proved that there is something for which you cannot construct an automaton; namely, you cannot construct an automaton which can predict*

> *in how many steps another automaton which can solve a certain problem will actually solve it. So, you can construct an [[universal]] automaton which can do anything any automaton can do, but you cannot construct an automaton which will predict the behaviour of any arbitrary automaton. In other words, you can build an organ which can do anything that can be done, but you cannot build an organ which tells you whether it can be done.*

The following three proofs by contradiction of the unsolvability of the halting problem use Cantor's diagonalization technique. They use algorithmic arguments, which, by the Church-Turing thesis, are valid for the class of recursive functions as well. (One might prefer not to make use of such "proofs by the Church-Turing thesis." Then it would be necessary to construct an explicit model, a concrete implementation, of these algorithms. — See, for example, A. Turing's article [446].)

Proof 1:

Consider an arbitrary algorithm $A(x)$ with input x. x is a string of symbols. Assume that there exists a "halting algorithm" HALT (whose existence should be disproved) which is able to decide whether A terminates on x or not. [I.e., HALT($A(x)$) $\downarrow$, i.e., the termination of HALT, is a property of this fictious "halting algorithm".]

Using HALT($A(x)$) it is easy to construct another algorithm, which will be denoted by B and which has as input any effective program A and which proceeds as follows: Upon reading the program A as input, B makes a copy of it. This can be readily achieved, since the program A is presented to B in some encoded form #(A), i.e., as a string of symbols. The code #(A) is used as input string for A itself; i.e., B forms A(#(A)), henceforth denoted by $A(A)$, and hands it over to its subroutine HALT. Then, B proceeds as follows:

case *(i)*: if HALT($A(A)$) decides that $A(A)$ halts, then B does not halt. This can for instance be realised by an infinite DO-loop.

case *(ii)*: if HALT($A(A)$) decides that $A(A)$ does *not* halt, then B halts.

We shall demonstrate that there is something wrong with B and, since a derivation of B from HALT is trivial (i.e., all manipulations of B despite the computation of HALT are obviously computable), that there must be something wrong with HALT: Recall that A is arbitrary and has not been specified yet. What happens if we substitute B for A, i.e., if we input B into itself and carry the argument through?

case *(i)*: Assume that $B(B)$ halts, then HALT($B(B)$) forces $B(B)$ not to halt;

case *(ii)*: assume that $B(B)$ does not halt, then HALT($B(B)$) steers $B(B)$ into the halting mode. In both cases one arrives at the contradiction *"B (does not) halt[s] if B [does not] halt(s)"*. This contradiction can only be consistently avoided by assuming the nonexistence of B and, since B is a trivial construction from HALT, the *impossibility of a halting algorithm* HALT.

This proof is important not only because of its relevance for the investigation of effectively computable processes but also due to its *method*, which is very similar to Cantor's *diagonalization* method described previously. The syntactic structure is essentially given by

$$\texttt{if } A(A)\downarrow \;\Rightarrow\; \texttt{define } B(\texttt{HALT}(A(A)))\uparrow$$
$$\texttt{if } A(A)\uparrow \;\Rightarrow\; \texttt{define } B(\texttt{HALT}(A(A)))\downarrow \quad ,$$

$$
\begin{aligned}
\texttt{substitute}\ B &\rightarrow A \ . \\
\texttt{contradiction}: B(B)\uparrow\downarrow &\Rightarrow B(\texttt{HALT}(B(B)))\downarrow\uparrow \\
&\Rightarrow B \texttt{ does not exist} \\
&\Rightarrow \texttt{HALT does not exist} \ ,
\end{aligned}
$$

which is structurally equivalent to Cantor's diagonalization method and which is employed for a proof of Gödel's undecidability theorem as well.

Proof 2: A very similar argument goes like this. Assume a function f (whose existence we shall disproof) which is defined as follows: $f(k)$ is either one more than the value of the kth computable algorithm applied to the number $k \in \mathbb{N}$ or zero if this algorithm does not halt on input k. f itself cannot be computable, for if the $k = n$th algorithm calculated it, one would obtain $f(n) = f(n)+1$, which is impossible. The only way that f can fail to be computable is because one cannot decide if the nth program ever halts on input n.

Proof 3: Consider algorithms for enumerating sets of natural numbers, and number these algorithms. Define a set of natural numbers associated with all algorithms which do not contain their own number in their output set. This set of natural numbers cannot be computable, for any algorithm which would attempt to compute it would end up with a dilemma known as Russel's paradox [400] of "`a male barber who shaves all those and only those who do not shave themselves`", who cannot shave himself nor avoid doing so (a more abstract version of Russel's paradox is obtained by considering "`the set of all sets which are not member of themselves`"). There is only way out od the dilemma: such a set cannot be computable because it is generally impossibile to decide whether or not a specific algorithm outputs a specific natural number. This is Church's variant of the halting problem.

Since there is a one-to-one translation between algorithmic entities and formal systems, one has proven a form of Gödel incompleteness as well. A statement about the convergence of an algorithm can be interpreted as a mathematical statement which sould, at least in principle, be decidable within a formal system. For, if it were always possible to prove a theorem whether or not a particular algorithm will halt, this would contradict the halting problem.

An immediate physical consequence of the recursive unsolvability of the halting problem is the following corollary:

C 9.3 (Unpredictable physical observables of mechanistic physical systems)
There exist mechanistic physical systems (whose laws and parameters are known) whose behaviour is unpredictable (undecidable).

Examples of physical systems are general-purpose computers, such as (insert your favourite brand here:) "...," denoted by U. These machines are *universal* up to finite computational resources. Therefore, the recursive undecidability of the halting problem applies to them. (The finiteness and discreteness of their state spaces limits the argument to some extend. Nevertheless, for all practical reasons these limitations are irrelevant: every megabit of memory space is worth $2^{1000000} \approx 10^{300000}$ possible states.) More explicitly, by implementing an arbitrary (legal) program p on U (assume an empty input list), a physical "halting" parameter $s_U(p)$ can be defined by

$$s_U(p) = \begin{cases} 0 & \text{if } U(p) \downarrow \\ 1 & \text{if } U(p) \uparrow \end{cases} \tag{9.2}$$

In general, by theorem 9.2 (i.e., due to the recursive unsolvability of the halting problem), $s_U(p)$ is unpredictable (undecidable).

Another, very similar, undecidable physical entity has been introduced by K. Popper [366], pp. 181-182. In two remarkable papers [366], K. Popper gives two other arguments for the impossibility of self-prediction, based on what Popper calls "Gödelian sentences" and the "Ödipus effect." In summary, he states,

> *"[[The arguments]] all agree that calculators are incapable of answering all questions which may be put to them about their own future, however completely they are informed; and the reason for this appears to lie in the fact that it is impossible for a calculator to possess completely up-to-date initial information about itself. ('Know thyself', it seems, is a contradictory ideal; we cannot fully know our limitations — at least not those to our knowledge.)"*

9.4 Maximum halting time

For *finite* program length, the halting problem may appear trivial to solve. Since computable systems are predictable on a step-by-step basis, one could in principle "empirically solve" the halting problem by just running a program and watch its performance. If a program does not terminate after "a very long time," it could be considered as non-terminating. One problem with this method is that one needs a criterion for the maximal time at which a program which halts has done so. However, in general this time is uncomputable; otherwise computability of the halting time would contradict the recursive unsolvability of the halting problem, theorem 9.2 (for details see chapter 7, p. 99).

Nevertheless, it is possible to give some estimate for the *longest* computation time associated with programs with smaller than or equal to a fixed complexity n which halt [109]. Recall theorem (8.9), p. 105, stating that *either a program p halts in time less than or equal to the* busy beaver function $\Sigma(H(p)+O(1))$, *or else it never halts.* Since, as has been argued before, computability of $\Sigma(H(p)+O(1))$ would again contradict the recursive unsolvability of the halting problem, theorem 9.2, the maximum time of computation for a program of length n exceeds any computable function of n. Indeed, if one would be willing to proceed empirically, one would have to wait longer than any computable time to decide whether an arbitrary program halts or what its output is! The empirical method of solving the halting problem is thus of no use for practical purposes even for programs of finite length.

It is tempting to exclude *"Zeno squeezed oracles"* (cf. section 1.5.2, p. 24) or, in A. Grünbaum's terminology ([219], p. 630), *"infinity machines"* by the requirement of consistency. Indeed, posession of such an oracle would make the TIME algorithm (cf. 100) and thus the "halting function"

$$\mathtt{HALT}(U(p) = x(n)) = \begin{cases} 1 & \text{if } \mathtt{TIME}(U(p) = x(n)) < \infty \\ 0 & \text{if } \mathtt{TIME}(U(p) = x(n)) = \infty \end{cases}$$

[cf. equation (8.1), p. 100] "mechanically" computable. (A warning: the term "mechanic" here refers to processes which are normally considered as not physically realisable.)

But then it would be possible to go through the algorithmic argument (based on G. Cantor's diagonalization method) against the recursive solvability of the halting problem. A seemingly paradoxical result, resulting in an inconsistency, would follow: an algorithm B could be "constructed" which halts if it does not halt and which does not halt if it halts.

However, at a second glance, such absurdities can be translated into somewhat more harmless statements. Consider again the above "halting function" `HALT` when applied to B. Let us assume that $B(B)$ halts, i.e., $\texttt{HALT}(B(B)) = 1$ and let us follow the *"actual construction"* of $B(B)$: B is constructed such that, with the knowledge that $B(B)$ halts, $B(B)$ would not halt; i.e., $\texttt{HALT}(B(B)) = 0$. But then, B is constructed such that, with this latter knowledge, $B(B)$ would halt; i.e., $\texttt{HALT}(B(B)) = 1$. But then, B is constructed such that, with this knowledge, $B(B)$ would not halt; i.e., $\texttt{HALT}(B(B)) = 0 \cdots$ *ad infinitum.* I.e., in this procedural sense, $\texttt{HALT}(B(B))$ does not converge to a "true value;" it is just fluctuating between the values 0 and 1, yielding an infinite sequence $1010101010101 \cdots$. This situation resembles very much the situation if one actually tries to evaluate the liar paradox "`this statement is false.`" Because if it is false, then it is in accordance with what it claims; therefore it is true. Yet, if it is true, then it is not in accordance with what it claims; therefore it is false. Yet if it is false, then it is in accordance with what it claims; therefore it is true. Yet, if it is true, then it is not in accordance with what it claims; therefore it is false $\cdots$ *ad infinitum.* One might say that in this procedural view, paradoxes give rise to very simple forms of evolution and change.

9.5 Gödel's incompleteness theorems

In what follows, a formal system will be called *inconsistent* if *both* a statement *and* its negation are provable therein. (A related version of consistency is the requirement that at least one statement is *not* provable therein.) Furtheremore, a formal system will be called *incomplete* if *neither* some statement *nor* its negation is provable therein.

The historical context of K. Gödel's incompleteness theorem was the *formalist program*, put forward, among others, by D. Hilbert [234]. This program aimed at reconstructing mathematics in purely syntactic terms. In particular, one of its goals was to "condense" *all* mathematical truth into a formal system which has a finite description; any proof should be a finite and *"mechanic" manipulation of lists of symbols* (in particular consistency proofs). Stated differently, such a formal system should be implementable as finite-size program on a computer, which should output all true theorems of mathematics. Anything that could be proven by non-constructive techniques should be derivable also by *finitistic* means of proof. (This statement is sometimes referred to as the "conservation program" [417].) The formalist hope was to be able to replace "truth" by a finite notion of "provability." Gödel showed in 1931 the impossibility of such a pursuit [204]. In particular he proved the following

T 9.4 (Incompleteness Theorems (K. Gödel [204])) *Let L be a recursively enumerable*

and consistent formal system containing (Peano) arithmetic.

(i) *There is a (metamathematically "true") statement* φ *which asserts its own unprovability and which cannot be decided within L; i.e., neither "*φ*" nor "*not φ*" are provable within L.*

(ii) Within *L it is impossible to derive the* consistency *of L.*

T 9.5 (Gödel's incompleteness theorems, pedestrian version)
No "reasonable" finite theory encompasses all of mathematics. It is impossible to decide "from within" a theory whether it is free of contradictions.

The term *"consistent"* means that at least one statement is not derivable within L. In particular, L should be free of contradictions, such as the statement "p and not p." (If such a form would occur then any statement would be derivable.)

Proof:

The following proof closely follows C. Smorynski's account [417] in *Handbook of Mathematical Logic* [27], p. 821. Let $\#(\varphi)$ denote the *code* of some statement φ. $L \vdash \varphi$ stands for "φ can be derived in L." φx is a formula with free variable x. Function symbols which are defined in L correspond to logical connectives and quantifiers. E.g., for all formulae φ, ψ, $L \vdash \mathtt{NEG}(\#(\varphi)) = \#(\neg\varphi)$, $L \vdash \mathtt{IMP}(\#(\varphi), \#(\psi)) = \#(\varphi \rightarrow \psi)$ and so on. For formulae φx and terms t, the *substitution operator* SUB is defined by

$$L \vdash \mathtt{SUB}(\#(\varphi x), \#(t)) = \#(\varphi t) \quad . \tag{9.3}$$

If one encodes derivations, binary relation DER can be defined such that for closed (no free variables) t_1, t_2,

$$L \vdash \mathtt{DER}(t_1, t_2) \tag{9.4}$$

(read "t_1 proves t_2" or "t_1 is a proof of t_2") iff t_1 is the code of a derivation in L of the formula with code t_2. It follows that $L \vdash \varphi$ iff $L \vdash \mathtt{DER}(t, \#(\varphi))$ for some closed term t. By defining

$$\mathtt{PROVE}(x) \leftrightarrow \exists y\, \mathtt{DER}(y, x) \quad , \tag{9.5}$$

one obtains a predicate defining provability ($\exists$ stands for "there exists").

The encoding can be carried out in such a way that the following *derivability conditions* are satisfied:

$$L \vdash \varphi \quad \text{implies} \quad \mathtt{PROVE}(\#(\varphi)) \quad , \tag{9.6}$$

$$L \vdash \mathtt{PROVE}(\#(\varphi)) \rightarrow \mathtt{PROVE}(\#(\mathtt{PROVE}(\#(\varphi)))) \quad , \tag{9.7}$$

$$L \vdash \mathtt{PROVE}(\#(\varphi)) \wedge \mathtt{PROVE}(\#(\varphi \rightarrow \psi)) \rightarrow \mathtt{PROVE}(\#(\psi)) \quad . \tag{9.8}$$

T 9.6 (Diagonalization Lemma) *Let* φx *in the language of L have one free variable as indicated. Then there is a sentence* ψ *such that*

$$L \vdash \psi \leftrightarrow \varphi(\#(\psi)) \quad . \tag{9.9}$$

Proof of diagonalization Lemma:

Given φx, let $\theta x \leftrightarrow \varphi(\mathtt{SUB}(x, x))$ be the diagonalization of φ. Let $m = \#(\theta x)$ and let $\psi = \theta m$. Then

$$\begin{aligned} \psi &\leftrightarrow \theta m \\ &\leftrightarrow \varphi(\mathtt{SUB}(m, m)) \\ &\leftrightarrow \varphi(\mathtt{SUB}(\#(\theta x), m)) \quad (\text{since } m = \#(\theta x)) \\ &\leftrightarrow \varphi(\#(\theta m)) \\ &\leftrightarrow \varphi(\#(\psi)) \quad . \end{aligned}$$

By substituting $\neg\mathtt{PROVE}(x)$ for φ, one obtains the

T 9.7 (First incompleteness theorem) *Let $L \vdash \psi \leftrightarrow \neg\mathtt{PROVE}(\#(\psi))$. Then:*
(i) $L \nvdash \psi$;
(ii) under an additional assumption, $L \nvdash \neg\psi$;

Proof of first incompleteness theorem:

(i) Observe that by (9.6), $L \vdash \psi$ implies $L \vdash \mathtt{PROVE}(\#(\psi))$, which implies $L \vdash \neg\psi$, contradicting the consistency of L.

(ii) The additional assumption is a strengthening of the converse to (9.6), namely

$$\mathtt{PROVE}(\#(\varphi)) \quad \text{implies} \quad L \vdash \varphi \quad .$$

With $L \vdash \neg\psi$, $L \vdash \neg\neg\mathtt{PROVE}(\#(\psi))$, so that $L \vdash \mathtt{PROVE}(\#(\psi))$ and, again by the additional assumption, $L \vdash \psi$, again contradicting the consistency of L.

T 9.8 (Second incompleteness theorem) *Let* $\mathtt{CON}$ *be* $\neg\mathtt{PROVE}(\#(\Lambda))$, *where Λ is any convenient contradictory statement. Then*

$$L \nvdash \mathtt{CON} \quad .$$

Proof of second incompleteness theorem: Let $L \vdash \psi \leftrightarrow \neg\mathtt{PROOF}(\#(\psi))$ again be as in theorem 9.7. It is shown that $L \vdash \psi \leftrightarrow \mathtt{CON}$.

$L \vdash \psi \rightarrow \neg\mathtt{PROOF}(\#(\psi))$ implies $L \vdash \psi \rightarrow \neg\mathtt{PROOF}(\#(\Lambda))$, since $L \vdash \Lambda \rightarrow \psi$ by (9.6), implies $L \vdash \mathtt{PROVE}(\#(\Lambda \rightarrow \psi))$, which by (9.8) implies $L \vdash \mathtt{PROVE}(\#(\Lambda)) \rightarrow \mathtt{PROVE}(\#(\psi))$.

Conversely, by (9.7), $L \vdash \mathtt{PROVE}(\#(\psi)) \rightarrow \mathtt{PROVE}(\#(\mathtt{PROVE}(\#(\psi))))$, which by (9.6) and (9.8) implies $L \vdash \mathtt{PROVE}(\#(\psi)) \rightarrow \mathtt{PROVE}(\#(\neg\psi))$, since $\varphi \leftrightarrow \neg\mathtt{PROVE}(\#(\psi))$. Then, $L \vdash \mathtt{PROVE}(\#(\psi)) \rightarrow \mathtt{PROVE}(\#(\psi \wedge \neg\psi))$ by (9.6) and (9.8), which implies $L \vdash \mathtt{PROVE}(\#(\psi)) \rightarrow \mathtt{PROVE}(\#(\Lambda))$. But $L \vdash \neg\mathtt{PROVE}(\#(\Lambda)) \rightarrow \neg\mathtt{PROVE}(\#(\psi))$, which is $L \vdash \mathtt{CON} \rightarrow \psi$ by definition.

Remarks:

(i) As no attempt is made to extensively discuss Gödel's findings here, the interested reader is referred to the literature, for example to C. Smorynski's review article [417], to *Volume I* of Gödel's *Collected Works* [205], M. Davis [136] and to G. Kreisel [277]. For more informal introductions, see R. Rucker [397], E. Nagel & J. R. Newman, [331, 332], D. R. Hofstadter [237], R. Penrose [351] and J. L. Casti [86, 87, 88, 89, 90, 91], among others.

(ii) Gödel's incompleteness theorems are again proved by the method of *diagonalization*, which has already been introduced previously (p. 113, 114). Diagonalization already manifests itself in the statement "`this statement is unprovable.`"

(iii) Since, as has been pointed out before, there is an equivalence between the notions of *formal system* and *effective computation* (p. 29), statements referring to the unsolvability of some computational task directly translate into statements referring to the incompleteness of formal systems. Another version of (the first) part *(i)* of Gödel's incompleteness theorem thus follows immediately as corollary of the recursive unsolvability of the halting problem. Assume it were always possible to prove the theorem associated with the problem of whether or not a particular algorithm will halt. This would contradict the recursive unsolvability of the halting problem.

(iv) As stated earlier (p. 111), Gödel himself seemed to have interpreted his findings as the impossibility to give a complete and finite formal description of the concept of truth of statements about a theory *within* that theory.

(v) Note that the set of code numbers of theorems of the formal system PT is a *recursively enumerable* set which is *not recursive* (i.e., $\mathbb{N} - PT$ is not recursively enumerable). This is due to the recursive unsolvability of the halting problem. Although one can systematically derive and list PT by an effective computation, one can never be sure whether some well formed formula which at some finite time is not in PT (as "short" as it may be) turns up at a later time.

(vi) A. Church, A. M. Turing, S. Feferman, G. Kreisel and others have pursuit the question of whether it is possible to pass on to a complete formal system by a transfinite limit. One could exploit the constructive feature of Gödel's incompleteness result, i.e., the construction of a true though unprovable theorem. Let us start with an initial system, say L_1, whose theorems concerning integers are correct. By Gödel's construction it is possible to effectively compute a true but unprovable theorem A_1, and to include this theorem as an axiom into a new system $L_2 = L_1 \cup \{A_1\}$ and so on. The iteration $L_n = L_1 \cup \{A_1, \ldots, A_{n-1}\}$ might be done mechanically by a process of finite description, associated with a formal system $\overline{L}$ which thus includes all L_n. Nevertheless, since $\overline{L}$ can be effectively generated, it is still incomplete, and one must proceed further into the transfinite to overcome this incompleteness. This is made possible by *a constructive notation of the ordinals* and by considering the limit of this process. In that way one could "overcome" Gödel incompleteness at the price of finding a constructive notation of ordinals. For a much more detailed discussion the reader is referred to S. Feferman's review article [168].

9.6 Recursive unsolvability of the rule inference problem

Assume two (universal) computers U and U'. The second computer U' is being programmed with an arbitrary algorithm p, of which the first computer U has no knowledge of. The task of U, which shall be called the *inference problem*, is to conjecture the "law" or algorithm p by analysing output "behaviour" of $U'(p)$. The following theorem states that this task cannot be performed by any effective computation / recursive function.

T 9.9 (Recursive unsolvability of the rule inference problem (E. M. Gold [207]))
Assume some particular (universal) machine U which is used as a "guessing device." Then there exist total recursive functions which cannot be "guessed" or inferred by U.

Proof:

The following proof follows M. Li and P. M. B. Vitányi [301] and uses diagonalization. Assume that there exists a "perfect guesser" U which can identify all total recursive functions (wrong). It is possible to construct a function $\varphi^* : \mathbb{N} \rightarrow \{0, 1\}$, such that the guesses of U are wrong infinitely often, thereby contradicting the above assumption. Define $\varphi^*(0) = 0$. One may construct φ^* by simulating U. Suppose the values $\varphi^*(0)$, $\varphi^*(1)$, $\varphi^*(2), \cdots, \varphi^*(n-1)$ have already been constructed. Then, on input n, simulate U, based on the previous series $\{0, \varphi^*(0)\}, \{1, \varphi^*(1)\}, \{2, \varphi^*(2)\}, \cdots, \{n-1, \varphi^*(n-1)\}$, and define $\varphi^*(n)$ equal to 1 plus the guess of U of $\varphi^*(n)$ mod 2. In this way, U can never guess φ^* correctly; i.e., make only a finite number of mistakes.

Informally, the idea of the proof is to take any sufficiently powerful rule inference method (guessing method) and to define from it a (total) recursive function which is not identified by it. Pointedly stated, this amounts to constructing an algorithm which (passively!) "fakes" the "guesser" U by simulating some particular recursive function φ corresponding to the algorithm p until U pretends to guess this function correctly. In the diagonalization step, the algorithm then switches to a different function $\varphi^* \neq \varphi$, such that U's guesses are incorrect. A "feedback" is not essential to the argument, for U' may switch algorithms independently of U. The generic example, "`The first output of` U' `is one and each subsequent output of` U' `is one greater than the guess of` U `of that output`," is mentioned in the review by D. Angluin and C. H. Smith [10].

One can also interpret this result in terms of the recursive unsolvability of the halting problem: there is no recursive bound on the time the guesser U has to wait in order to make sure that his guess is correct.

For the original proof, as well as a discussion and related topics, the reader is referred to appendix I, pages 470-473, of E. M. Gold [207], as well as to the reviews by D. Angluin and C. H. Smith [10] and M. Li and P. M. B. Vitányi [301], and to an article by L. M. Adleman and M. Blum [4].

An immediate consequence of the recursive unsolvability of the rule inference problem is the following corollary:

C 9.10 (Recursive unsolvability of inference for mechanistic physical systems)
There exist mechanistic physical systems whose laws and parameters cannot be determined recursively (by any effective computation).

9.7 True > provable

Let again $O(1)$ denote a bounded function whose absolute value is less than or equal to an unspecified constant. We shall first state a theorem by G. Chaitin (reviewed, among other references, by G. Chaitin in [110, 112], see also M. van Lambalgen [287]) concerning the computability of bounds on H:

T 9.11 *Let L be an axiomatic theory containing (Peano) arithmetic whose arithmetical consequences are true.*

(i) *There exists a constant c_L such that within L no statement of the form "$H(n) > c_L$" is provable.*

(ii) *Let #(L) be a recursively enumerable index of L. Then there exists some constant c' independent of L such that within L no statement of the form "$H(n) > H(\#(L))+c'$" is provable.*

(iii) *There exist particular theories whose axioms have information content $H(\text{axioms}) = m + O(1)$ in which it is possible to establish all true propositions of the form "$H(x) = k$" with $k < H(\text{axioms})+O(1)$ and "$H(x) \geq H(\text{axioms})+O(1)$".*

We shall give only a rather informal idea of the proof [101, 112], which is based on "Berry's paradox," For the historical background, see, for instance, R. Rucker [397], p. 100. One form of it is given by "`the smallest positive integer that cannot be specified in less than a hundred words`", which, if it would exist, would have just been specified by fourteen words. As a consequence, this number cannot exist. In a rephrasing closer to algorithmic information theory, this paradox can be expressed as, "`the first positive integer` x `such that` $H(x) > 100$ `words.`" More generally, "`consider the least natural number that can't be defined in less than` n `bits.`" — Although, by equation (7.34), p. 95, this phrase is not longer than $\log_2^*(n)+O(1)$ bits, it purports to specify a number whose definition needs a phrase which is *at least* n bits long. This yields a contradiction for sufficiently great numbers n, more specifically for all $n > \log_2^*(n) + O(1)$. As a consequence, the above statement is consistent only for $n \leq \log_2^*(n) + O(1)$. A compact formal proof of can be found in G. J. Chaitin's book *Information-Theoretic Incompleteness* [112].

So far, the *computational complexity*, i.e., the minimal time consumption for calculating proofs of theorems in *(ii)* has been ignored. It has been proven by G. Chaitin [101] that the *time of computation* for propositions of the form "$H(x) = k$" with $k < m = H(\text{axioms}) + O(1)$ and "$H(x) \geq m = H(\text{axioms}) + O(1)$" is uncomputable. Informally stated, any derivation of such theorems amounts to their "mechanised" computation. Due to the recursive unsolvability of the halting problem or, more precisely, due to the maximal halting time, which is roughly proportial to $\Sigma(m)$, there does not exist any bound from above on the halting time which is recursive in m.

The ability of computable processes to produce objects with higher algorithmic information content than themselves will be reviewed next. — Indeed, a program which, for example, counts and outputs integers in successive order will eventually end up with an integer of higher algorithmic information content than its length.

In the spirit of the incompleteness theorem (9.11) for lower bounds on algorithmic information, one may be tempted to formulate the following erroneous statement, which might be a source of misconception:

> *"Consider a consistent theory, representable as formal system. Within such a theory no theorem representable as string x can be generated which has algorithmic information more than $O(1)$ greater than the information content of the axioms of the theory. I.e., x is a theorem only if $H(x) \leq H(\text{axioms}) +$*

$O(1)$.

Or, in pedestrian version,

You cannot generate a 20-pound theorem with a 10-pound theory if pound is the unit of algorithmic information."

As has been pointed out by G. Chaitin, this statement is false. G. Chaitin stated [114]:

"I've said that "You can't prove a 10 pound theorem from a 5 pound set of axioms." Many people erroneously take this to mean that I am asserting the following theorem: If A is a set of axioms that proves theorem T, then $H(T) < H(A) + O(1)$. I most certainly do not: this theorem is obviously false. Let me give two proofs. "Deep" proof: if the axioms A yield infinitely many theorems T, then some T are arbitrarily complex, because there are only finitely many simple strings. Proof via natural counter-example: Consider Peano arithmetic and all theorems of the form "n = n", where n is the numeral for a very complex natural number.

A more technical statement is "You can't prove a 10 pound theorem from a 5 pound set of axioms AND KNOW THAT YOU'VE DONE IT." (I.e., know that you've proven a theorem with a particular complexity that substantially exceeds that of the axioms.) Restated in this slightly more careful fashion, it is now obvious that my assertion is an immediate corollary of my basic theorem that one can prove "$H(S) > n$" only if $n < H(A) + O(1)$."

A simple counterexample is the theorem "$n = n$", where n is some numeral of a "very complex" natural number. The algorithmic information of "$n = n$" can *exceed* the information content of the axioms of the theory, since there always exists n's such that $H(n) \gg H(\text{axioms}) + O(1)$. Notice, however, that the above statement is true if one specifies *a particular theorem* for instance by specifying the place t in the enumeration of provable theorems. Then such a derivation corresponds to the *application* of a recursive function (see theorem (7.6), p. 93). The trouble occurs if one considers infinite, boundless computations, such as derivation of "all" formulas or the enumeration of "all" naural numbers in $\mathbb{N}$ — for a few more details, see section 7.3, p. 97. Another way of expressing this fact is the following

T 9.12 (Creation of algorithmic information) *There exist formal systems whose axioms contain finite algorithmic information, in which it is possible to generate theorems of arbitrary high algorithmic information.*

T 9.13 (Pedestrian version) *Eventually you can generate a 20-pound theorem with a 10-pound theory if pound is the unit of algorithmic information.*

More examples:

(i) Consider an algorithm which outputs the natural numbers, one after the other. Such an algorithm can be "very short," since, by starting from 0, in the n'th iteration step it needs only to add 1 to the previous number and output it. Eventually it will output a natural number corresponding to a bit string with higher algorithmic information than the length of the program, which is $O(1)$.

(ii) Consider *Peano's axioms*, representable as finite string *PA*; *PA* has a finite algorithmic information content $H(PA) = O(1)$; nevertheless, since the Peano axioms allow *count-*

ing, it is possible to derive and enumerate *all* natural numbers. By theorem 7.30, p. 95, "many" binary numerals of length k have algorithmic information $H(k) = k+H(k)+O(1)$, which may exceed $H(PA)$. Some of these numbers $\{n_L\} \subset \mathbb{N}$ will even be Gödel numbers or representations of formal systems L whose algorithmic information exceeds the algorithmic information of the original Peano axioms, i.e., $H(n_L) > H(PA)$. Notice, however, that there exist "extremely" large numbers with a "very" small algorithmic information content: take, for example, $2^{2^{2^{2^{2^{2^{2^{2}}}}}}}$ (cf. section 7.3, p. 97).

An immediate consequence is the fact that the algorithmic information necessary to specify some set S may be *smaller* than the algorithmic information of some of its constituents, i.e., $H(S) < \max_{s_i \in S} H(s_i)$. An example for such a set is the set of natural numbers $\mathbb{N}$, for which $H_\infty(\mathbb{N}) < \max_{n \in \mathbb{N}} H(n) = \infty$.

This can be combined with theorem 9.11 by the following scenario:

C 9.14 (Pedestrian version) *Eventually you can generate a 20-pound theorem from a 10-pound theory if pound is the unit of algorithmic information, but intrinsically you wont be able to prove that.*

9.8 Paradoxical combinator

In the *type-free* representation of the recursive functions by *combinatory logic* and, more specifically, by the *lambda calculus*, no difference is made between a "function" and its "argument" and "value." This approach emphasises the procedural character of effectively computable systems. It "unifies" several paradoxical constructions based on the method of diagonalization by introducing an abstract *fixed point combinator*, or *paradoxical combinator Y*. Informally, the fixed point combinator Y, which has been originally obtained from an analysis of the Russell paradox, is a procedural representation for self-reference. Its existence is assured by the following theorem. Let "$\succeq$" be a relation, as defined in the reviews by H. B Curry and R. Feys [129], p. 151, 178, and H. P. Barendregt [22].

T 9.15 (Fixed point combinator) *Given any combinator x, there is a combinator y which is obtained from the* fixed point operator *Y by*

$$Yx \succeq y \succeq xy \quad .$$

Proof: Let f, g be two functions of one argument, and h be a function of two arguments. Then define the operator $Bfgx \succeq f(g(x))$, called *elementary compositor*, and the operator $Whx \succeq h(x,x)$, called *elementary duplicator*. Assume an arbitrary combinator x, then $BWBx(BWBx)$ is such a combinator Y, which can be verified by evaluating $Yx \succeq y \succeq xy$.

It follows from the definition that all terms Yx cannot be reduced to a form which is not further reducible; further reductions render a regress which is infinite if not stopped: $Yx \succeq y \succeq xy \succeq xxy \succeq xxxy \succeq xxxxy \succeq \cdots$ *ad infinitum*. In what follows we describe applications of the fixed point operator. Notice that the identification of the combinator x with an object is purely *syntactical*; the *semantic* aspect of "meaning" belongs to an interpretative metalevel.

Russell paradox: Let $F(f)$ be a property of properties f defined by $F(f) = \neg f(f)$, where $\neg$ is the symbol for negation. On substitution of F for f, one arrives at $F(F) = \neg F(F)$. If one considers the statement "$F(F) = \neg F(F)$" as a proposition, one arrives at a contradiction. The resolution of this "paradox" is the *exclusion* of such propositions, *restricting* the class of allowed propositions.

When creating classical set theory, one of G. Cantor's main objectives was to cope with infinities in a suitable mathematically form (and not to cause schoolchildren headache by drilling them in useless formalism). This program fails just because of a contradiction of the above type: For Cantor, *"a set is a Many which allows itself to be thought of as One"* [80]. Stated differently, *"a set is the representation of a thought"*. In this sense every "thought" corresponds to a set, and set theory encompasses all forms of thoughts, not necessarily of mathematical origin.

If we allow most general (also nonrational) processes to stand for the term "thought" in the above definition, we are lead to inconsistencies. Assume a "set $\mathfrak{V}$ of all sets that are no members of themselves", i.e., $\mathfrak{V} = \{x \mid x \not\in x\}$. Insertion of $\mathfrak{V}$ for x yields a contradiction, since if $\mathfrak{V} \in \mathfrak{V}$, then by definition $\mathfrak{V} \notin \mathfrak{V}$. As a consequence of this type of "diagonalization" argument, $\mathfrak{V}$ cannot be a set although it is "thinkable".

One solution is to restrict the term "thought" to "thoughts" which are effectively computable from specific "axioms" or "basic premises". One such "system of thoughts" is the Zermelo-Fraenkel system. The above contradiction is resolved by excluding "pathological thoughts," thereby avoiding contradiction in this restricted universe.

Epimenides' paradox: if $x = \neg$`True` is identified with the statement that a statement is false, then the existence of a fixed point operator expresses Epimenides' paradox by $Yx = y = xy = \neg$`True`y.

Gödel incompleteness: if $x = \neg$`PROVE` is identified with the statement that a theorem is unprovable in some specified formal system, then the existence of a fixed point operator expresses Gödel's incompleteness theorem by $Yx = y = xy = \neg$`PROVE`y.

The above examples resemble K. Popper's *"paradox of Tristram Shandy"* (p. 187), an absurd attempt of complete self-comprehension "going wild," or Zeno's "paradox of Achilles an the Tortoise" (p. 24).

Chapter 10

Complementarity

In this chapter, the quantum mechanical concept of *complementarity* will be modelled by computational complementarity. One advantage of computational complementarity over quantum complementarity, if you like, is the use of elementary primitives from the theory of (finite) automata. There is no need for introducing the sort of 'WODOO' magic which is sometimes encountered in discussions on the epistemology of quantum theory. To put automaton models for quantum systems in the proper perspective: Strictly speaking, they correspond to nonlocal hidden variable models. The "hidden" physical entities are the "true" initial states of the automata.

Computational complementarity has been first investigated by E. F. Moore in his article *Gedanken-Experiments on Sequential Machines* [328]. The name *computational complementarity* is due to D. Finkelstein [177, 178], who also made the first attempt to construct logics from experimentally obtained propositions about automata; see also the more recent investigation by A. A. Grib and R. R. Zapatrin [216, 217]. The following investigation has been carried out independently. Although the goals are very similar, the methods and techniques used here differ from the ones used by previous authors. Therefore, no attempt is made to discuss these previous results in detail.

10.1 Historical review of quantum complementarity

The following paragraphs contain a very brief, incomplete and personal review of quantum complementarity. For a more comprehensive treatment, see, for instance, the monography *The Philosophy of Quantum Mechanics* by M. Jammer [243].

Informally speaking, quantum complementarity states that there are complementary observables such that it is impossible to simultaneously and independently measure all of them with arbitrary accuracy. The experimenter has then to choose which one of the complementary observables should actually be measured. The actual measurement process modifies measurement(s) of other observable(s) which are complementary to the measured observable. To put it pointedly: the experimenter has to decide which one of the many possible observables shall be measured. Measurement of this observables restricts or makes impossible measurement of other, complementary, observables.

This is also the case for entangled subsystems which are spatially separated. (The terminology *"entanglement"* has been used by E. Schrödinger; the German original is *"Verschränkung"* [406].) Events associated with one subsystem depend on events associated with the other subsystem. Both subsystems may be space-like separated, a property which is often referred to as *"nonseparability"* or *"nonlocality."* For a stunning consequence of these quantum mechanical features, the reader is referred to the *delayed-choice experiment* envisioned by J. A. Wheeler [457].

A. Einstein, B. Podolsky and N. Rosen (EPR) have attempted to point to what they called the *incompleteness of quantum theory* [154], and have sparked a very lively debate ever since [458]: Assume two spatially separated subsystems which are entangled. I.e., as the EPR argument goes, a precise measurement of some observable, say, spin or polarisation, in one subsystem could be interpreted as an indirect precise measurement of a corresponding "observable" in the other subsystem as well. (This conjecture is usually supported by conservation or symmetry arguments.) EPR associate elements of physical reality with the event which has been actually measured *as well as* with the indirectly (!) inferred event. (The notation used in this book differs from the one used by EPR; see 4.2, p. 45.) Consider two such observables, which are complementary if they are measured in one and the same subsystem. The EPR argument asserts that if one of these observables is measured in one subsystem and the other observable is measured in the other subsystem, then one can infer both — albeit in the quantum mechanical point of view, complementary — observables in one and the same subsystem with arbitrary accuracy. Since quantum mechanics declares that complementary observables cannot be measured simultaneously with arbitrary accuracy, this, the EPR argument closes, demonstrates the incompleteness of quantum theory. Stated pointedly: the EPR argument claims that quantum mechanics *cannot* predict what the experimenter *can* "measure (& infer)." Finally, the authors conjectured that it should be possible to invent alternative theories which are "more complete" than quantum mechanics.

Such theories have indeed been proposed. They can be subsumed under the title *hidden variable theories.* I believe it is not unfair to state that most of the more serious models share two common features: they do not predict more observable phenomena than quantum mechanics and they are nonlocal. Yet, they provide a sort of "classical arena" hidden to the experimenter.

The Einstein-Podolsky-Rosen argument has been extended in various forms. J. S. Bell gave a statistical analysis of the *correlation function* of a system consisting of two entangled subsystems [29]. For a very clear introduction, see A. Peres [352], A. Shimony [413] and J. F. Clauser and A. Shimony [121], among others. D. M. Greenberger, M. Horne and A. Zeilinger presented a stronger result by an analysis of the correlation function of a system consisting of *three* entangled subsystems [215]. While Bell's original argument is statistical, i.e., requires observation of a great number of events, the Greenberger-Horn-Zeilinger setup is demonstrated, a least in principle, by a *single* measurement. For a short review, see also N. D. Mermin's article [318]. One conclusion of all the so-far accumulated evidence is that quantum systems indeed do not allow the independent co-existence of complementary observables. (In nonlocal hidden variable models co-existing complementary observables are no longer independent.)

In a theorem by S. Kochen and E. P. Specker, the impossibility of an independent co-existence of complementary observables is interpreted algebraically [269]. These authors have demonstrated that in general it is impossible to "enrich" the quantum description by additional (hidden) variables such that in this "more complete" theory the logical operations `and`, `or` and `not` are defined as in the classical propositional calculus. Stated differently, no Hilbert lattice (of dimension > 2) can be embedded in a Boolean lattice such that the lattice operations of meet, join and orthocomplementation survive in their original context, i.e., as operations on independent elements. See also A. Peres [358] and I. Pitowsky [364], chapter 4.

10.2 Automaton propositional calculus

We shall deal, informally speaking, with the intrinsic perception of a computer-generated universe. The investigation is based on the construction of primitive *experimental statements* or *propositions.* Then the *structure* of these propositions will be discussed, thereby defining algebraic relations and operations between the propositions. It will be shown that certain features known from experimental (quantum) logics, in particular complementarity, can be modelled by the experimental logics of finite automata.

The methods introduced here somewhat resemble those invented by J. von Neumann and G. Birkhoff [42], C. H. Randall [381], G. W. Mackey [306], J. Jauch [244], C. Piron [362] and others in the context of quantum and empirical logic.

Logical structures which have been invented abstractly (such as Boolean logic) may not be suitable *empirically.* An "empirically adequate" logical formalism should be obtained [not (only) from "reasonable" first principles, but] by studying the experimentally decidable propositions and their experimentally determined interrelations. — This epistemologic approach might be compared to the program of non-Euclidean geometry, put forward by Gauss, Lobachevsky, Bolyai (among others) and operationalised by Einstein.

Lattice theory (see chapter 5, p. 51) provides an effective formal representation of such logical structures. As no attempt is made here to intensively discuss these methods here, the reader is referred to the literature, in particular to G. Birkhoff's *Lattice Theory* [43], J. Jauch's *Foundations of Quantum Mechanics* [244], G. Kalmbach's *Orthomodular Lattices* [251], P. Pták and S. Pulmannová's [376] *Orthomodular Structures as Quantum Logics*, R. Giuntini's *Quantum Logic and Hidden Variables* [201] and A. Dvurečenskij's *Gleason's Theorem and Its Applications* [144], among others.

G. Birkhoff and J. von Neumann originally obtained an "experimental logic" by a *"top-down"* approach. It is based on von Neumann's Hilbert space formalism of quantum mechanics. Physical properties corresponding to experimental propositions are idendified with projection operators on the Hilbert space; or equivalently with the associated subspaces. The lattice of closed subspaces of a Hilbert space, henceforth called *Hilbert lattice,* corresponds to a lattice of experimental propositions. called the *calculus of propositions.* At best, given suitable correspondence rules, quantum mechanical Hilbert lattices and propositional calculi should be one-to-one translatable. Subsequently, J. Jauch [244] C. Piron [362], among others, proposed a "bottom-up" approach by considering the experimental structure of propositions as the primary entities. For a historical introduction, see M. Jammer's *The Philosophy of Quantum Mechanics* [243].

A complete lattice theoretic characterisation of Hilbert lattices has been given by W.

J. Wilbur [460]; see also R. Piziak [365]. Wilbur's axioms has also been shortly reviewed in chapter 5, p. 68. A purely algebraic characterisation of infinite dimensional Hilbert lattices has not (yet?) been given. It is for instance known that infinite dimensional Hilbert lattices are atomic, irreducible, complete, satisfy the orthomodular and the orthoarguesian laws and the exchange axiom [251]. Yet, these features do not sufficiently define a Hilbert lattice: As has been proven by H. A. Keller [262, 253], there are spaces which are not Hilbert spaces but have atomic, irreducible, complete lattices satisfying the orthomodular law and the exchange axiom. One may conjecture that additional axioms characterising Hilbert lattices algebraically may be necessary. Indeed, one could speculate that such a characterisation is not recursively enumerable [290]. In an actual proof one might try to embed arithmetic in Hilbert lattices.

In what follows, a method similar to quantum logics will be introduced for the intrinsic logics of automaton universes. Although specific classes of finite automata will be analysed, these considerations apply to universal computers as well. (Finite automata can be simulated on universal computers.) A *"bottom-up"* approach is pursued. First, the possible experimental propositions and their experimentally determined interrelations are studied. Then, these can be compared with and (for specific automaton universes even) modelled after the experimental logic of quantized systems. The first step requires the construction of an "automaton propositional calculus." The latter step requires an algorithmic feature which has first been discussed by E. F. Moore in his article *Gedanken-Experiments on Sequential Machines* [328] and which in D. Finkelstein's words [178] is called *"computational complementarity."* Yet, for the same reasons as before, such a correspondence between an automaton universe and a quantized system may only be weakly defined.

10.2.1 Statement of the problem

An (i,k,n)-automaton A will be considered whose transition and output tables are known beforehand and which has i internal states, k input symbols and n output symbols. Recall the single-copy and multi-copy *"Gedankenexperiments."*[1] In a sense, an automaton is treated as a "black box," whose transition and output tables (i.e., informally speaking, its "intrinsic machinery") are given in advance but *whose initial state is unknown.* The automaton is "feeded" with certain input sequences from the experimenter and responds with certain output sequences. We shall be interested in the *distinguishing problem*: *"identify an unknown initial state."*

10.2.2 Propositions, relation and operations

Consider propositions of the form

[1]Formally, one may define [125] an experiment on an automaton as a function $e : O^* \rightarrow I \cup \mathfrak{A}$, where O^* is the set of all output sequences generated by the automaton, I is the set of all input symbols, and $\mathfrak{A}$ is the set of all "conclusions," or "answers," or propositions. This can be understood as follows. The effect of some output sequence is *either* some "conclusion" or "answer" in $\mathfrak{A}$ about the automaton, *or* the further input of a symbol in an attempt to reach a "conclusion" or "answer" in $\mathfrak{A}$. I.e., the experimenter concludes from a particular sequence of outputs certain *propositions*, or "statements" about the automaton.

"`the automaton is in state` a_j"

with $(1 \leq j \leq i)$. Propositions can be composed to form expressions of the form

"`the automaton is in state` a_j `or in state` a_m `or in state` $a_l \cdots$"

which are again propositions, where the "`or`" is understood here as the usual "or" operation of the classical propositional calculus. Any proposition composed by propositions can be represented by a set. E.g., the above statement "`the automaton is in state` a_j `or in state` a_m `or in state` $a_l \cdots$" represents the set $\{j, m, l, \ldots\}$.

The element **1** is given by the set of *all* states $\{1, 2, \ldots, i\}$. (Recall that we are considering an (i,k,n)-automaton with i internal states, k input symbols and n output symbols.) This corresponds to a proposition which is always satisfied:

"`the automaton is in some internal state`"

The element **0** is given by the *empty* set $\emptyset$ (or $\{\}$). This corresponds to a proposition which is false (by definition the automaton has to be in *some* internal state):

"`the automaton is in no internal state`"

The class of all propositions and their relations will be called *automaton propositional calculus*. In what follows, automaton propositional calculi are denoted by $\mathfrak{A}$ or $\tilde{\mathfrak{A}}$ if they are defined intrinsically or extrinsically, respectively. (In this context, intrinsic and extrinsic stands for the one-automaton and multi-automaton configuration, respectively. For a definition, see chapter 6, p. 73.) Each particular outcome which, if defined, has the value `TRUE` or `FALSE`, shall be called "event." In this sense, an automaton propositional calculus, just as the quantum propositional calculus, is obtained *experimentally*. It consists of all potentially measurable *elements of the automaton reality* and their logical structure, with the implication as order relation.

Not all of the propositions and their corresponding subsets may be accessible by experiments. — This is where the considerations become nontrivial. An experiment is said to "identify" a proposition of the form "`the automaton is in state` a_j `or in state` a_m `or in state` $a_l \cdots$" if the proposition can be decided by performing the experiment (yielding either one of the results `TRUE` or `FALSE`). Lattice relations of the form "$p_1 \rightarrow p_2$" as well as the lattice operations "$p_1 \vee p_2$," "$p_1 \wedge p_2$" and "$\neg p$" have the usual meaning "implies," "or," "and," "not" only if an experiment properly identifies the propositions resulting from these operations. I.e., only if these operations are operational and can be defined experimentally, they are in $\mathfrak{A}$.

A reminder: Let A be a set. A *partition* is a family $\{a_i\}$ of nonempty subsets of A with the following properties:

(i) $a_i \cap a_j = \emptyset$ or $a_i = a_j$;

(ii) $A = \bigcup_i a_i$.

E.g., a partition $\{\{1\}, \{2, 3\}\}$ generates the set $\{1, 2, 3\}$.

The elementary propositions can be conveniently constructed by a partitioning of automaton states generated from the input/output analysis of the automaton as follows:

D 10.1 (Partition of propositions, version I)

Let V be the set of partitions *($\emptyset$ is the empty input sequence)*

$$V = \bigcup_w v(w) = \{v(\emptyset), v(s_1), \ldots, v(s_k), v(s_1 s_2), \ldots\} \quad . \tag{10.1}$$

Each $v(s_j \cdots s_m)$ is a partition of the propositions which can be identified from experiments using the input sequence "$s_j \cdots s_m$." I.e.,

$$
\begin{aligned}
v(\emptyset) &= \{\text{partition of propositions identifyable with no input}\} \quad , \\
v(s_j) &= \{\text{partition of propositions identifyable with input } s_j\} \quad , \\
&\vdots \\
v(s_j \cdots s_m) &= \{\text{partition of propositions identifyable with input } s_j \cdots s_m\} \quad , \\
&\vdots
\end{aligned}
$$

A proposition p_j is identified by a partition if p_j is an element of that partition or the union of elements thereof.

D 10.2 (Partition of propositions, version II)

Let $w = s_1 s_2 \cdots s_k$ be a sequence of input symbols,

$$a_{i,w} = a_i \delta_{s_1}(a_i)\delta_{s_2}(\delta_{s_1}(a_i)) \cdots \delta_{s_k}(\cdots \delta_{s_1}(a_i)\cdots) \tag{10.2}$$

and

$$z = o(a_{i,w}) = o(a_i)o(\delta_{s_1}(a_i))o(\delta_{s_2}(\delta_{s_1}(a_i))) \cdots o(\delta_{s_k}(\cdots \delta_{s_1}(a_i)\cdots)) \quad . \tag{10.3}$$

Let

$$\alpha_z^w = \{a_i \mid o(a_{i,w}) = z\} \tag{10.4}$$

be the set of initial states which, on some fixed input sequence w, yield some fixed output sequence $z = t_0 t_1 t_2 \cdots t_k$. I.e., α_z^w is the equivalence class of propositions identifyable by input w and output z. The elements $\{\alpha_z^w\}$ of the partition

$$v(w) = \bigcup_z \{\alpha_z^w\} \tag{10.5}$$

define the equivalence classes of propositions identifiable by input w and output z.

$$V = \bigcup_w v(w) = \{v(\emptyset), v(s_1), \ldots, v(s_k), v(s_1 s_2), \ldots\} \tag{10.6}$$

is the set of partitions.

Remarks:

(i) $\{1, 2, 3\}$ is interpreted as "$1 \vee 2 \vee 3$."

(ii) A *Mathematica* package for computing $V = \cup_w v(w)$ is listed in appendix A.3, p. 250.

Examples:

(i) Let p_1 be the proposition that the automaton is in state 1. Then p_1 is identified by a partition of the form $\{\cdots, \{1\}, \cdots\}$. p_1 is *not* identified by a partition of the form $\{\cdots, \{1, 2\}, \cdots\}$.

(ii) Let $p_1 \vee p_2$ be the proposition that the automaton is in state 1 or in state 2. Then $p_1 \vee p_2$ is identified by a partition of the form $\{\cdots, \{1\}, \{2\}, \cdots\}$ or $\{\cdots, \{1, 2\}, \cdots\}$.

D 10.3 ("Or" and "and" operations)

Let p_i be propositions of the form "`the automaton is in state` *a_i." The proposition*

$$p_1 \vee p_2 \tag{10.7}$$

(interpretable as "p_1 or p_2") defines a proposition of the form "`the automaton is in state` *a_1* `or`[2] `in state` *a_2" (or the set theoretic union "$p_1 \cup p_2$") if and only if there exist input sequences $s_j \cdots s_m$ (including the empty sequence $\emptyset$), such that $p_1 \vee p_2$ is identified by the partition $v(s_j \cdots s_m)$. (For arbitrary composite statements p_j, p_m, a definition of the composition "$p_j \vee p_m$" is straightforward.)*

The proposition

$$p_j \wedge p_m \tag{10.8}$$

(interpretable as "p_j and p_m") defines a proposition of the form "p_j `and`[3] *p_m" (or the set theoretic intersection "$p_j \cap p_m$") if and only if there exist input sequences $s_j \cdots s_m$ (including the empty sequence $\emptyset$), such that $p_1 \wedge p_2$ is identified by the partition $v(s_j \cdots s_m)$.*

D 10.4 (Complement operation)

The complement

$$\neg p_1 \tag{10.9}$$

(or p_1') of a proposition p_1 (has the meaning of "not p_1" and) is defined if and only if

$$\begin{aligned} p_1 \wedge \neg p_1 &= \mathbf{0} \\ p_1 \vee \neg p_1 &= \mathbf{1} \end{aligned}$$

(or, with the propositions p_1 and $\neg p_1 = p_j$ expressed as sets, $p_1 \cap p_j = \mathbf{0} = \emptyset$ and $p_1 \cup p_j = \mathbf{1} = \{1, 2, \ldots, i\}$), and there exist input sequences $s_j \cdots s_m$ (including the empty sequence $\emptyset$), such that $\neg p_1$ is a proposition identified by the partition $v(s_j \cdots s_m)$.

Examples:

(i) If $\mathbf{1} = \{1, 2, 3, 4, 5\}$ and $p_1 = \{1, 2\}$, then $\neg p_1 = \{3, 4, 5\}$.

(ii) If $\mathbf{1} = \{1, 2, 3, 4, 5\}$ and $p_1 = \{5\}$, then $\neg p_1 = \{1, 2, 3, 4\}$.

The order relation "$\preceq$" can be defined by a properly defined *implication* as follows.

D 10.5 (Partial order)

A partial order relation $p_j \preceq p_m$, *or*

$$p_j \rightarrow p_m \tag{10.10}$$

(with the interpretation "p_j implies p_m," or with "whenever p_j is true it follows that p_m is true, too") is defined if and only if p_j `implies`[4] *p_m (set theoretically, this is equivalent to requiring that $p_j \subset p_m$; i.e., p_j is subset of p_m), and there exist input sequences $s_j \cdots s_m$ (including the empty sequence $\emptyset$), such that p_j and p_m are propositions identified by the partition $v(s_j \cdots s_m)$.*

---

[2] The "`or`" is understood here as the usual "or" operation of the classical propositional calculus.

[3] The "`and`" is understood here as the usual "and" operation of the classical propositional calculus.

[4] The "`implies`" is understood here as the usual "implies" operation of the classical propositional calculus.

Remarks:

(i) The above definition of implication (order relation) *cannot* be consistently applied to *all* automata. An automaton (counter-) example will be given (p. 137) whose automaton propositional calculus is not a lattice and whose "implication" is not transitive [402].

(ii) Not all classes of partitions correspond to Moore-type automata. A more general automaton model are Mealy-type automata.

(iii) The partial order relation can be conveniently represented by drawing the Hasse diagram thereof. This can be done by proceeding in two steps. First, the Boolean lattices of propositional structures based on all relevant state partitions $v(w)$ are constructed. [This can be done by generating the set of all subsets of $v(w)$ and identifying the subset relation "$a \subset b$" with the implication "$a \rightarrow b$."] Then, the union of all the Boolean subalgebras generated in that way renders the complete partial order of the automaton propositional calculus. This can also be understood graph theoretically [225, 344]. A *Mathematica* package for computing the graphical representation of the Hasse diagram of automaton propositional calculi is listed in appendix A.3, p. 250.

Examples

10.2.3 Trivial automaton

Consider a two-state (2,2,1)-automaton represented in table 10.1. If the two intrinsic states are labelled a and b, the two inputs 0, 1 and the output 0, this automaton represents a machine which constantly outputs 0, regardless of the input & regardless of its present and past internal states. No experiment distinguishes between the two states a and b, and no conclusion can be drawn from the phenomenology upon the "hidden machinery" of the automaton, in particular of the existence of two separate internal states. The isomorphic minimal-state automaton of the trivial automaton is just a one-state machine. Its intrinsic world is monotonous, and its `TRUE` propositions are of the form, "`the digit '0' appears`," or equivalently, "`the trivial automaton is in state 'a' or in state 'b'`." All state partitions corresponding to all input sequences coincide:

$$v(\emptyset) = v(0) = v(1) = v(00) = \cdots = \{\{a, b\}\} \quad . \tag{10.11}$$

Extrinsically, i.e., in the multi-automaton configuration, one could not extract more information from the trivial automaton than by the single-automaton configuration. Extrinsic and intrinsic propositional calculi $\tilde{\mathfrak{A}}$ and $\mathfrak{A}$ are identical, i.e., $\mathfrak{A} = \tilde{\mathfrak{A}}$ and can be represented by the lattice drawn in Fig. 10.1. $\mathfrak{A}$ is isomorphic to the Boolean lattice 2^1.

10.2.4 "Eating" automaton

Consider a (3,2,3)-automaton represented in table 10.2. The resulting output sequences are repetitions of "$\cdots$ *eat* $\cdots$," regardless of the input. The output is a one-to-one map of the state space of the automaton. All state partitions corresponding to all input sequences coincide:

$$v(\emptyset) = v(0) = v(1) = v(00) = \cdots = \{\{1\}, \{2\}, \{3\}\} \quad . \tag{10.12}$$

	a	b
δ_0	b	a
δ_1	a	a

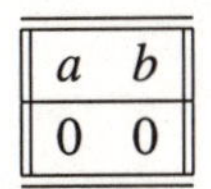

a	b
0	0

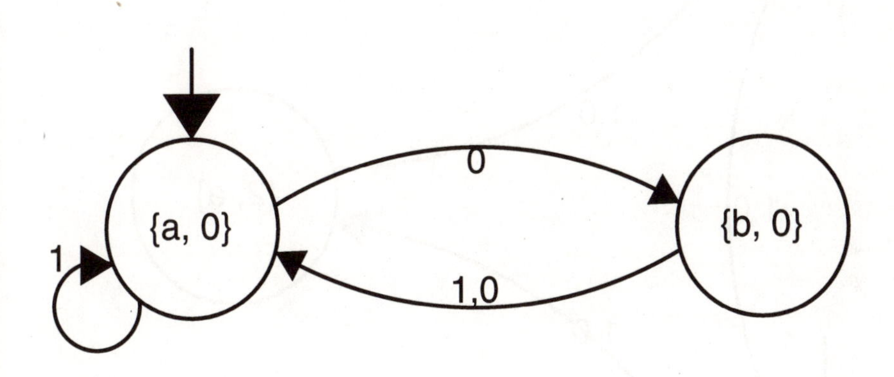

Table 10.1: Transition and output tables and figure of a trivial 2-state (2,2,1)-automaton. Internal states are labelled by *a*, *b*, input symbols by 0, 1 and the output symbol by 0.

	1	2	3
δ_0	2	3	1
δ_1	2	3	1

1	2	3
e	a	t

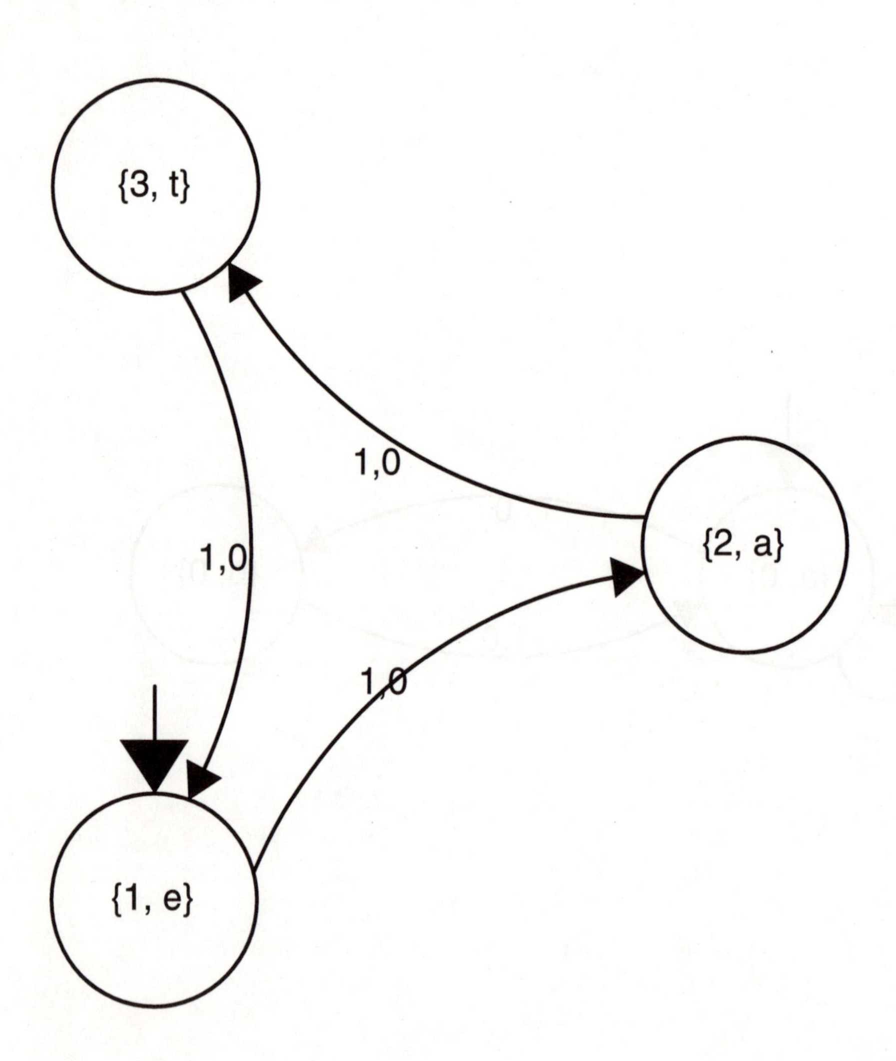

Table 10.2: Transition and output tables and figure of a (3,2,3)-automaton. Internal states are labelled by 1, 2, 3, input symbols by 0, 1 and the output symbols by *e, a, t*.

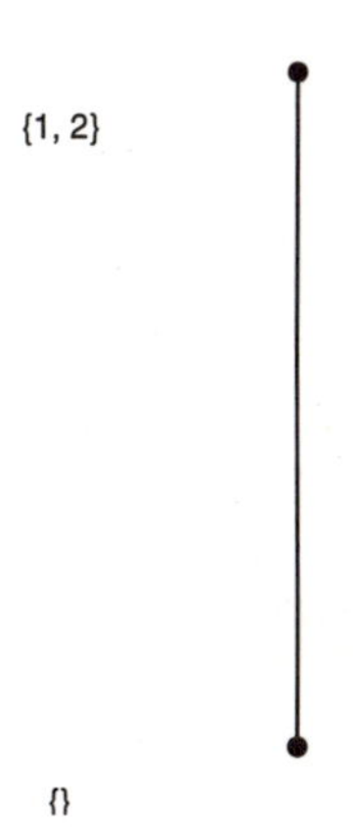

Figure 10.1: Lattice of the extrinsic and intrinsic propositional calculus of the trivial automaton.

Once again, for the eating automaton, the extrinsic and intrinsic propositional calculi $\tilde{\mathfrak{A}}$ and $\mathfrak{A}$ coincide, i.e., $\mathfrak{A} = \tilde{\mathfrak{A}}$. The eating automaton propositional calculus can be represented by the lattice drawn in Fig. 10.2. It is isomorphic to the Boolean lattice 2^3.

10.2.5 "Counterexample"

Not all automata render a "propositional calculus" which is a lattice and whose "implication" ("order relation") is transitive [402]. Consider the (4,2,2)-automaton defined in table 10.3. The "intrinsic propositional calculus" is defined by the partitions

$$v(\emptyset) = \{\{1 \vee 2\}, \{3 \vee 4\}\} \quad , \tag{10.13}$$
$$v(0) = \{\{1\}, \{2\}, \{3 \vee 4\}\} \quad , \tag{10.14}$$
$$v(1) = \{\{1 \vee 2\}, \{3\}, \{4\}\} \quad . \tag{10.15}$$

The resulting order structure is drawn in Fig. 10.3. It is obviously no lattice. The "implication" is not transitive either, because $1 \rightarrow 1 \vee 2$ requires input "0" and $1 \vee 2 \rightarrow 1 \vee 2 \vee 3$ requires input "1," but $1 \rightarrow 1 \vee 2 \vee 3$ cannot be realised by any experiment. This example shows that in general there is no guarantee that to every automaton there corresponds a properly defined automata propositional calculus which is a lattice and whose implication is a partial order relation.

One way of approaching the problem is to consider *adaptive* experiments instead of preset experiments, i.e., to analyse each output symbol separately, beginning with the first output (which, in the case of Moore-type atomata, comes for free) and reacting accordingly. In the case of the automaton discussed, the first output (empty input string) of an adaptive experiment would be "0" or "1," depending on whether the automaton is

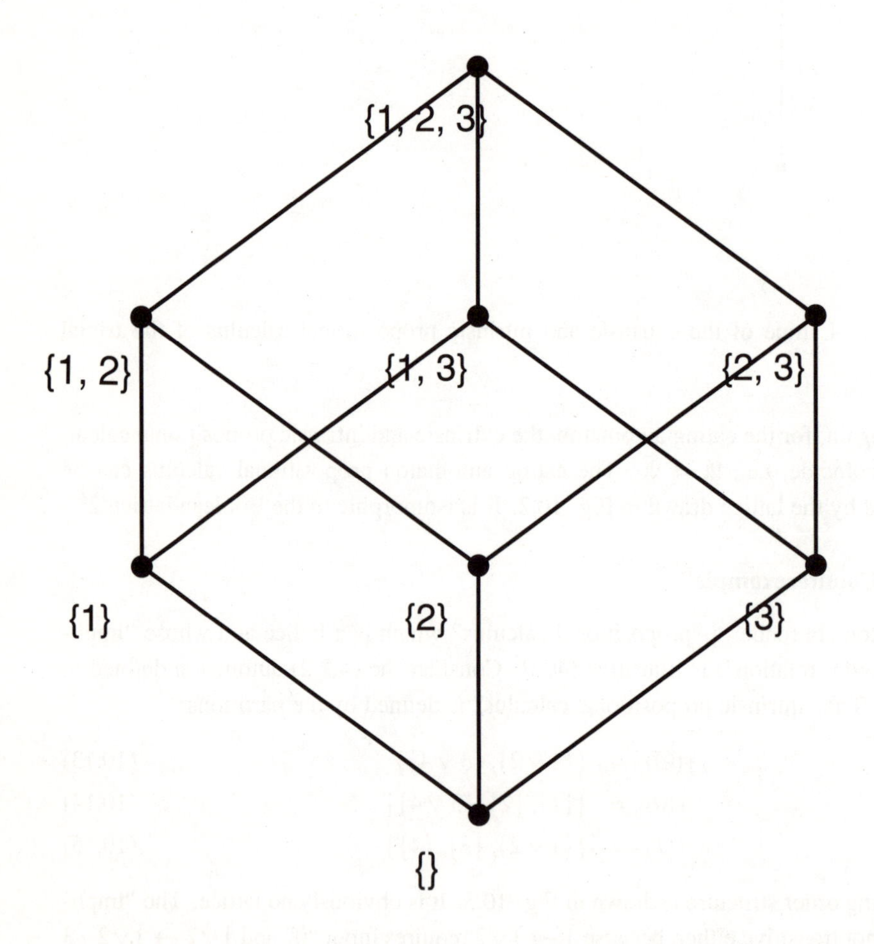

Figure 10.2: Boolean lattice of the extrinsic and intrinsic propositional calculus of the eating automaton.

	1	2	3	4
δ_0	2	3	4	4
δ_1	2	2	4	1

1	2	3	4
0	0	1	1

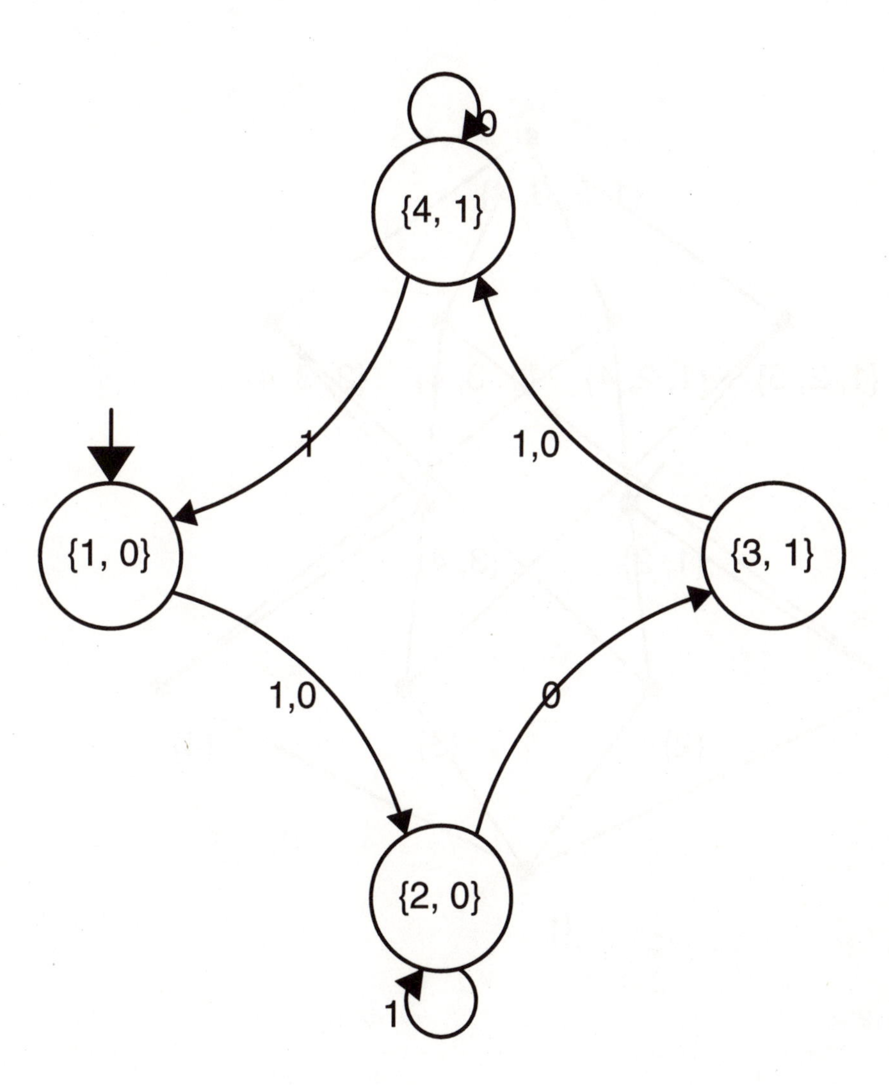

Table 10.3: Transition and output tables and figure of an automaton with 4 internal states, labelled 1, 2, 3, 4 and two input and output symbols, labelled 0, 1.

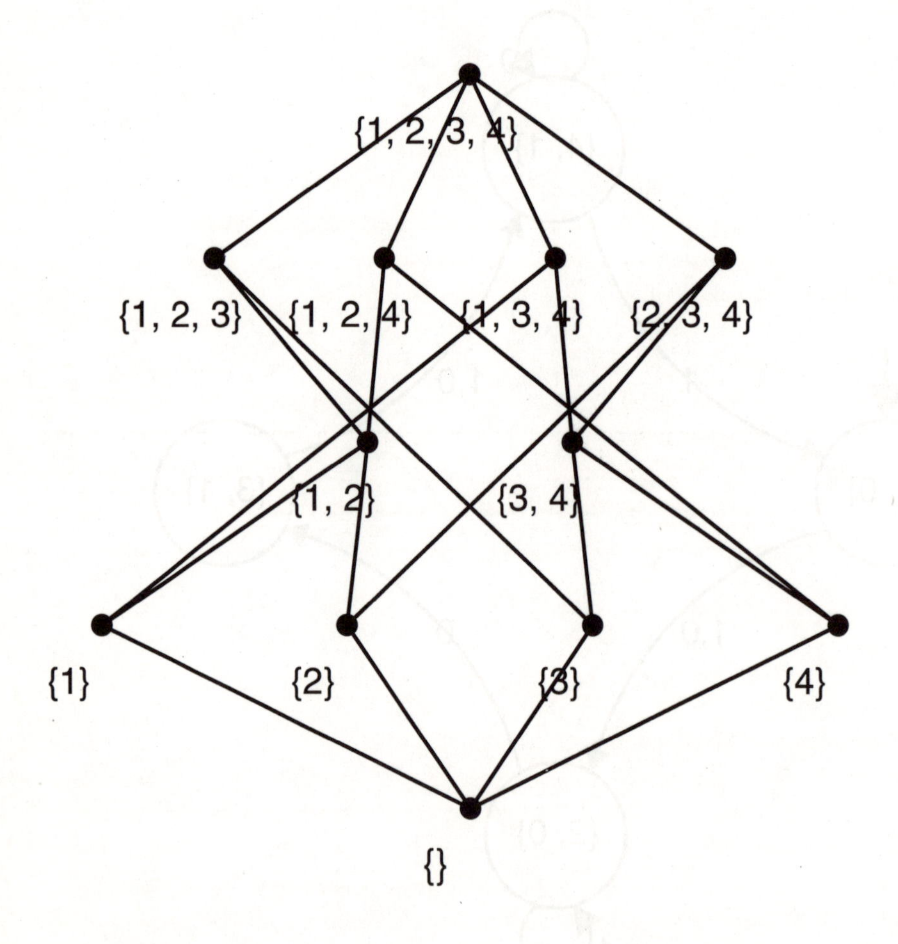

Figure 10.3: Example of a structure of propositions which is no lattice.

in states $\{1 \vee 2\}$ or $\{3 \vee 4\}$, respectively. Input "0" or "1" distinguishes between the remaining states, and the intrinsic automaton propositional calculus becomes Boolean.

10.3 Computational complementarity

In quantum theory, *complementarity* is often stated as follows ([319], p. 154):

> *The description of the properties of microscopic objects in classical terms requires pairs of complementary variables; the accuracy in one member of the pair cannot be improved without a corresponding loss in the accuracy of the other member.*

Or, stated explicitly:

> *It is impossible to perform measurements of position x and momentum p with uncertainties (defined by the root-mean square deviations) Δx and Δp such that the product of $\Delta x \Delta p$ is smaller than a constant unit of action $\frac{\hbar}{2}$.*

A historical review of the physical concept of *complementarity* is for instance given M. Jammer's book *The Philosophy of Quantum mechanics* [243], chapter 4.

Computational complementarity is a very similar structure which is based on finite automata. There, measurement of one aspect of an automaton makes impossible the measurement of another aspect. E.g., in the case of the Moore automaton (see below) measurement of the proposition $\{1 \vee 2\}$ makes impossible measurement of the proposition $\{1 \vee 3\}$. In a sense, computational complementarity is a "poor man's version" of diagonalization for finite systems: whereas diagonalization changes an infinite number of entities, in the case of finite automata, computational complementarity affects only a finite number of observables.

10.3.1 Moore automaton

Consider the *Moore automaton* [328], defined in table 10.4. The Moore automaton is an example of a type of automata with the following remarkable feature.

T 10.6 (Computational complementarity (E. F. Moore [328]))

There exists a (finite) automaton such that any pair of its states are distinguishable, but there is no experiment which can determine in what state the automaton was at the beginning of the experiment.

The term *"computational complementarity"* has been introduced by D. Finkelstein [178]. J. Conway calls this phenomenon *"Moore's uncertainty principle"* [125]. Computational complementarity is introduced here with Moore's original example. Readers may find the Mealy automaton defined in 10.5, p. 10.5 more comprehendible.

Proof:

Consider the Moore automaton in an arbitrary initial state. Recall that there is only a *single* copy to play with.

First note that any experiment will distinguish between the state 4 and any other state, since if the automaton is in state 4, any such experiment will begin with "1" at the first position of the output sequence. What remains to be looked at are the pairs $(1, 2)$, $(1, 3)$ and $(2, 3)$.

	1	2	3	4
δ_0	4	1	4	2
δ_1	3	3	4	2

1	2	3	4
0	0	0	1

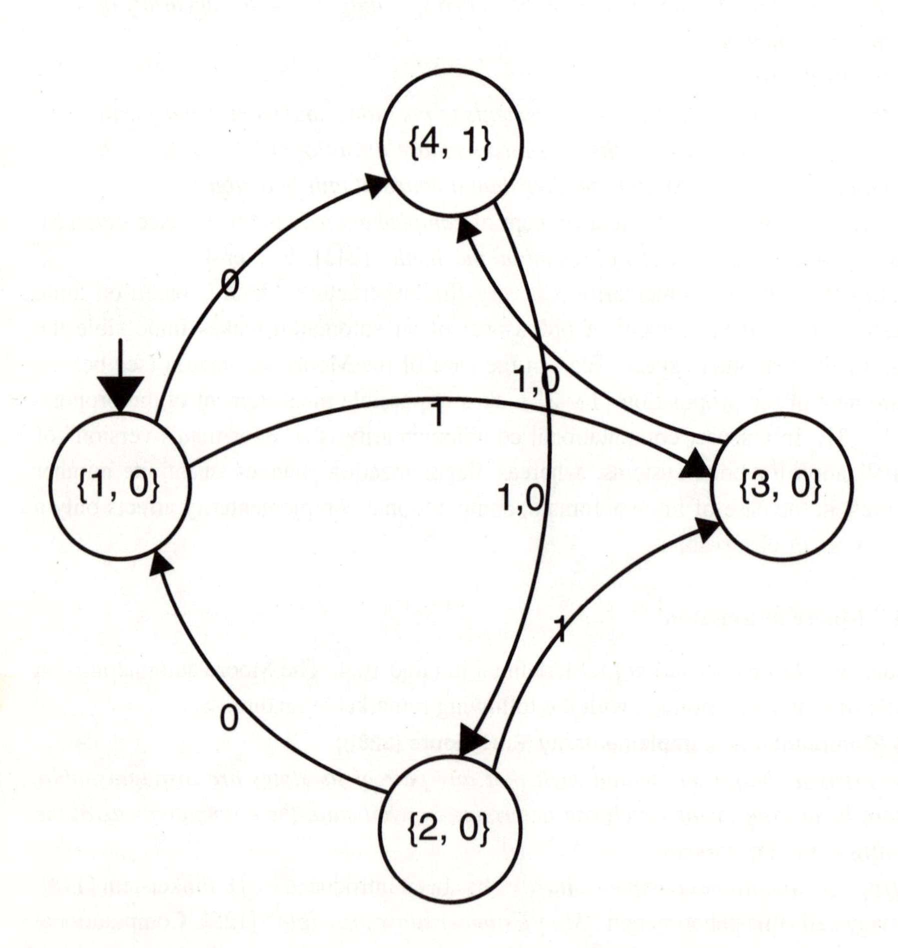

Table 10.4: Transition and output tables and figure of the Moore automaton with 4 internal states, labelled 1, 2, 3, 4 and two input and output symbols, labelled 0, 1.

Lets look at the pair $(2, 3)$. Any input of length one (i.e., either 0 or 1) will distinguish between these states. I.e., if the Moore automaton were either in state 2 or in state 3 initially, one could see this immediately by reading the *second* bit of the experimental sequence.

What remains to be looked at are the pairs $(1, 2)$ and $(1, 3)$. *(i)* Lets assume we start our experiment with input "1." If the automaton would be either in state 1 or in state 3, this would induce a transition according to

input	1	···	
state	1	3	···
output	0	0	···

or

input	1	···	
state	3	4	···
output	0	1	···

.

This means that we could distinguish $(1, 3)$. However, if the automaton would be either in state 1 or in state 2, this would induce a transition to state 3, i.e.,

input	1	···	
state	1	3	···
output	0	0	···

or

input	1	···	
state	2	3	···
output	0	0	···

.

This means that we would get identical output and we could *not* distinguish $(1, 2)$.

(ii) Lets assume we start our experiment with input "0." If the automaton would be either in state 1 or in state 2, this would induce a transition according to

input	0	···	
state	1	4	···
output	0	1	···

or

input	0	···	
state	2	1	···
output	0	0	···

.

This means that we could distinguish $(1, 2)$. However, if the automaton would be either in state 1 or in state 3, this would induce a transition to state 3, i.e.,

input	0	···	
state	1	4	···
output	0	1	···

or

input	0	···	
state	3	4	···
output	0	1	···

.

This means that we would get identical output and we could *not* distinguish $(1, 3)$.

In other words, by a *single-copy* experiment, it may only be possible to obtain *partial* information about the Moore automaton's internal state. In particular one has to decide whether one would like to distinguish between the states 1 and 2 — and thus start the experiment with input "0" — *(exclusive) or* between the states 1 and 3 — and thus start the experiment with input "1". Hence, stated differently, if the initial state is chosen at random, the states 1,2 and 3 cannot be distinguished from one another at the same time, since any distiction between the states 1 and 2 makes impossible a distinction between the states 1 and 3, and any distiction between the states 1 and 3 makes impossible a distinction between the states 1 and 2. Based on this observation, let us consider the automaton propositional calculus next.

Extrinsic propositional calculus

Again the following terminology is used: the proposition "`the Moore automaton is in state` 1 `or` 2 `or` 3" is denoted by $1 \vee 2 \vee 3$, or by $\{1, 2, 3\}$.

Consider the following propositions:

1: "`the Moore automaton is in state` 1"

2: "`the Moore automaton is in state` 2"

3: "`the Moore automaton is in state` 3"

4: "`the Moore automaton is in state` 4"

The *extrinsic propositional calculus* of the Moore automaton can be obtained from multi-automaton distinguishing experiments; i.e., by allowing experiments on an arbitrary number of identical copies of an automaton. All of the states are pairwise distinguishable. Therefore, $\tilde{\mathfrak{A}}$ is trivially Boolean, and hence it is distributive and thus modular and orthomodular. It is represented by Fig. 10.4.

Intrinsic propositional calculus

Consider the *intrinsic propositional calculus* $\mathfrak{A}$, obtained from the single-automaton setup. Because the states $1, 2, 3$ cannot be identified simultaneously and separately, the propositional calculus is not trivial (Boolean). Since the empty (i.e., no) input identifies states 4 and $1 \vee 2 \vee 3$, the partition $v(\emptyset)$ is

$$v(\emptyset) = \{\{4\}, \{1, 2, 3\}\} \quad . \tag{10.16}$$

Since the input string 0 (and all other strings beginning with 0) identifies states 2, $1 \vee 3$ and 4, the partition $v(0)$ is

$$v(0) = \{\{2\}, \{1, 3\}, \{4\}\} \quad . \tag{10.17}$$

Since the input string 1 (and all other strings beginning with 1) identifies states 3, $1 \vee 2$ and 4, the partition $v(1)$ is

$$v(1) = \{\{3\}, \{1, 2\}, \{4\}\} \quad . \tag{10.18}$$

The intrinsic Moore automaton propositional calculus can be derived according to definitions 10.3 and 10.5. It is the pasting of two blocks 2^3. The atoms of the blocks are the elements of $v(0)$ and $v(1)$. Both blocks share one common element, i.e., $\{4\}$. The intrinsic Moore automaton propositional calculus is represented in Fig. 10.5. It is an orthocomplemented lattice: the complements are given by:

$$\begin{aligned}
\neg\mathbf{0} &= \mathbf{1} \quad ; \\
\neg 2 &= 1 \vee 3 \vee 4 \quad ; \\
\neg 3 &= 1 \vee 2 \vee 4 \quad ; \\
\neg 4 &= 1 \vee 2 \vee 3 \quad ; \\
\neg(1 \vee 2) &= 3 \vee 4 \quad ; \\
\neg(1 \vee 3) &= 2 \vee 4 \quad ; \\
\neg(2 \vee 4) &= 1 \vee 3 \quad ; \\
\neg(3 \vee 4) &= 1 \vee 2 \quad ; \\
\neg(1 \vee 2 \vee 3) &= 4 \quad ; \\
\neg(1 \vee 2 \vee 4) &= 3 \quad ; \\
\neg(1 \vee 3 \vee 4) &= 2 \quad ; \\
\neg\mathbf{1} &= \mathbf{0} \quad .
\end{aligned}$$

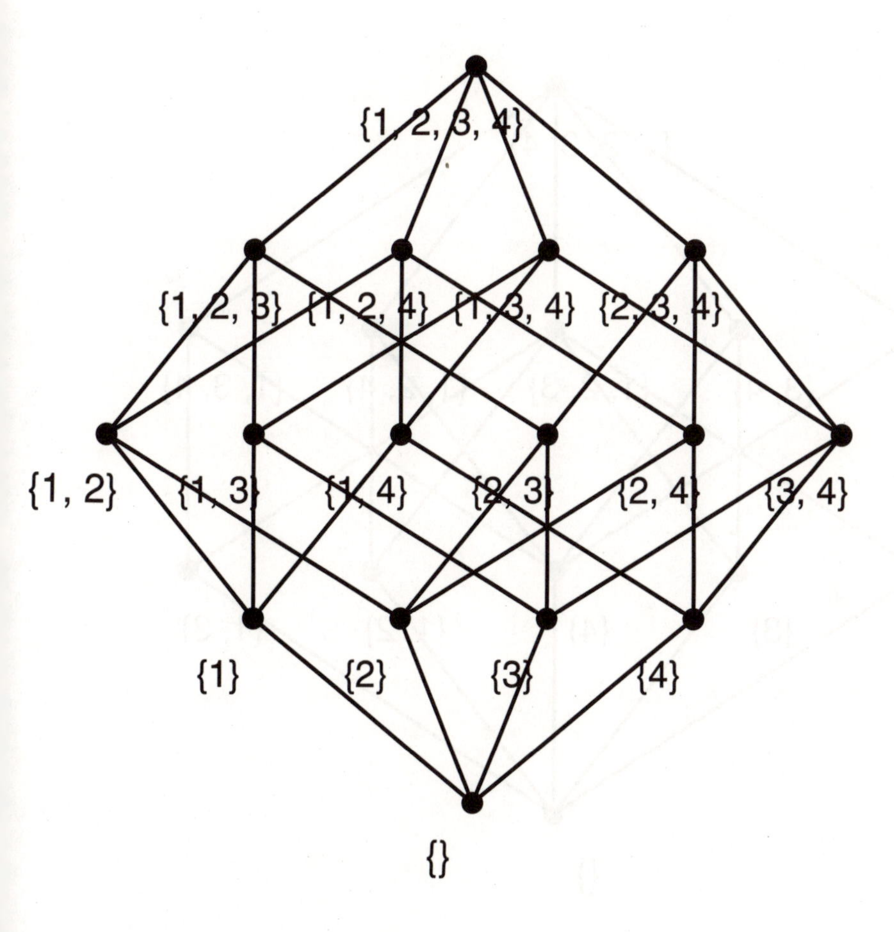

Figure 10.4: Boolean lattice of the extrinsic propositional calculus of the Moore automaton.

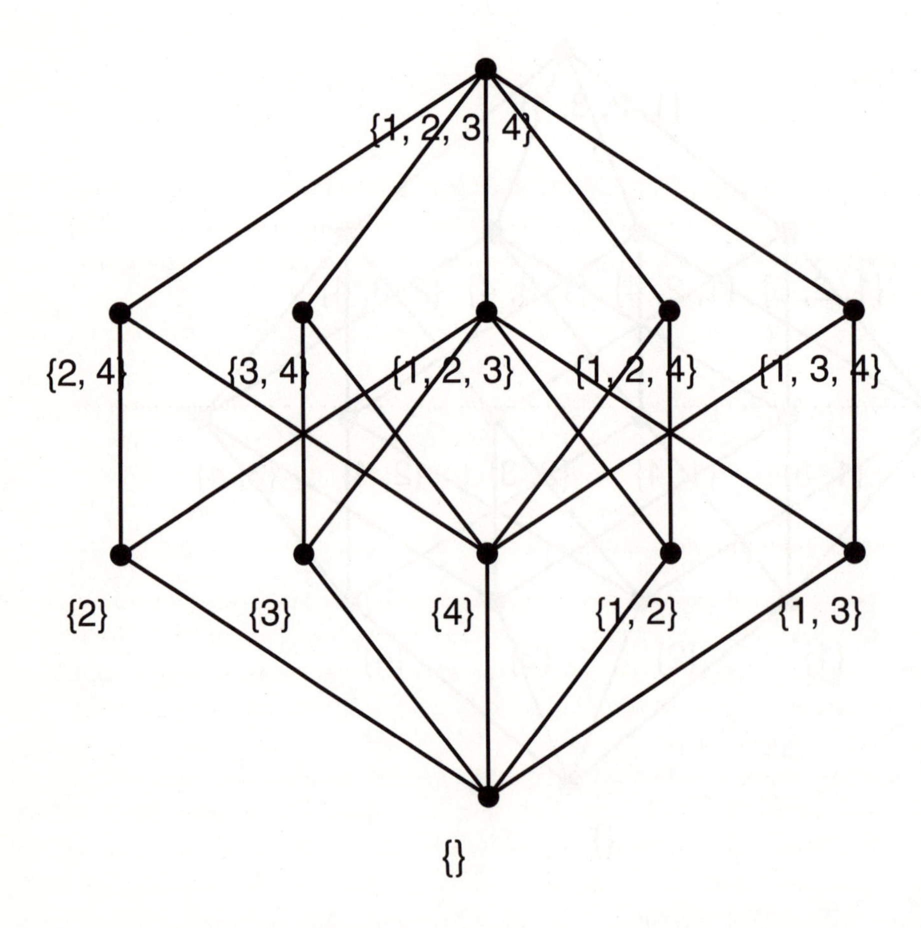

Figure 10.5: Lattice of the intrinsic propositional calculus of the Moore automaton.

The intrinsic propositional calculus of the Moore automaton is *not a distributive* and thus *not a Boolean* lattice. The intrinsic propositional calculus of the Moore automaton is a *modular* lattice, since it does not contain the lattice of Fig. 5.2, p. 57 as a subalgebra. Since modularity implies orthomodularity, the intrinsic propositional calculus of the Moore automaton is an *orthomodular* lattice. — It does not contain the subalgebra O_6 drawn in Fig. 5.3, p. 57.

A physical interpretation of $\mathfrak{A}$ has been suggested by R. Giuntini [201], p. 161.

In summary, by comparing the extrinsic and intrinsic Moore automaton propositional calculi, one obtains

$$\mathfrak{A} \neq \tilde{\mathfrak{A}} \quad . \tag{10.19}$$

J. H. Conway and W. Brauer, among others, have studied computational complementarity of the Moore type in some detail [125, 28]. In what follows, some results are reviewed.

T 10.7 ((Conway [125], Brauer [28])) *Two distinguishable states of an (i,k,n)-automaton of the Moore type can be distinguished by some input word of length at most* $i-n$. *[Two distinguishable states of any (i,k,2)-automaton of the Moore type can be distinguished by an input word of length at most* $i-2$.*]*

Moore uncertainties cannot occur in Moore type automata with fewer than four internal states.

An arbitrary number of pairwise distinguishable states of any (i,k,n)-automaton of the Moore type can be distinguished by an input word of length at most $(i-n+1)i^i$. *(If* $n > i$, *the trivial experiment with no input distinguishes between the states.)*

10.3.2 Simple "quantum-like" automaton

Another explicit model of a non distributive and modular (orthomodular) finite automaton propositional calculus which is a pasting of three Boolean algebras 2^2 is not of the Moore but of the Mealy type. A Mealy-type automaton presents no "free" output at the beginning (no input) and responds with an output as some function of its internal state and of the input.

Consider the transition and output tables 10.5 of a (3,3,2)-automaton. Input of 1, 2 or 3 steers the automaton into the states 1, 2 or 3, respectively. At the same time, the output of the automaton is 1 only if the guess is a "hit," i.e., if the automaton was in that state. Otherwise the output is 0. Hence, after the measurement, the automaton is in a definite state, but if the guess is no "hit," the information about the initial automaton state is lost. Therefore, the experimenter has to decide before the actual measurement which one of the following hypotheses should be tested (in short-hand notation, "$\{1\}$" stands for "`the automaton is in state` 1" *etc.*): $\{1\} = \neg\{2,3\}$, $\{2\} = \neg\{1,3\}$, $\{3\} = \neg\{1,2\}$. Measurement of either one of these three hypotheses (or their complement) makes impossible measurement of the other two hypotheses.

No input, i.e., the empty input string $\varnothing$, identifies all three internal automaton states. This corresponds to the trivial information that the automaton is in *some* internal state. Input of the symbol 1 (and all sequences of symbols starting with 1) distinguishes between the hypothesis $\{1\}$ (output "1") and the hypothesis $\{2,3\}$ (output "0"). Input of

	1	2	3
δ_1	1	1	1
δ_2	2	2	2
δ_3	3	3	3

	1	2	3
o_1	1	0	0
o_2	0	1	0
o_3	0	0	1

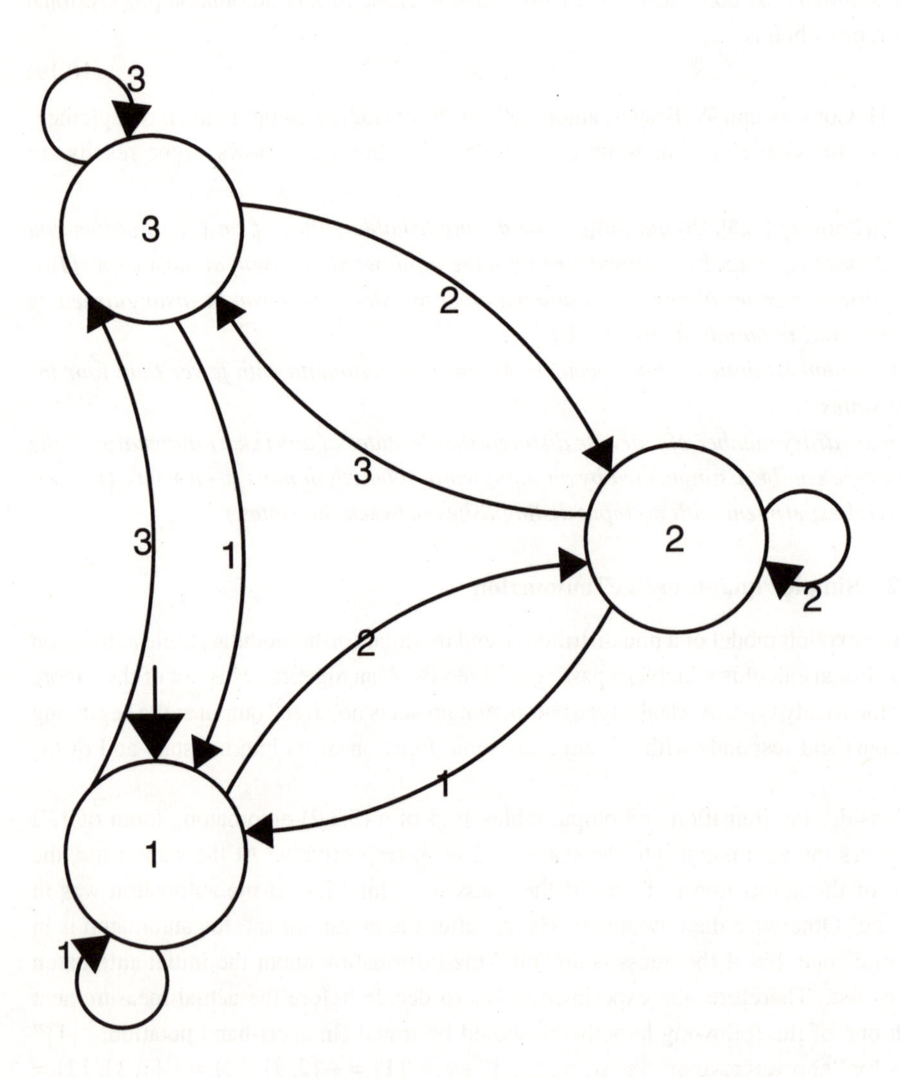

Table 10.5: Transition and output tables and figure of a (3,2,2)-automaton of the Mealy type.

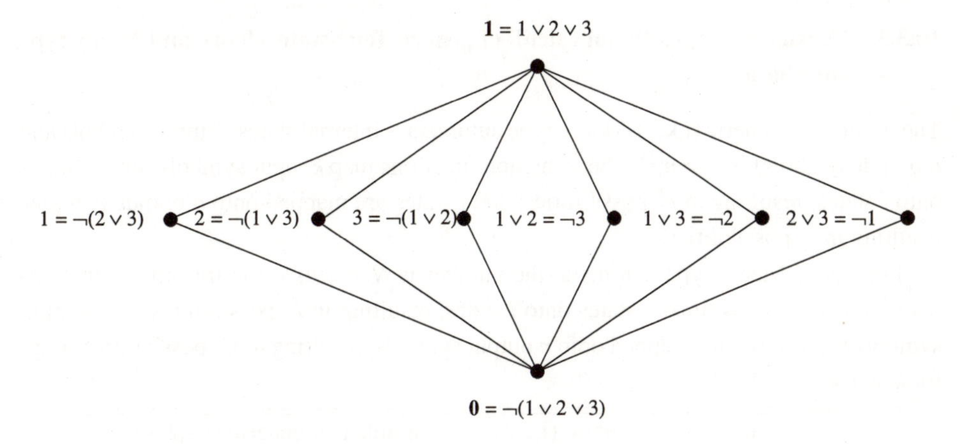

Figure 10.6: Lattice *MO*3 of intrinsic propositional calculus of a (3,2,2)-automaton of the Mealy type.

the symbol 2 (and all sequences of symbols starting with 1) distinguishes between the hypothesis $\{2\}$ (output "1") and the hypothesis $\{1,3\}$ (output "0"). Input of the symbol 3 (and all sequences of symbols starting with 1) distinguishes between the hypothesis $\{3\}$ (output "1") and the hypothesis $\{1,2\}$ (output "0"). The intrinsic propositional calculus is thus defined by the partitions

$$v(\emptyset) = \{\{1,2,3\}\} \quad , \tag{10.20}$$
$$v(1) = \{\{1\},\{2,3\}\} \quad , \tag{10.21}$$
$$v(2) = \{\{2\},\{1,3\}\} \quad , \tag{10.22}$$
$$v(3) = \{\{3\},\{1,2\}\} \quad . \tag{10.23}$$

It can be represented by the lattice drawn in Fig. 10.6. This lattice is of the "Chinese latern" *MO*3 form. It is non distributive because it contains the lattice of Fig. 5.5(d), p. 63 as sublattice. It is modular, since it does not contain a subalgebra of the form drawn in Fig. 5.2, p. 57, and hence it is orthomodular.

The obtained intrinsic propositional calculus in many ways resembles the lattice obtained from photon polarisation experiments or from other incompatible quantum measurements. Consider an experiment measuring photon polarisation. Three propositions of the form

"`the photon has polarisation` p_{ϕ_1},"
"`the photon has polarisation` p_{ϕ_2},"
"`the photon has polarisation` p_{ϕ_3}"

cannot be measured simultaneously for the angles $\phi_1 \neq \phi_2 \neq \phi_3 \neq \phi_1 (\bmod \pi)$. An irreversible measurement of one direction of polarisation would result in a state preparation, making impossible measurement of the other directions of polarisation, and resulting in a propositional calculus of the "Chinese latern" form *MO*3. [The *extrinsic propositional calculus* of this automaton is trivial (Boolean).]

10.3.3 Census of propositional calculi of generic four-state Moore and Mealy type automata

The number of generic (i,k,n)-Moore type automata (i internal states, k input symbols and n output symbols) is $N = n^i i^{ik}$: the transition functions map k input symbols times i states onto i states, resulting in i^{ik} possibilities, and i states are mapped onto n output symbols, resulting in n^i possibilities.

For (i,k,n)-Mealy type automata, the number is $N = (in)^{ik}$: the transition functions map k input symbols times i states onto i states, resulting in i^{ik} possibilities, and k input symbols and i states are mapped onto n output symbols, resulting in n^{ik} possibilities. E.g., for $k = n = 2, i \geq 1$,

	number of generic (i,2,2)-Moore automata	number of generic (i,2,2)-Mealy automata
$N(1,2,2)$	2	4
$N(2,2,2)$	64	256
$N(3,2,2)$	5832	46656
$N(4,2,2)$	1048576	16777216
$N(5,2,2)$	312500000	10000000000
$N(6,2,2)$	139314069504	8916100448256
$N(7,2,2)$	86812553324672	11112006825558016
$N(8,2,2)$	72057594037927936	18446744073709551616
$N(9,2,2)$	76848453272063549952	39346408075296537575424
$N(10,2,2)$	102400000000000000000000	104857600000000000000000000

The number of non isomorphic automata grows substantially slower: for Mealy-type automata, M. A. Harrison showed that the number of classes of non isomorphic automata is asymptotic (i.e., for $i \to \infty$) to $(in)^{ik}/i!$; i.e., non isomorphism reduces the number of automata by a factor of $1/i!$ (cf. [228], theorem 6.7).

With regards to the structure of their (intrinsic) propositional calculi, one expects an increasingly complex behaviour as the number of states i is increased. One may indeed speculate that the limit of an unbounded number of internal states yields a model of computation which is at least as powerful as a universal computer. — After all, for example, a Cellular Automaton is nothing but an infinite array of interconnected finite automata. (Using cardinal numbers, one may speculate that there are $\aleph_1$ many infinite-state automata but only $\aleph_0$ many Turing machines.)

Generic Moore automaton logics

Extrinsic automaton propositional calculi, i.e., ones obtained by multi-automata configurations, are always Boolean and thus, in a sense, trivial. If there are i pairwise distinguishable states, then $\widetilde{\mathfrak{A}(i)} = 2^i$. If not mentioned otherwise, in what follows we shall therefore concentrate on the *intrinsic* automaton propositional calculi.

The Hasse diagrams of the automaton propositional calculi for all (1,x,x)-, (2,2,2)-, (3,2,2)-, (4,2,2)- Moore type automata and for some (5,2,2)- Moore type automata are listed in Fig. 10.7. They can be obtained by "brute force," i.e., without utilising isomorphism, in four steps: *(i)* generate all automata; *(ii)* generate the set of all state partitions for these automata; *(iii)* generate the graphs (of the Hasse diagram) of the automaton propositional calculi; *(iv)* apply `Compress` (for a listing, see the appendix, p. 261) to reduce the set to all non isomorphic graphs. The last step includes the problem of graph isomorphism, in particular the use of St. Skiena's function `IsomorphicQ[g_Graph,h_Graph]`, to generate the set of graphs which are not isomorphic to each other. Graph isomorphism is conjectured NP-hard [416].

Examples:

(i) The set of state partitions of all (2,2,2)-Moore type automata is given by
$\{\{\{\{1,2\}\}\},\{\{\{1\},\{2\}\}\}\}$.

(ii) The set of state partitions of all (3,2,2)-Moore type automata is given by
$\{\{\{\{1,2,3\}\}\},$
$\{\{\{1\},\{2,3\}\}\},$
$\{\{\{2\},\{1,3\}\}\},$
$\{\{\{3\},\{1,2\}\}\},$
$\{\{\{1\},\{2\},\{3\}\}\},$
$\{\{\{1\},\{2,3\}\},\{\{1\},\{2\},\{3\}\}\},$
$\{\{\{2\},\{1,3\}\},\{\{1\},\{2\},\{3\}\}\},$
$\{\{\{3\},\{1,2\}\},\{\{1\},\{2\},\{3\}\}\}\}$.
The Hasse diagrams occurring in the places 2,3,4 , as well as the elements in the places 5,6,7,8, are isomorphic.

(iii) These are some (computer-generated) sets of state partitions of (4,2,2)- Moore type automata:
$\{\{\{\{1,2,3,4\}\}\},$
$\{\{\{1\},\{2,3,4\}\}\},$
$\{\{\{2\},\{1,3,4\}\}\},$
$\{\{\{3\},\{1,2,4\}\}\},$
$\{\{\{4\},\{1,2,3\}\}\},$
$\{\{\{1,2\},\{3,4\}\}\},$
$\{\{\{1,3\},\{2,4\}\}\},$
$\{\{\{1,4\},\{2,3\}\}\},$
$\{\{\{1\},\{2\},\{3,4\}\}\},$
$\{\{\{1\},\{3\},\{2,4\}\}\},$
$\{\{\{1\},\{4\},\{2,3\}\}\},$
$\{\{\{1\},\{2,3,4\}\},\{\{1\},\{2\},\{3,4\}\}\},$
$\{\{\{2\},\{1,3,4\}\},\{\{1\},\{2\},\{3,4\}\}\},$
$\{\{\{3\},\{1,2,4\}\},\{\{1\},\{3\},\{2,4\}\}\},$
$\{\{\{3\},\{1,2,4\}\},\{\{2\},\{3\},\{1,4\}\}\},$
$\{\{\{3\},\{1,2,4\}\},\{\{3\},\{4\},\{1,2\}\}\},$
$\{\{\{4\},\{1,2,3\}\},\{\{1\},\{4\},\{2,3\}\}\},$
$\{\{\{4\},\{1,2,3\}\},\{\{2\},\{4\},\{1,3\}\}\},$
$\{\{\{4\},\{1,2,3\}\},\{\{3\},\{4\},\{1,2\}\}\},$

$$
\begin{array}{l}
\{\{\{1,2\},\{3,4\}\},\{\{1\},\{2\},\{3,4\}\}\},\\
\{\{\{1,3\},\{2,4\}\},\{\{1\},\{3\},\{2,4\}\}\},\\
\{\{\{1,3\},\{2,4\}\},\{\{2\},\{4\},\{1,3\}\}\},\\
\{\{\{1,3\},\{2,4\}\},\{\{1\},\{2\},\{3\},\{4\}\}\},\\
\{\{\{1,4\},\{2,3\}\},\{\{1\},\{4\},\{2,3\}\}\},\\
\{\{\{1,4\},\{2,3\}\},\{\{2\},\{3\},\{1,4\}\}\},\\
\{\{\{1,4\},\{2,3\}\},\{\{1\},\{2\},\{3\},\{4\}\}\},\\
\{\{\{1\},\{3\},\{2,4\}\},\{\{2\},\{3\},\{1,4\}\}\},\\
\{\{\{1\},\{3\},\{2,4\}\},\{\{2\},\{4\},\{1,3\}\}\},\\
\{\{\{1\},\{3\},\{2,4\}\},\{\{3\},\{4\},\{1,2\}\}\},\\
\{\{\{1\},\{3\},\{2,4\}\},\{\{1\},\{2\},\{3\},\{4\}\}\},\\
\{\{\{1\},\{4\},\{2,3\}\},\{\{2\},\{3\},\{1,4\}\}\},\\
\{\{\{1\},\{4\},\{2,3\}\},\{\{2\},\{4\},\{1,3\}\}\},\\
\{\{\{1\},\{4\},\{2,3\}\},\{\{3\},\{4\},\{1,2\}\}\},\\
\{\{\{1\},\{4\},\{2,3\}\},\{\{1\},\{2\},\{3\},\{4\}\}\},\\
\{\{\{3\},\{1,2,4\}\},\{\{1\},\{3\},\{2,4\}\},\{\{2\},\{3\},\{1,4\}\}\},\\
\{\{\{3\},\{1,2,4\}\},\{\{1\},\{3\},\{2,4\}\},\{\{3\},\{4\},\{1,2\}\}\},\\
\{\{\{3\},\{1,2,4\}\},\{\{1\},\{3\},\{2,4\}\},\{\{1\},\{2\},\{3\},\{4\}\}\},\\
\{\{\{3\},\{1,2,4\}\},\{\{2\},\{3\},\{1,4\}\},\{\{1\},\{2\},\{3\},\{4\}\}\},\\
\{\{\{4\},\{1,2,3\}\},\{\{1\},\{4\},\{2,3\}\},\{\{2\},\{4\},\{1,3\}\}\},\\
\{\{\{4\},\{1,2,3\}\},\{\{1\},\{4\},\{2,3\}\},\{\{3\},\{4\},\{1,2\}\}\},\\
\{\{\{4\},\{1,2,3\}\},\{\{1\},\{4\},\{2,3\}\},\{\{1\},\{2\},\{3\},\{4\}\}\},\\
\{\{\{4\},\{1,2,3\}\},\{\{2\},\{4\},\{1,3\}\},\{\{1\},\{2\},\{3\},\{4\}\}\},\\
\{\{\{1,3\},\{2,4\}\},\{\{1\},\{3\},\{2,4\}\},\{\{2\},\{4\},\{1,3\}\}\},\\
\{\{\{1,3\},\{2,4\}\},\{\{1\},\{3\},\{2,4\}\},\{\{1\},\{2\},\{3\},\{4\}\}\},\\
\{\{\{1,4\},\{2,3\}\},\{\{1\},\{4\},\{2,3\}\},\{\{2\},\{3\},\{1,4\}\}\},\\
\{\{\{1,4\},\{2,3\}\},\{\{1\},\{4\},\{2,3\}\},\{\{1\},\{2\},\{3\},\{4\}\}\},\\
\{\{\{1\},\{3\},\{2,4\}\},\{\{2\},\{3\},\{1,4\}\},\{\{1\},\{2\},\{3\},\{4\}\}\},\\
\{\{\{1\},\{3\},\{2,4\}\},\{\{2\},\{4\},\{1,3\}\},\{\{1\},\{2\},\{3\},\{4\}\}\},\\
\{\{\{1\},\{4\},\{2,3\}\},\{\{2\},\{3\},\{1,4\}\},\{\{1\},\{2\},\{3\},\{4\}\}\},\\
\{\{\{1\},\{4\},\{2,3\}\},\{\{2\},\{4\},\{1,3\}\},\{\{1\},\{2\},\{3\},\{4\}\}\},\\
\{\{\{3\},\{1,2,4\}\},\{\{1\},\{3\},\{2,4\}\},\{\{2\},\{3\},\{1,4\}\},\{\{1\},\{2\},\{3\},\{4\}\}\},\\
\{\{\{4\},\{1,2,3\}\},\{\{1\},\{4\},\{2,3\}\},\{\{2\},\{4\},\{1,3\}\},\{\{1\},\{2\},\{3\},\{4\}\}\},\cdots\}
\end{array}
$$

Generic Mealy automaton logics

Moore type automata are too restricted universes to represent a sufficiently comprehensive model for automata logic. In particular, it is not possible with them to generate state partitions corresponding to pure "Chinese latern" *MOn* forms. This can be traced back to the first "free" bit of output of information, which is supplied to the experimenter even without any input. This deficiency can be overcome by introducing more general models such as Mealy automata.

Fig. 10.8 shows $\mathfrak{F}_4$, the Hasse diagrams of generic intrinsic propositional calculi of Mealy automata up to 4 states. Fig. 10.9 shows only those propositional calculi

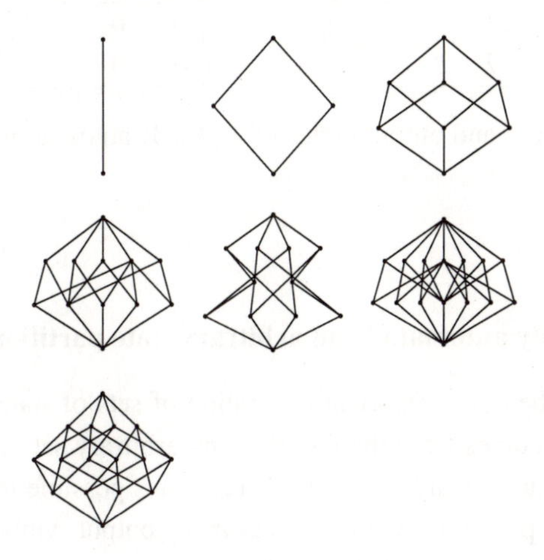

Figure 10.7: Non isomorphic Hasse diagrams of the intrinsic propositional calculi of all (1-4,2,2)- and of one (5,2,2)- Moore type automata.

which are *lattices*. The state partitions can be generated not by analysing automata but by permutation (for a *Mathematica* program, see A.3, p. 261).

Examples:

(i) The set of state partitions of all (2,2,2)-Mealy type automata is given by
$\{\{\{\{1\},\{2\}\}\},\{\{\{1\},\{2\}\},\{\{2,1\}\}\},\{\{\{2,1\}\}\}\}$.

(ii) The set of state partitions of all (3,2,2)-Mealy type automata is given by
$\{\{\{\{1,2,3\}\}\},$
$\{\{\{1\},\{2,3\}\}\},$
$\{\{\{1\},\{2,3\}\},\{\{1,2\},\{3\}\}\},$
$\{\{\{1,2\},\{3\}\}\},$
$\{\{\{1,2\},\{3\}\},\{\{1,3\},\{2\}\}\},$
$\{\{\{1\},\{2,3\}\},\{\{1,2\},\{3\}\},\{\{1,3\},\{2\}\}\},$
$\{\{\{1\},\{2,3\}\},\{\{1,3\},\{2\}\}\},$
$\{\{\{1,3\},\{2\}\}\},$
$\{\{\{1,3\},\{2\}\},\{\{1\},\{2\},\{3\}\}\},$
$\{\{\{1\},\{2,3\}\},\{\{1,3\},\{2\}\},\{\{1\},\{2\},\{3\}\}\},$
$\{\{\{1\},\{2,3\}\},\{\{1,2\},\{3\}\},\{\{1,3\},\{2\}\},\{\{1\},\{2\},\{3\}\}\},$
$\{\{\{1,2\},\{3\}\},\{\{1,3\},\{2\}\},\{\{1\},\{2\},\{3\}\}\},$
$\{\{\{1,2\},\{3\}\},\{\{1\},\{2\},\{3\}\}\},$
$\{\{\{1\},\{2,3\}\},\{\{1,2\},\{3\}\},\{\{1\},\{2\},\{3\}\}\},$
$\{\{\{1\},\{2,3\}\},\{\{1\},\{2\},\{3\}\}\},$

	1	2	3	4
δ_0	1	1	1	1
δ_1	2	2	2	2

	1	2	3	4
o_0	0	1	1	1
o_1	0	1	2	2

Table 10.6: Transition and output tables of a (4,2,3)-automaton of the Mealy type.

$$\{\{\{1\},\{2\},\{3\}\}\}\}\quad .$$

Construction of Mealy automata from arbitrary state partitions

The justification for the combinatorical generation of sets of state partitions for generic Mealy type automata comes from the fact that any arbitrary set of state partitions corresponds to infinitely many Mealy automata. It is always possible to assume (at least) one input symbol per state partition and (at least) as many output symbols as there are distinct elements in the state partition. In this case, one can construct a Mealy automaton by associating the same output symbol with all the states contained in the same element of the state partition (per input symbol), regardless of the initial symbol. As a consequence, after the first input symbol, it is possible to distinguish between its states according to the elements in the state partition associated with that input symbol. Furthermore — in order to destroy information about the automaton's initial state after one input symbol — the transition function should be defined such that every input yields a transition of the Mealy automaton into a state purely depending on the input, regardless of its previous state.

Take, for instance, the set

$$V = \{\{\{1\},\{2,3,4\}\},\{\{1\},\{2\},\{3,4\}\}\}.$$

This set of state partitions requires two input symbols, say "0" and "1," corresponding to the two state partitions $\{\{1\},\{2,3,4\}\}$ and $\{\{1\},\{2\},\{3,4\}\}$, respectively. It requires (at least) three output symbols, say "0", "1" and "2" for the states 1 and 2 and 3, 4, respectively; and "0", "1" for the states 1 and 2, 3, 4, respectively. The transition and output table of a Mealy automaton realising the above state partitions is therefore given in table 10.6.

In order to make all internal automaton states accessible, i.e., reachable by sequences of input symbols from arbitrary internal states, one has to add additional input symbols whose associated output is redundant. This can be demonstrated in another example, a Mealy-type automaton realising the propositional calculus of Fig. 10.5, p. 146. It is given in tables 10.7.

More generally: Assume an arbitrary set of state partitions V with m elements $\{v_1,\dots,v_m\}$; i.e., $|V| = m$ ["$|x|$" stands for the *cardinality* (number of elements) of a set x]. Consider an arbitrary state partition $v_j = \{\alpha_l\}$, then the set

$$A = \bigcup_{a_l\in v_j} \alpha_l \tag{10.24}$$

	1	2	3	4
δ_1	1	1	1	1
δ_2	2	2	2	2
δ_3	3	3	3	3
δ_4	4	4	4	4

	1	2	3	4
o_1	0	0	0	1
o_2	0	1	0	2
o_3	0	0	1	2
o_4	0	0	0	1

Table 10.7: Transition and output tables of a (4,4,3)-automaton of the Mealy type.

is associated with the set of internal states of the automaton to be constructed. Without loss of generality one may assume that the internal states are labelled by successive positive integers, beginning with 1; i.e., $A = \{1, 2, 3, \cdots, i\} \subset \mathbb{N}$. The number of states is given by

$$i = |A| \quad . \tag{10.25}$$

In order for the internal states to be accessible (see argument above), one needs at least i input symbols, where i stands for the number of internal states. In order to obtain as many state partitions as there are in V, one needs at least $m = |V|$ input symbols. By combining both limits from below on the number of input symbols yields the requirement that there are at least

$$k = \sup\{m, i\} = \sup\{|V|, i\} \tag{10.26}$$

input symbols. Each of the k input symbols is associated with one state partition; i.e., if $I = \{s_1, s_2, s_3, \ldots, s_k\}$ is the input alphabet, then

$$s_j \rightarrow v_l = v(s_j) \quad , \tag{10.27}$$

where $s_j \in I$, $v_l \in V$.

The requirement that there are at least as many output symbols as there are elements in the state partition with the highest number of elements yields a bound from below on the number of output symbols

$$n = \sup_{1 \leq j \leq m} |v_j| \quad . \tag{10.28}$$

The output functions are chosen to match the elements of the state partition. I.e., each element of a partition corresponds to exactly one output symbol. (There may be more output symbols as there are elements in the partitions.)

With a transition, all information of the past state of the automaton should be destroyed. As stated before, in order to make the internal states accessible, there have to be at least i input symbols to steer the automaton into its i internal states. Assume that the internal states as well as the input symbols are labelled by successive positive integers, beginning with 1. Let q denote an arbitrary internal state. Then, the transition function

$$\delta(q, s_j) = 1 + s_j \bmod i \tag{10.29}$$

guarantees that the transition of the Mealy automaton depends solely on the input symbol and not on the previous internal state. Therefore, if $m < i$, i.e., if the number of state partitions is smaller than the number of internal states, then some input symbols will correspond to the same state partition. If $m > i$, i.e., if the number of state partitions is greater than the number of internal states, then some input symbols will induce a transition into the same state. If $m = i$, i.e., if the number of state partitions is equal the number of internal states, then the correspondence is 1-1 and the transition function is also simple: every input symbol induces a transition into one internal state.

Note that any propositional calculus of Moore-type automata can be obtained by Mealy-type automata as well (but not *vice versa*). If the first output of Moore-type automata is omitted, then both automaton types realise the same class of propositional calculi.

Figure 10.8: The class $\mathfrak{F}_4$ of non isomorphic Hasse diagrams of the intrinsic propositional calculi of generic 4-state automata of the Mealy type.

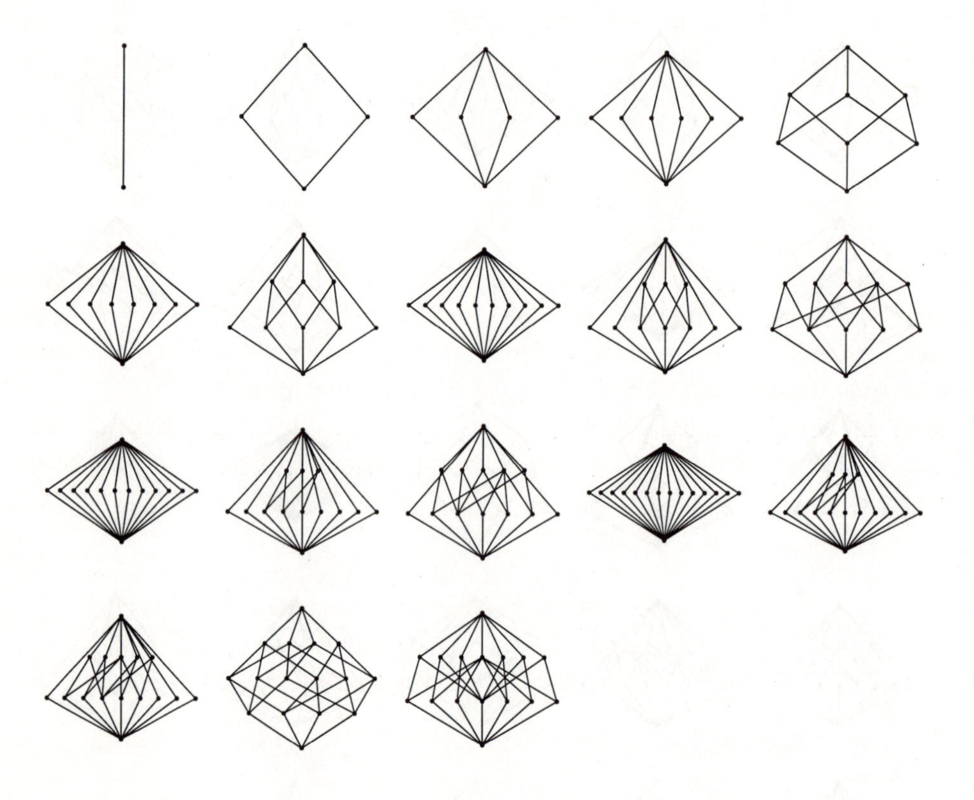

Figure 10.9: Non isomorphic Hasse diagrams of *lattice* propositional calculi of generic 4-state automata of the Mealy type.

10.4 The inverse problem

The previous sections concentrated on the construction of a suitable propositional calculus from the input/output analysis of an automaton. What is called the inverse problem here is the construction of suitable automata which correspond to particular (orthomodular) lattices, in particular to subalgebras of Hilbert lattices. One could speculate that, stated pointedly, similar to the "induction" of the Hilbert space formalism of quantum mechanics from an experimental quantum propositional calculus, a correspondence between a certain class of automaton propositional calculi and subalgebras of Hilbert lattices could be postulated. Stated differently: *"given an arbitrary orthomodular (subalgebra of a Hilbert) lattice $\mathfrak{L}$; is it possible to construct an automaton propositional calculus $\mathfrak{A}$ realising $\mathfrak{L}$?"*

In developing an answer to this question, the notion of "generic automaton propositional calculus" has to be made precise. This can be done by the following definition of a *partition logic* (cf. section 5.4.2, p. 62).

D 10.8 (Partition logic) *Consider a set M and a set $\mathfrak{P}$ of partitions of M. Every partition $P \in \mathfrak{P}$ generates a Boolean algebra of the subsets in the partition P. As for Boolean algebras, the partial order relation is identified with the subset relation (set theoretic inclusion) and the complement is identified with the set theoretic complement. The pasting of an arbitrary number of these Boolean algebras is called a* partition logic.

Since any set of partitions can be realised by some automaton propositional calculus and *vice versa* (cf. p. 154), the class of partition logics and the class of automaton propositional calculi are equivalent. We are now in the position to state the following result (for the notion of prime, see definition 5.30, p. 58):

T 10.9 ([402]) *Any orthomodular lattice $\mathfrak{L}$ is isomorphic (1-1 translatable) to some partition logic (automaton propositional calculus) iff $\mathfrak{L}$ is prime. I.e.,*

$$\textit{prime orthomodular lattice} \begin{array}{c}\Rightarrow\\ \not\Leftarrow\end{array} \left\{ \begin{array}{c} \text{partition logic} \\ \text{(automaton propositional calculus)} \end{array} \right\} .$$

For a proof, see M. Schaller and K. Svozil [402].

Remark:

The notions of two-valued states and prime ideals are equivalent (cf. p. 58). Therefore, prime orthomodular lattices admit two-valued states.

Examples:

For elementary orthomodular lattices, theorem 10.9 can be verified easily. Recall theorem 5.34, p. 59, stating that every orthomodular lattice is a pasting of its blocks. (Block decomposition may be *NP*-hard.) "At face value," every automaton state partition $v(\cdots)$ with n elements generates a Boolean algebra 2^n. If these Boolean algebras are identified with blocks, the set of automaton state partitions V represents a complete family of blocks of the automaton propositional calculus.

Some concrete examples of the construction of an automaton state partition from a prime orthomodular lattice will be considered next. — In general there will be infinitely many automata whose propositional calculi are isomorphic to $\mathfrak{L}$.

For an easy start, consider the pasting of two *disjoint* blocks; e.g. of two 2^3. Label the atoms of these blocks by $1,2,3$ and $4,5,6$. Imagine two separate automata A_1 and A_2 whose internal states are labelled by $1,2,3$ and $4,5,6$, respectively. Assume that all internal automata states can be distinguished separately, yielding two state partitions $\{\{\{1\},\{2\},\{3\}\}\}$ and $\{\{\{4\},\{5\},\{6\}\}\}$. — A (nonunique) construction technique for Mealy automata from arbitrary state partitions has already been described above (p. 154). The pasting of these subalgebras can for instance be achieved by substituting the union of the first atom of one algebra *and* all atoms from the other algebra for the first atom of one algebra and *vice versa*. E.g., in the above example,

$$\begin{aligned}\{\{\{1\},\{2\},\{3\}\}\} &\longrightarrow \{\{\{1,4,5,6\},\{2\},\{3\}\}\} \quad ,\\ \{\{\{4\},\{5\},\{6\}\}\} &\longrightarrow \{\{\{1,2,3,4\},\{5\},\{6\}\}\} \quad .\end{aligned}$$

The set of state partitions associated with the new automaton is

$$\{\{\{1,4,5,6\},\{2\},\{3\}\},\{\{1,2,3,4\},\{5\},\{6\}\}\} \quad .$$

The Greechie diagram (see 5.39, p. 60) and the Hasse diagram corresponding to this state partition is drawn in Fig. 10.10.

In a similar way, a pasting of two almost disjoint Boolean algebras (a Greechie logic) with one common atom (and its complement) could be obtained by additionally substituting the union of one atom of one algebra *and* one atom from the other algebra for the first atom of one algebra and *vice versa*. The two respective atoms should no longer occur in other elements of the partition. E.g., in the above example, states 3 & 6 are identified:

$$\begin{aligned}\{\{\{1\},\{2\},\{3\}\}\} &\longrightarrow \{\{\{1,4,5\},\{2\},\{3,6\}\}\} \quad ,\\ \{\{\{4\},\{5\},\{6\}\}\} &\longrightarrow \{\{\{1,2,4\},\{5\},\{3,6\}\}\} \quad ,\end{aligned}$$

yielding the set of partitions

$$\{\{\{1,4,5\},\{2\},\{3,6\}\},\{\{1,2,4\},\{5\},\{3,6\}\}\} \quad .$$

The Greechie and Hasse diagrams of this type of pasting is drawn in Fig. 10.11. A generalisation to arbitrary pastings of an arbitrary number of blocks is straightforward. For example, it is relatively straightforward to construct an automaton which is pasting of two 2^4 with *two* common elements; see Fig. 10.12. A possible realisation by state partitions is

$$\{\{\{1,7,8\},\{2\},\{3,5\},\{4,6\}\},\{\{3,5\},\{4,6\},\{7\},\{1,2,8\}\}\} \quad .$$

Another "all-time favourite" pasting is represented in Fig. 10.13. A possible realisation by state partitions is

$$\begin{aligned}\{\{\{1,5,6,7,8,9\},\{2\},\{3,4\}\} \quad ,\\ \{\{3,4\},\{1,2,5,8,9\},\{6,7\}\} \quad ,\\ \{\{6,7\},\{8\},\{1,2,3,4,5,9\}\}\} \quad .\end{aligned}$$

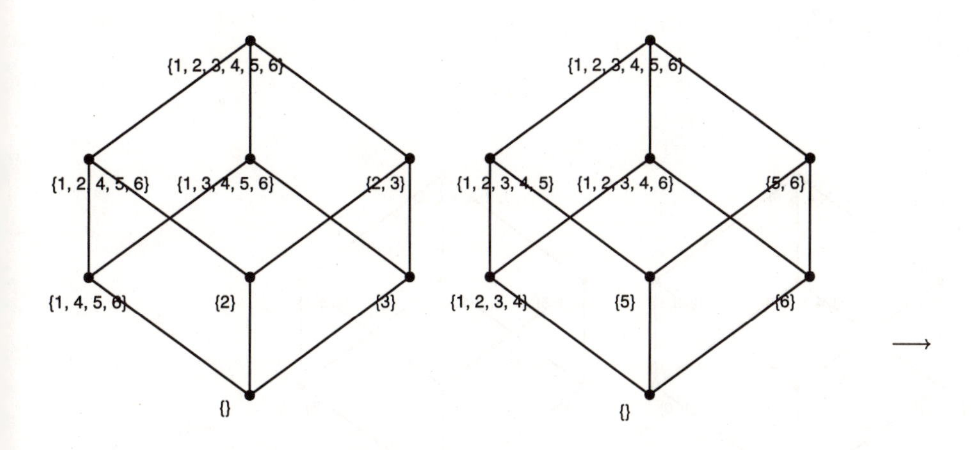

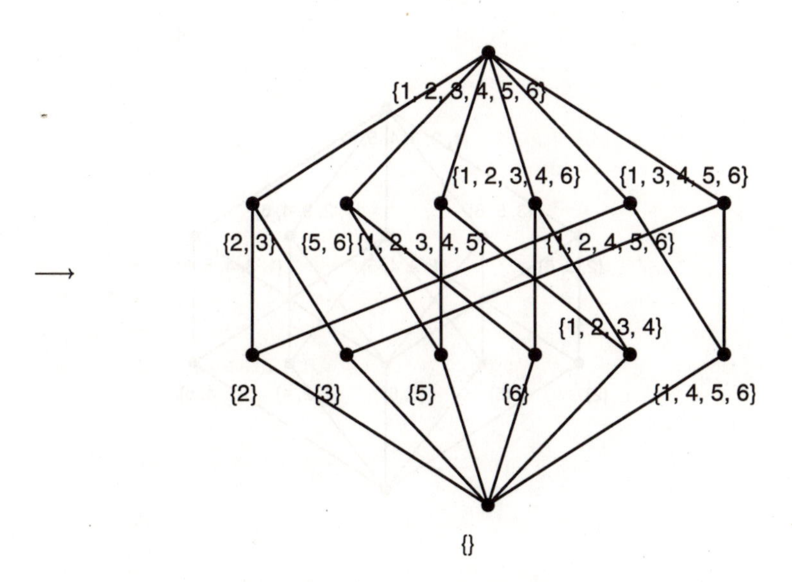

Figure 10.10: Greechie and Hasse diagrams of a pasting of two disjoint blocks.

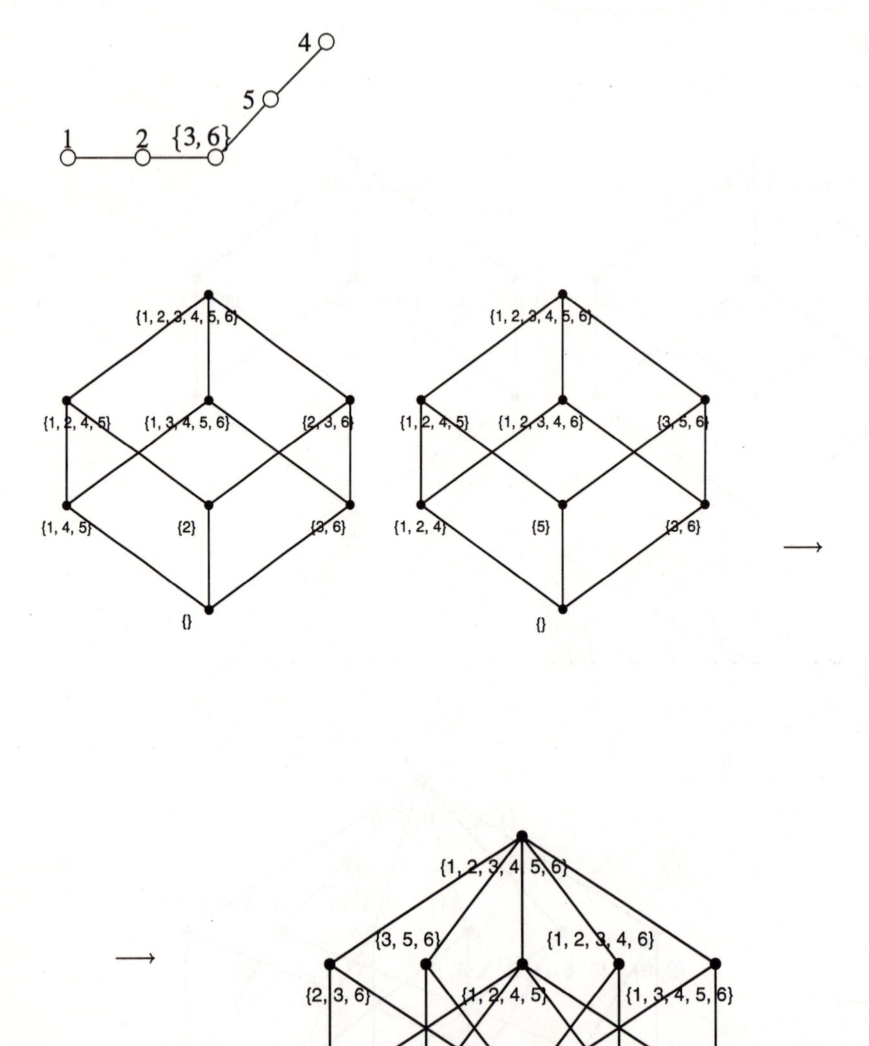

Figure 10.11: Greechie and Hasse diagrams of a pasting of two almost disjoint blocks.

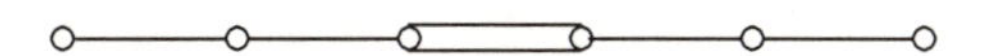

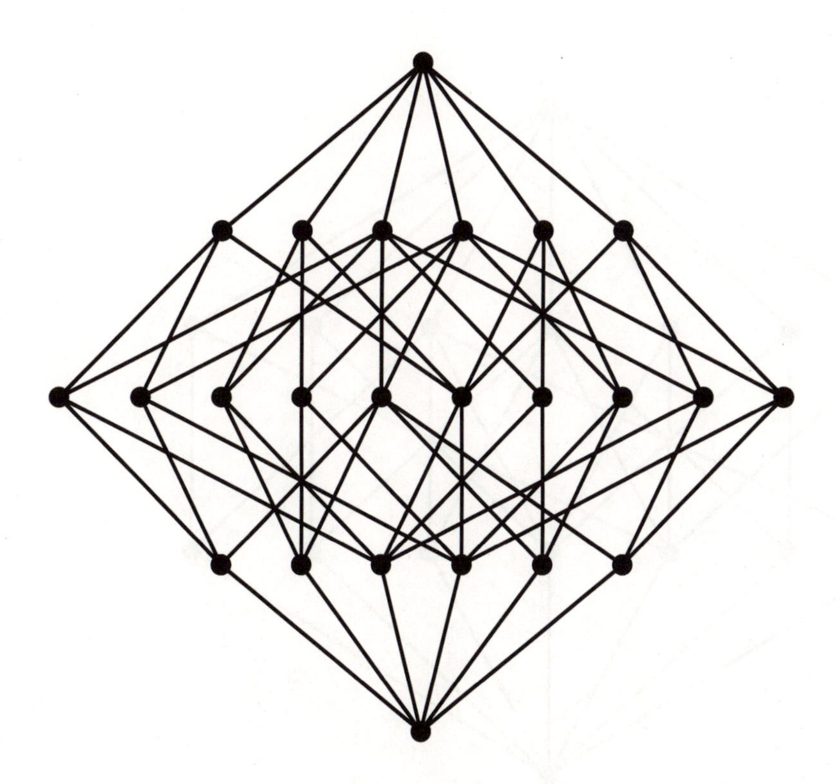

Figure 10.12: Greechie and Hasse diagrams of a pasting of two 2^4 with two common atoms.

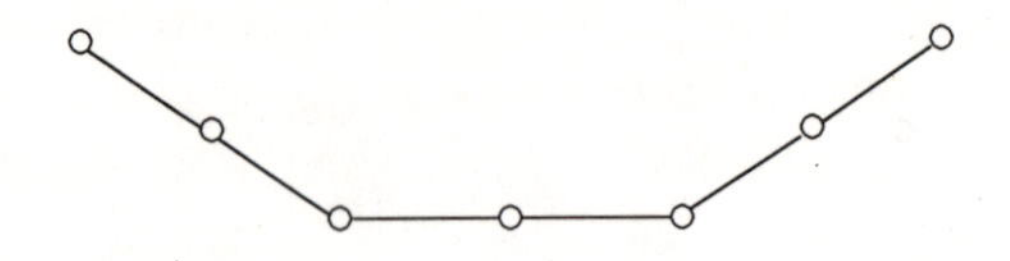

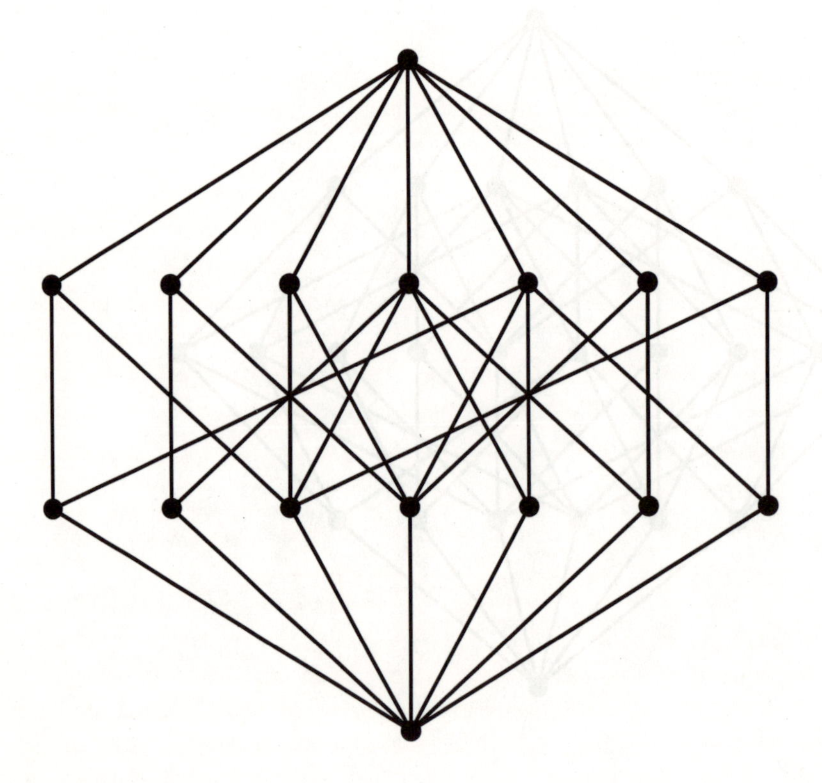

Figure 10.13: Greechie and Hasse diagrams of a pasting of three almost disjoint 2^3 without loop.

In summary: One interpretation of pasting is the generation of a new "product" automaton A from a family of automata $A^* = \{A_1, A_2, \ldots, A_l\}$. Let all A_i's have disjoint input alphabets of exactly one symbol per automaton. (More symbols per automaton would do as well.) Let all A_i's have disjoint internal states. Let all A_i's have identical output alphabets (some output symbols may not be needed). The input alphabet of A consists of the union of the disjoint input alphabets from the automata in A^*. The internal states of A consist of the union of the disjoint internal states from the automata in A^*, with certain "pasted" elements identified.

To realise a pasting (in the form described here) of l blocks of at most of order 2^j (i.e., l Boolean algebras with at most j atoms), one needs at least l input symbols and j output symbols. These numbers are not optimal, since the construction method employed here is not optimal.

Not all automata propositional calculi correspond to orthomodular lattices. By the loop lemma 5.38, p. 60, any pasting of almost disjoint Boolean subalgebras which contains a loop of order 3 or 4 is no orthomodular lattice. As an example, consider the following partition:

$$\{\{\{1\},\{2\},\{3,4,5,6\}\},\{\{2\},\{4\},\{1,3,5,6\}\},\{\{1\},\{4\},\{2,3,5,6\}\}\} \quad .$$

The structure is obtained by a pasting of three algebras 2^3; see Fig. 10.14. The Greechie diagram reveals a loop of order 3. A loop of order 4 is contained in the propositional calculi drawn in Figs. 10.15 and 10.16, which contain the subalgebra O_6 drawn in Fig. 5.3, p. 57 and which are therefore no orthomodular lattices. For a review of more general results, see G. Kalmbach, *Orthomodular Lattices* [250], chapter 4.

T 10.10 (J. R. Greechie [210]) *The orthomodular poset given in Fig. 10.17 does not correspond to any partition logic. It admits no states.*

Proof by contradiction: Assume that there exists a partition logic corresponding to the orthomodular poset drawn in Fig. 10.17 (wrong). The set of all internal automaton states is denoted by A. Consider an arbitrary element $a_i \in A$. This element has to be contained in exactly one of the atoms of the block x. Without loss of generality one can assume that a_i is contained in the atom denoted by 1. a_i has also to be contained in exactly one of the atoms of the block y. Without loss of generality one can assume that a_i is contained in the atom denoted by 2. a_i has also to be contained in exactly one of the atoms of the block z. Without loss of generality one can assume that a_i is contained in the atom denoted by 3. Any attempt to associate a_i with one of the atoms of the block d yields a contradiction with the assumption that a_i has to be contained in one *and only one* of the atoms of the blocks x, y and z.

This can be traced back to the feature that there exists two disjoint coverings $\{a, b, c, d\}$ and $\{x, y, z\}$ of the orthomodular poset by its blocks such that the number of blocks in the two coverings is different. An orthomodular lattice which does not correspond to a partition logic is given in P. Pták and S. Pulmannová [376], p. 37.

The following corollary is a direct consequence of theorem 10.9:

C 10.11 *Any prime orthomodular subalgebra of a Hilbert lattice is isomorphic (1-1 translatable) to some finite automaton propositional calculus. I.e.,*

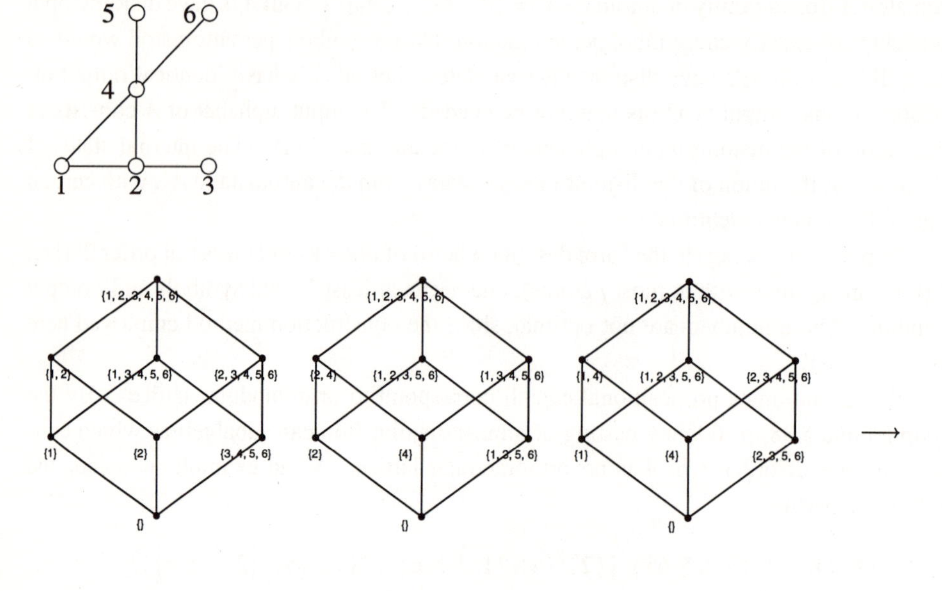

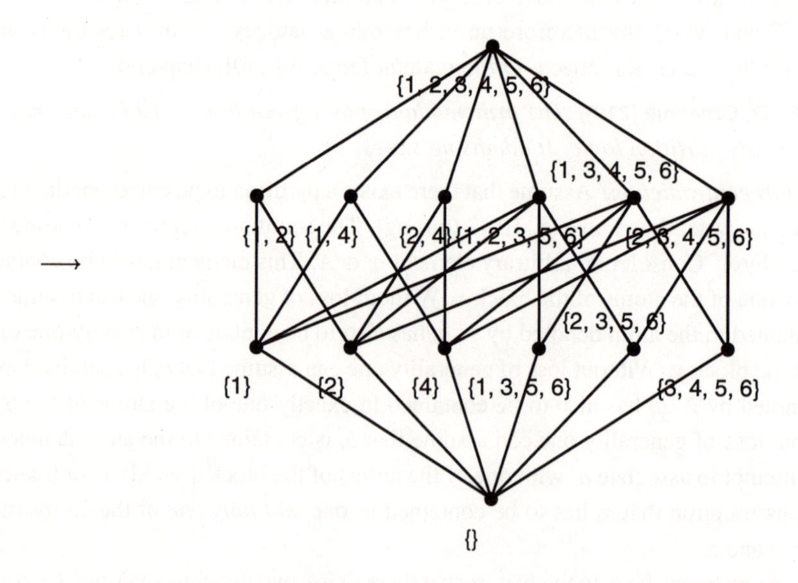

Figure 10.14: Greechie and Hasse diagrams of a pasting of three almost disjoint blocks which is no orthomodular partially ordered set.

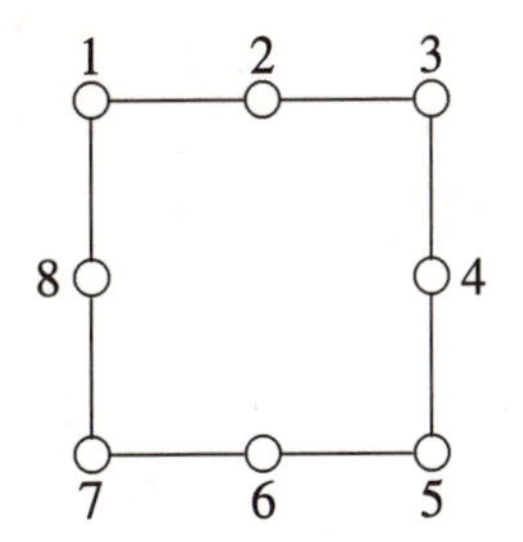

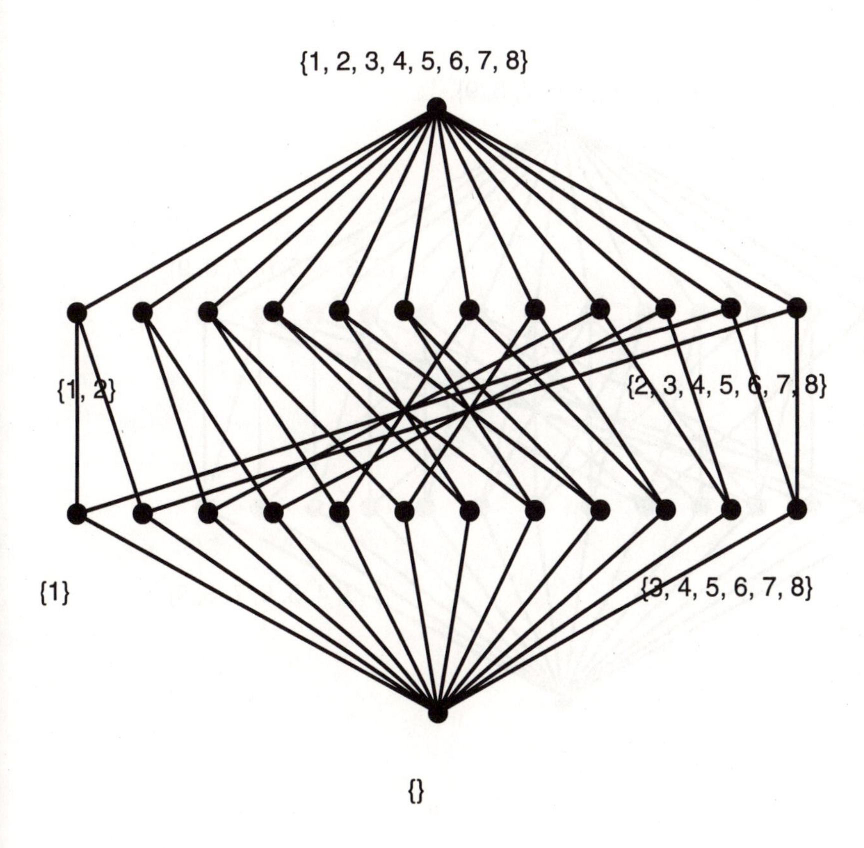

Figure 10.15: Greechie and Hasse diagrams of a non orthomodular lattice resulting from a block pasting with loop of order four.

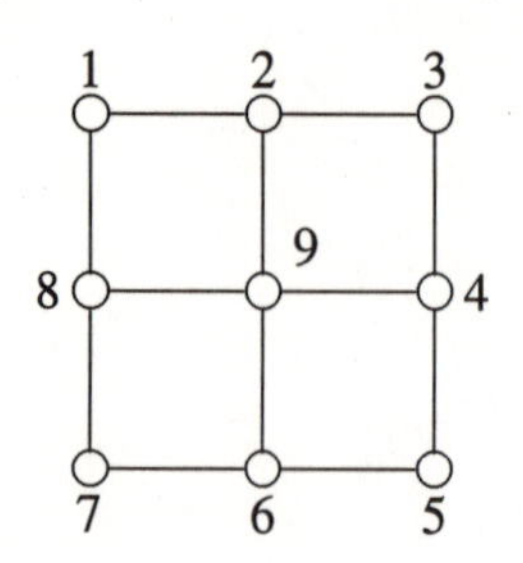

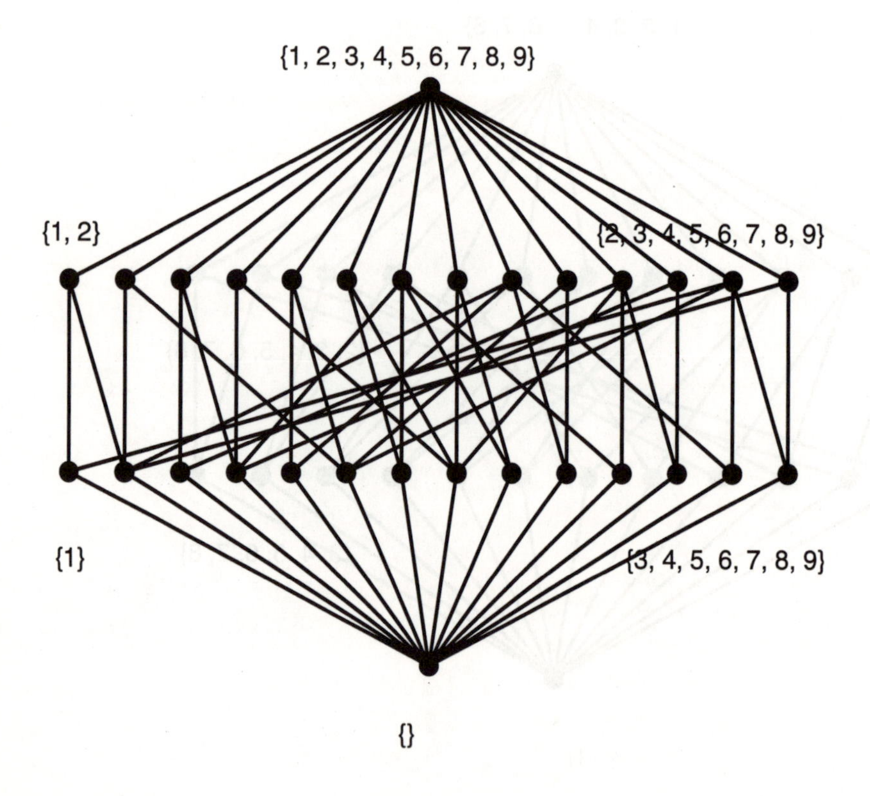

Figure 10.16: Greechie and Hasse diagrams of another non orthomodular lattice resulting from a block pasting with loop of order four.

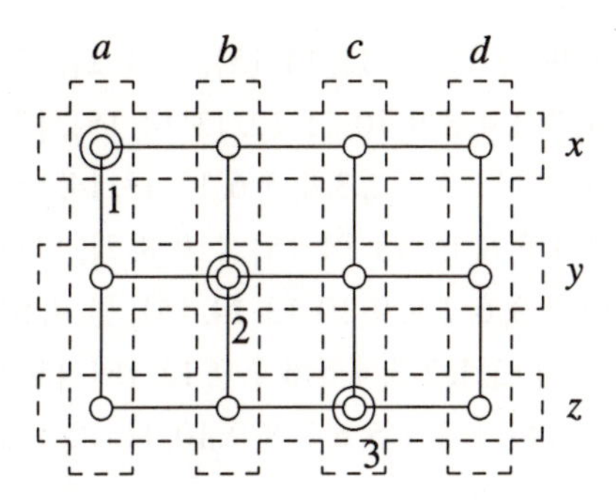

Figure 10.17: Greechie diagram of an orthomodular poset which does not correspond to any partition logic. Blocks are denoted by dashed boxes. There are two disjoint coverings consisting of the set of partitions $\{a, b, c, d\}$ and the set of partitions $\{x, y, z\}$ such that the number of blocks in the two coverings is different.

$$\left\{ \begin{array}{c} \text{prime orthomodular subalgebra} \\ \text{of Hilbert lattice (quantum logic)} \end{array} \right\} \begin{array}{c} \Rightarrow \\ \not\Leftarrow \end{array} \left\{ \begin{array}{c} \text{partition logic} \\ \text{(automaton propositional calculus)} \end{array} \right\}$$

Every Hilbert lattice is orthomodular, but not every orthomodular lattice is the subalgebra of some Hilbert lattice. For example, consider the Greechie and Hasse diagrams drawn in Fig. 10.18 of an orthomodular lattice introduced by R. J. Greechie and reviewed by R. Giuntini ([201], chapter 15, p. 139). The *orthoarguesian law* does not hold in this lattice. Yet, by the loop lemma 5.38, p. 60, it is an orthomodular lattice. With the transformations

$$\begin{aligned}
1 &\rightarrow \{1,4,5\},\\
2 &\rightarrow \{2\},\\
3 &\rightarrow \{3,6,7,8,9,10,11,12,13,14,15\},\\
4 &\rightarrow \{4,1,2\},\\
5 &\rightarrow \{5\},\\
6 &\rightarrow \{6,1,2,3,4,8,9,10,11,12,13,14,15\},\\
7 &\rightarrow \{7\},\\
8 &\rightarrow \{8,10,11\},\\
9 &\rightarrow \{9,1,2,3,4,5,6,12,13,14,15\},\\
10 &\rightarrow \{10,8,7\},\\
11 &\rightarrow \{11\},\\
12 &\rightarrow \{12,2,3,6,7,8,9,10,13,14,15\},\\
13 &\rightarrow \{13,1,2,3,4,5,6,7,9,12,15\},\\
14 &\rightarrow \{14\},\\
15 &\rightarrow \{15,1,3,4,5,6,7,8,9,10,11,12,13\} \quad ,
\end{aligned}$$

the set of state partitions is given by

$$\{\{\{1,4,5\},\{2\},\{3,6,7,8,9,10,11,12,13,14,15\}\} \quad ,$$

$$\begin{aligned}
&\{\{3,6,7,8,9,10,11,12,13,14,15\},\{4,1,2\},\{5\}\} \quad ,\\
&\{\{5\},\{6,1,2,3,4,8,9,10,11,12,13,14,15\},\{7\}\} \quad ,\\
&\{\{7\},\{8,10,11\},\{9,1,2,3,4,5,6,12,13,14,15\}\} \quad ,\\
&\{\{9,1,2,3,4,5,6,12,13,14,15\},\{10,8,7\},\{11\}\} \quad ,\\
&\{\{11\},\{12,2,3,6,7,8,9,10,13,14,15\},\{1,4,5\}\} \quad ,\\
&\{\{8,10,11\},\{13,1,2,3,4,5,6,7,9,12,15\},\{14\}\} \quad ,\\
&\{\{14\},\{2\},\{15,1,3,4,5,6,7,8,9,10,11,12,13\}\} \quad \} \quad .
\end{aligned}$$

The transition and output tables of a Mealy-type automaton realising the set of state partitions and thus yielding an orthomodular lattice which is not a subalgebra of some Hilbert lattice is enumerated in table 10.8. It remains to be seen whether microscopic phenomena indeed can be represented by Hilbert lattices or, alternatively, by the larger class of orthomodular lattices, or even by the larger class $\mathfrak{F}_n$ (see below. So far, no experiment has been performed in order to test, e.g., the orthoarguesian property of quantum mechanics in Hilbert space.

Having established a similarity between types of lattices of the subspaces of a Hilbert space (or more general geometric spaces) — i.e., those which are prime (which implies that they admit states) — and certain automaton propositional calculi, one might then proceed by considering *(probability) measures* on such non distributive structures. It can be expected that by Gleason-type theorems a Hilbert space formalism similar to quantum mechanics is recovered.

10.5 Features of automaton universes

Having considered the family $\mathfrak{F}_4$ of all intrinsic propositional calculi of generic four-state automata (of the Mealy type), and having in mind the correspondence of certain automaton propositional calculi with certain quantized systems, it is not entirely unreasonable to speculate about the logico-algebraic structure of automaton universes in general. To put it pointedly, one could ask *"how would creatures embedded in a universal computer perceive their universe?"*

The lattice-theoretic answer might be sketched as follows. Let $\mathfrak{F}_i$ stand for the family of all intrinsic propositional calculi of automata with i states. From the point of view of logic, the intrinsic propositional calculi of a universe generated by universal computation is the limiting class $\lim_{n\rightarrow\infty}\mathfrak{F}_n$ of all automata with $n\rightarrow\infty$ states. Since $\mathfrak{F}_1\subset\mathfrak{F}_2\subset\mathfrak{F}_3\subset\cdots\subset\mathfrak{F}_i\subset\mathfrak{F}_{i+1}\subset\cdots$, this class "starts with" the propositional calculi represented by Fig. 10.8, p. 157.

It is tempting to speculate that we live in a computer generated universe. But then, if the "underlying" computing agent were universal, *there is no a priori reason to exclude propositional calculi even if they do not correspond to an orthomodular subalgebra of a Hilbert lattice.* I.e., to test the speculation that we live in a universe created by universal computation, we would have to look for phenomena which correspond to automaton propositional calculi not contained in the subalgebras of some $\mathfrak{C}(\mathfrak{H})$ — such as, for instance, the one represented by Fig. 10.18, p. 171.

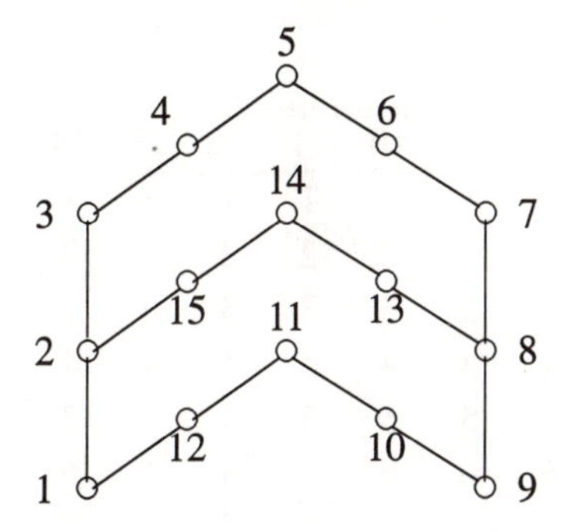

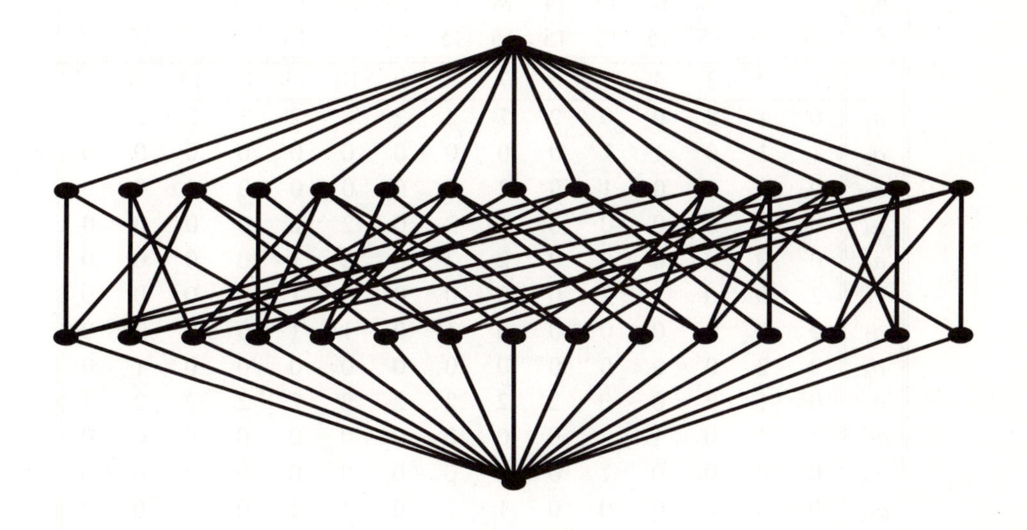

Figure 10.18: Greechie and Hasse diagram of an orthomodular lattice which is no subalgebra of a Hilbert lattice.

	1	2	3	4	5	6	7	8	9	10	11	12	13	14	15
δ_1	1	1	1	1	1	1	1	1	1	1	1	1	1	1	1
δ_2	2	2	2	2	2	2	2	2	2	2	2	2	2	2	2
δ_3	3	3	3	3	3	3	3	3	3	3	3	3	3	3	3
δ_4	4	4	4	4	4	4	4	4	4	4	4	4	4	4	4
δ_5	5	5	5	5	5	5	5	5	5	5	5	5	5	5	5
δ_6	6	6	6	6	6	6	6	6	6	6	6	6	6	6	6
δ_7	7	7	7	7	7	7	7	7	7	7	7	7	7	7	7
δ_8	8	8	8	8	8	8	8	8	8	8	8	8	8	8	8
δ_9	9	9	9	9	9	9	9	9	9	9	9	9	9	9	9
δ_{10}	10	10	10	10	10	10	10	10	10	10	10	10	10	10	10
δ_{11}	11	11	11	11	11	11	11	11	11	11	11	11	11	11	11
δ_{12}	12	12	12	12	12	12	12	12	12	12	12	12	12	12	12
δ_{13}	13	13	13	13	13	13	13	13	13	13	13	13	13	13	13
δ_{14}	14	14	14	14	14	14	14	14	14	14	14	14	14	14	14
δ_{15}	15	15	15	15	15	15	15	15	15	15	15	15	15	15	15

	1	2	3	4	5	6	7	8	9	10	11	12	13	14	15
o_1	0	1	2	0	0	2	2	2	2	2	2	2	2	2	2
o_2	1	1	0	1	2	0	0	0	0	0	0	0	0	0	0
o_3	0	0	0	0	1	0	2	0	0	0	0	0	0	0	0
o_4	0	0	0	0	0	0	1	2	0	2	2	0	0	0	0
o_5	0	0	0	0	0	0	2	2	0	2	1	0	0	0	0
o_6	2	0	0	2	2	0	0	0	0	0	1	0	0	0	0
o_7	0	0	0	0	0	0	0	1	0	1	1	0	0	2	0
o_8	0	2	0	0	0	0	0	0	0	0	0	0	0	1	0
o_9	0	1	2	0	0	2	2	2	2	2	2	2	2	2	2
o_{10}	1	1	0	1	2	0	0	0	0	0	0	0	0	0	0
o_{11}	0	0	0	0	1	0	2	0	0	0	0	0	0	0	0
o_{12}	0	0	0	0	0	0	1	2	0	2	2	0	0	0	0
o_{13}	0	0	0	0	0	0	2	2	0	2	1	0	0	0	0
o_{14}	2	0	0	2	2	0	0	0	0	0	1	0	0	0	0
o_{15}	0	0	0	0	0	0	0	1	0	1	1	0	0	2	0

Table 10.8: Transition and output tables of an automaton corresponding to an orthomodular lattice which is not a subalgebra of some Hilbert lattice, since it does not satisfy the orthoarguesian law.

10.6 Bell-type inequalities for automata

The violation of Bell-type inequalities [29, 30] in quantum mechanics has sparked a lively debate about their meaning ever since. — For an elementary introduction, see, for instance, A. Peres [352]. See also the review by J. H. Clauser and A. Shimony [121] and the GHZ-paper [215]. It is therefore tempting to construct Bell-type inequalities from correlation functions of an automaton. Thereby it will be assumed that, in a slightly modified version of the distinguishing problem, an automaton is presented to *two* experimenters. Its initial state is chosen at random. Besides its (initial) state, every specification of the automaton (i.e., its number of states, its output function, its transition rules) is known to the experimenters. The problem is to find correlations between the two automaton experiments.

There is again a difference between the extrinsic and the intrinsic configuration: whereas in the extrinsic setup, the experimenters posess an unlimited number of identical copies of the automaton, in the intrinsic setup, both experimenters share a single automaton copy. In the latter case, if the automaton features computational complementarity, the automaton's response to one experimenter may affect its response to the other experimenter. It is therefore suggestive to identify the correlation function of the extrinsic experiment with the "classical" correlation functions. The intrinsically obtained correlation functions will be "quantum-like."

Assume that the automaton outputs "suitable" numbers $s_j \in \mathbb{R}$. The automaton's initial state is chosen at random. A correlation function (expectation value) can be defined by the *average product of outputs* of two (successive) experiments; i.e., by counting the frequency of occurrences for N experiments (here, "$\times$" stands for scalar multiplication):

$$C(j,m) = \langle s_j s_m \rangle = \lim_{N\to\infty} \frac{1}{N} \sum_{\text{experiments}} s_j \times s_m \quad . \tag{10.30}$$

For example, take the Mealy-type automaton defined in section 10.3.2, p. 147, but with a different output table: the symbol "0" (no hit) is substituted by the number "–1" (other output functions may do as well). The output table then is

s_i	1	2	3
o_1	1	–1	–1
o_2	–1	1	–1
o_3	–1	–1	1

Assume further that the initial automaton state which has to be guessed is presented at random, i.e., for a large number of experiments the extrinsic frequency of occurrence of the internal states 1, 2 and 3 is $\frac{1}{3}$. In order to compute the extrinsic correlation function $\widetilde{C}(j,m)$, it then suffices to consider table 10.9. The correlation function is symmetric in the arguments, i.e., $C(j,m) = C(m,j)$. One can therefore consider $C(|j-m|)$.

It is not difficult to see that the following "Bell-type" inequality holds:

$$-2 \le \widetilde{C}(j,m) + \widetilde{C}(j,m') + \widetilde{C}(j',m) - \widetilde{C}(j',m') \le 2 \quad , \tag{10.31}$$

	1	2	3
1	1	−1	−1
1	1	−1	−1
$\widetilde{C}(1,1)=$	$\frac{1}{3}(1+$	$1+$	$1)=1$

	1	2	3
1	1	−1	−1
¬1	−1	1	1
$\widetilde{C}(1,\neg 1)=$	$\frac{1}{3}(-1-$	$1-$	$1)=-1$

	1	2	3
1	1	−1	−1
2	−1	1	−1
$\widetilde{C}(1,2)=$	$\frac{1}{3}(-1-$	$1+$	$1)=-\frac{1}{3}$

et cetera .

Table 10.9: Evaluation of the extrinsic correlation function $\widetilde{C}$ of an automaton of the Mealy type ($\{2,3\}=\neg 1$).

where the bound from above is derived by setting $j=m=m'$ and $j'\neq m'$, the bound from below is derived by setting $j\neq m\neq m'$ and $j'=m'$.

In order to evaluate the intrinsic correlation function (expectation value) $C(j,m)$ properly, one would have to evaluate a properly defined probability measure on the non distributive lattice represented in Fig. 10.6. One way of doing this is to establish a link to the lattice of subspaces of a Hilbert space, and then using the scalar product of Hilbert space for the definition of probability. This would give a correlation function similar to the quantum mechanical one for the spin-$\frac{1}{2}$ system, resulting in a violation of the Bell inequality (10.31).

One may conceive that the two experimenters are located at (in what amounts to in their intrinsic model) "two spacelike separated points." Yet, since actually the experimenters perform measurements on one and the same automaton copy, they could recognise this "entanglement" or "inseparability," or "nonlocality." The obvious contradiction with their separateness might cause much puzzlement.

10.7 Modelling classical measurements

The above considerations on computational complementarity were motivated by modelling automata with "quantum-like" behaviour, at least with respect to the quantum mechanical feature of complementarity. These considerations shall now be extended to measurement processes in mechanistic systems of a general kind.

Assume two propositions (or features) of a (mechanistic) system, denoted by F_1 and F_2. Successive measurements of F_1 and F_2 do not necessarily "disturb" each other, but we are not interested in this trivial case. If, on the other hand, measurement of F_1 — caused by some kind of "interaction" of the system and the measurement apparatus — makes impossible a successive measurement of F_2, then the resulting propositional structure will reflect "complementarity." In particular, the propositional calculus will not be distributive and Boolean. Stated differently, the paradoxical attempt of certain measurement procedures to measure the object while at the same time changing its state, yields an experimental propositional calculus which, in some aspects, resembles the quantum mechanical one. — Indeed, one may speculate that quantum theory is the only theory so far which implicitly takes into account this kind of "complementarity."

For example, imagine a dark room with a tennis ball (or a frog!) moving in it. Assume further an experimenter therein, who is not allowed to "turn on some light," and has to touch the ball (frog) in order to measure its position and velocity. As the experimenter touches the ball (frog) in order to measure its position, it will change its velocity.

The above consideration seems to necessitate a non distributive experimental propositional calculus for classical physics as well; at least for those circumstances where, intuitively speaking, the measurement processes causes "disturbances" which are "of the same order" than the measured effect. Classical physics pretends that the interaction can be made arbitrarily small, although in many circumstances this requirement is of little operational relevance. Quantum mechanics postulates a bound from below — Planck's constant — for a transfer of action between a measurement apparatus and the measured object. Similar arguments have already been put forward and formalised in reference [432]. For earlier formalisations in the context of an "operational logic," see C. H. Randall [381] and G. W. Mackey [306].

Chapter 11

Extrinsic indeterminism

A system "forecast" or "prediction" will be decomposed into two distinct requirements or phases: the *algorithmic representation* or *description* of a system, succeeded by the actual *computation* of a prediction, based on that representation. It will be assumed that the system is "sufficiently complex" to support universal computation. One appropriate *representation* or *description* of a computable system is an *algorithm* or *program code* which, implemented (extrinsically) on different computers or (intrinsically) within the same system, yields a replica of that system. A mechanistic system can also be considered as formal system, with the *index* or *Gödel number* of the axioms serving as description. In this sense the terms "(algorithmic) description" and "program code," as well as "index" or "Gödel number" are synonyms. (Historically, such descriptions have been interpreted as "natural laws" governing the system.)

11.1 Extrinsic algorithmic description

We shall now be concerned with the question of inferring "laws," i.e., algorithmic descriptions, from the experimental input/output analysis of a system. Considerations will be restricted to *mechanistic*, i.e., *computable* systems. — By definition, for systems which are non mechanistic, no reasonable concept of "recursive law" can be given. (Probabilistic laws may still apply.) The system is perceived as a "black box" on which experiments are allowed only *via* some kind of input/output terminal. Classical results reviewed in chapter 9 apply.

11.1.1 Active mode

In the *active mode* (which is relevant for applications, where intuition or oracles are rarely encountered), some experimenter examines an arbitrary number of copies of a system and, based on these experiments, attempts to construct an algorithmic description of it. As a consequence of the recursive unsolvability of the rule inference problem ([207]; see also theorem 9.9, p. 121) this is generally impossible. There does not exist any systematic, i.e., recursive (effectively computable), method of deriving an algorithmic description from the input/output analysis of an arbitrary mechanistic system.

11.1.2 Passive mode

As no effective computation is in general able to perform the task to find a "law" of a mechanistic system, it is assumed that some oracle provides it. I.e., by some "higher" insight, the oracle outputs an algorithm which computes a step-by-step enumeration of the system evolution. As has been pointed out before, such an algorithm corresponds to a theory.

11.2 Prediction by simulation

Imagine statements of the form, *"feeded with program x and input y my computer will output z,"* or *"at time t the system will be in state xyz,"* or, *"on May 2nd of next year there will be sunshine in Vienna; a wind will blow from northwest at 5 km/hour."* As a consequence of the *recursive unsolvability of the halting problem,* such statements are undecidable. Indeed, there exist uncomputable observables even for computable systems whose "laws" and "input parameters" are completely determined. In particular, no effective computation can predict the behaviour of an arbitrary computable system in any "reasonable" (i.e., computable) time. Stated pointedly, in general there does not exist any "computational shortcut," no optimisation with respect to time, which would allow a forecast of the "distant future." — A "speedup" of a calculation is generally impossible.

Nevertheless, any universal computer U can simulate a mechanistic system once the recursive "law" governing the system is known. This can be done by encoding the mechanistic system (as a program) on U and performing the simulation of the system evolution on a "step-by-step" basis. In that way it is possible to simulate a mechanistic physical system *completely* and *in finite time.* I.e., any entity in the mechanistic system can be brought into a one-to-one correspondence with its simulation. In many cases, this simulation will take longer than the evolution of the original system, since the latter one is, stated pointedly, "the perfect computing agent to simulate itself." — Think, for instance, of systems which consist of a great number of processes which are "local" in the sense that they are not affected by other processes taking place in "great (space, time, ...) distance." Simulation of such systems on a (universal) computer with only one processor unit would be inefficient. [Performance could be improved by the use of (universal) computers which are adapted to the system; in the case of "local" processes, by the use of Cellular Automata instead of sequential machines such as the Turing machine.] Therefore, externally, given an algorithmic description, a prediction of a mechanistic physical system is not so much a problem of principle — let alone the recursive unsolvability of the rule inference problem — but a problem of *performance,* i.e., of time and other computational resources.

Of course, there exist trivial possibilities of forecast. E.g., one may think of a universal computer simulating the mechanistic system with *higher speed of computation* than the mechanistic system itself. A speedup can also be realised for periodic systems.

11.3 Examples

Consider the following *Gedankenexperiments*.

Example 1: Consider a physical system S producing numbers on a display. Assume an observer A, for whom S for all practical purposes is a "black box"; i.e., despite the input & output terminals (e.g., keyboard, display), A has no knowledge of S. Assume a second observer B, who, by intuition or other insight, knows that S calculates the digits of π, displays them, and in doing so, has arrived at a specific n'th digit. In this case one may ask the following questions.

(i) (Rule inference problem) How does A without communicating with B learn about the "meaning" of S, i.e., how could A find out that S outputs the digits of π?

(ii) (Halting problem) To what extend is the predictive power of B restricted by finite computational resources? — E.g., what sense makes any "knowledge" claimed by B that S has arrived at the $n = 10^{200}$'th decimal place of π, which is the digit "7," according to B? For even if A uses a whole galaxy as computer, and even if A is willing to wait for the result of the computation for a time comparable to the age of the universe, at least with present-day mathematical means, it is impossible to confirm this statement and to predict the 10^{200}'th decimal place of π [193, 194]. Although ideally π can be calculated deterministically to arbitrary precision, one is forced to apply a probabilistic description by restrictions in computational resources and intuition. (This was, after all, the perception of Laplace's "old" probability theory.) To put it pointedly: Although A may have no access to a CRAY 2 computer, A might be willing to believe D. H. Bailey's claim [19] (see also J. von Neumann *et al.* [338] for a false conjecture) that the next ten digits following the 29 359 000'th digit in the decimal expansion of π are 3, 4, 1, 9, 2, 8, 4, 1, 7, 8, but, on rational grounds, A cannot accept a claim such as *"with a probability greater than* 1/10, *the* 10^{200}*'th digit in a decimal expansion of* π *is* 7*"*.

Example 2: Assume the same setup as for example 1, with the difference that, for B, S produces Champernowne's sequence [115]; i.e., the enumeration of the set of all finite "decimal" words, i.e., words using the alphabet 0, 1, 2, 3, $\cdots$ 9, in lexicographic order ($9 > 8 > 7 > \cdots > 1 > 0$) "0 1 2 3 4 5 6 7 8 9 10 11 12 13 ...." Assume A performs some statistical tests in order to find some hint on how the sequence is generated. In particular, let us assume that A counts the frequencies of arbitrary fixed sequences such as "1", or "79", or "16$\cdots$0"— D. G. Champernowne has shown that the enumeration is a Bernoulli-sequence [115], i.e., any arbitrary partial sequence occurs with the expected limiting frequency. The same has been demonstrated numerically [19, 338] for the decimal expansion of π up to 26 million places and for partial sequences up to length 6.

The above *Gedankenexperiments* are no exceptions. Due to the recursive unsolvability of the rule inference problem, there are rather few physical systems whose laws can be determined exactly. But even if an initial value and the law governing the system were known precisely it would in general be impossible to forecast the future behaviour of the system: As has already been pointed out, due to the recursive unsolvability of the halting problem and the non recursive enumerability of the maximal halting time, no "computational shortcut" would be feasible.

The moral of these examples is the fact that *sequences "looking random" may stem from a low-complex but hidden and unknown deterministic evolution.* Indeed, due to a possible "creation" of algorithmic information, in the limit of infinite computation time, a low-complex dynamical system could produce a sequence which is random in the Martin-Löf/Solovay/Chaitin sense (see 14.1, p. 195), although no recursive bound could be given that would "guarantee" that, at any particular moment, the system outputs objects with incompressible algorithmic information content.

Presently, in physics no distinction is being made between *randomness*, as for instance postulated by quantum mechanics and continuum physics, and *undecidability*; the difference being that undecidability occurs even for computable systems (i.e., halting problem *etc.*), whereas randomness implies uncomputability. One may even speculate that, what is presently perceived as "randomness" in physics will eventually turn out to be the feature of "undecidability" of computable systems. Stated pointedly, *"insofar randomness is identified with unpredictability, deterministic systems are random, and (as will be argued later,) insofar randomness is defined by indeterminism, it is not operationalisable."*

Chapter 12

Intrinsic indeterminism

Since already externally the problems of induction and forecast are generally recursively unsolvable, one may suspect that the situation might not get worse in the intrinsic case. However, whereas in the external setup the experimenter is allowed to "set aside" a complete simulation of the observed system without altering the original system, this is not the case in the intrinsic setup. There, any model simulation of the system is necessarily part of that same system. This results in a worsening of the speedup theorem; i.e., of the unpredictability of the system phenomenology.

12.1 Intrinsic algorithmic description

The question of a description within the same system or process is related to the question of whether an agent can posess an intrinsic "theory," "description" or "blueprint" of itself. This has been analysed in the context of *self-reproduction*. Consider the question, *"Can a (universal) computer reproduce itself?"* According to J. von Neumann, there are (at least) two distinct meanings or interpretations of this question ([340], pp. 118-126). The answer will therefore be "yes" or "no," depending on the way in which self-reproduction is realised.

12.1.1 Passive mode

In the first mode of self-reproduction, which will be called the *"passive mode"* (quotation from A. W. Burk's editorial comments [340], pp. 125-126), *"the self-reproducing automaton contains with itself a passive description of itself and reads this description in such a way that the description cannot interfere with the automaton's operations."* Exactly *how* the description inducing self-reproduction is obtained is irrelevant to the argument. It comes from a source *external* to the automaton, presumably from an *oracle*. (As has been discussed already in chapter 11, p. 177), any external source trying to obtain the automaton's description has to cope with the recursive unsolvability of the rule inference problem.) See also remark *(iii)* below.

It is indeed possible to construct a program which *includes* its own description and, through that description, is able to reproduce itself. The affirmative answer can be intuitively envisioned as follows (see A. W. Burks, [340], p. 55): A Gödel number may be regarded as a description of a formula. At least in some cases, the Gödel number of a statement may be described in fewer symbols than the statement, else Gödel's self-referring undecidable statement (which is central for a proof of Gödel's incompleteness theorems) could not exist.

John von Neumann was one of the first who has put forward an explicit formal (Cellular Automaton) model of a universal self-reproducing automaton [340]. A formal proof of the existence of self-reproducing automata is too technical to be included here. See, for instance, von Neumann's original work [340], or H. Rogers' *Theory of Recursive Functions and Effective Computability* [390], p. 188, which also contains the following informal proof. It uses a scheme of a self-reproducing machine which has the structure $M = (D, C, E, (b, i))$, where D is a *"blueprint realiser"* (that can build an object from a given "blueprint program"), C is a *"program copier,"* E is some supplementary equipment for handling inputs and outputs for C and D, and (b, i) is a *"program"* consisting of b, which is the "blueprint program" for D, C, E; i is a set of supplementary instructions. The machine takes its orders from i and operates as follows. b is placed in D, and replicas D', C', E' of D, C, E are produced. Then (b, i) is placed in C, and a copy (b', i') is made. The reproduction $M' = (D', C', E', (b', i'))$ is then assembled. (The index " $'$ " stands for the copy here.)

Remarks:

(i) There is a difference between a *possible self-description* and the *impossibility of self-prediction.* Whereas certain finite algorithms can reproduce a copy of their own *past*, they are unable to "catch up" with their immediate presence and predict their future. This will be discussed below.

(ii) The reproduction circle as it has been conceived does *not contain any process step similar to diagonalization.* Paradoxical reproduction attempts which originate from diagonalization will be considered below.

(iii) It has been assumed that the "blueprint program" b (interpretable as the "algorithmic description" of M in terms of "laws" and "system parameters") is known beforehand, presumably from an *oracle*. Oracle computation is necessary here because due to the recursive unsolvability of the rule inference problem, no effective computation exists which in general outputs the "laws" governing a mechanistic system.

(iv) Note, however, that there exist computer states or configurations, in the Cellular Automaton context called the "Garden of Eden" configuration, which cannot be produced by any other configuration. Therefore, such configurations cannon reproduce themselves.

Example:

J. H. Conway's *Life Cellular Automaton* [38, 368] is a universal computer which may exhibit self-reproducing configurations.

The following theorem summarises the possibility of self-reproduction in the passive mode.

T 12.1 *There exist complete intrinsic theories (algorithmic descriptions) of computable systems which are defined passively, i.e., without self-examination.*

For a proof, see J. von Neumann [340].

12.1.2 Active mode

The second mode of description, called the *"active mode,"* is more relevant for application, where intuition or oracles are rarely encountered. In the active mode, the *"self-reproducing automaton examines itself and thereby constructs a description of itself"* (quotation again from A. W. Burk's editorial comments [340], p. 126).

Unfortunately, this characterisation may give rise to misunderstandings: if a self-reproducing automaton is specified as a device consisting of elements which can be *analysed, identified,* and, after this analysis, *restored* to their previous state, then self-reproduction by self-inspection can indeed be performed. As has been pointed out by R. Laing [282], one of many possible strategies is, informally speaking, to divide the automaton into two distinct parts. Each part contains, in some form, an analysing and a constructing element. Initially the first part analyses the second part, which is assumed to be passive, and constructs a copy of it. Then the first part activates the second string and becomes passive. The second part analyses the first part and constructs a copy of it. In that way, a copy of the original automaton is obtained.

Nevertheless, while feasible with specific assumptions, strategies of this type are not generally applicable. One reason why the above strategy fails for a more general type of automata is the feature of *computational complementarity* encountered in single-automaton experiments (see chapter 10, p. 127). A "diagnostic" stimulus analysing (part of) some automaton in general cannot be made "nondestructive:" Assume some (universal) computer consisting of parts which feature computational complementarity. Self-reproduction by self-inspection in the active mode would require that the act of observation of the initial state (distinguishing problem) would not destroy the possibility to measure other aspects of this part and would not make impossible the reconstruction of the original state. As has been pointed out by E. F. Moore [328], this is impossible in the single-system configuration; i.e., in a setup where only one copy of the part is available.

In the general case, self-reproduction by self-inspection is not feasible. Even before the publication of E. M. Gold's findings [207], J. von Neumann suspected that an argument utilising diagonalization in the form of the Richard paradox may give a hint on this question. In J. von Neumann's own words ([340], pp. 121-122):

> *... In order to copy a group of cells [[J. von Neumann uses a Cellular Automaton model]] ... it is necessary to "explore" that group to ascertain the state of each one of its cells and to induce the same state in the corresponding cell in the area where the copy is to be placed. This exploration implies, of course, affecting each cell of this group successively with suitable stimuli and observing the reactions. This is clearly the way in which the copying automaton **B** can be expected to operate, i.e., to take the appropriate actions on the basis of what is found in each case. If the object under observation consists of "quasi-quiescent" cells[1] ..., then these stimulations can be so arranged as to produce the reactions that **B** needs for its diagnostic purposes,*

[1] *... let these cells lie "outside", in the external, otherwise quiescent, region of the [[CA]] crystal ... such that they will normally not disturb (stimulate or otherwise transform) each other or the surrounding quiescent cells.*

> ***but no reactions that will affect other parts of the area which has to be explored. If an assembly G, which may itself be an active automaton, were to be investigated by such methods, one would have to expect trouble. The stimulations conveyed to it, as discussed above, for "diagnostic" purposes, might actually stimulate various parts of G in such a manner that other regions could also get involved, i.e., have the states of their cells altered. Thus G would be disturbed; it could change in ways that are difficult to foresee, and, in any case, likely to be incompatible with the purpose of observation; indeed, observing and copying presuppose an unchanging original.***
>
> ...
>
> ***If one considers the existing studies concerning the relationship of automata and logic, it appears very likely that any procedure for the direct copying of a given automaton G, without the possession of a description $\mathbf{L_G}$, will fail; otherwise one would probably get involved in logical antinomies of the Richard type.***

What is the *Richard paradox?* Published in 1905 by Jules Richard [384], it is one of the first "classical" paradoxa of (meta-)mathematics. Reviews can be found in St. C. Kleene's *Introduction to Metamathematics* [266], p. 38,39, as well as in J. von Neumann's *Theory of Self-Reproducing Automata* [340], pp. 123-125. Richard's paradox resembles the Berry paradox which uses a language to describe numbers. Let $E_1, E_2, E_3, \ldots$ be an enumeration of all expressions of the language which define functions of one natural number variable with two values 0 and 1. For example, the expression "`n is odd`" can be used to define a function which has the value 1 if n is odd and has the value 0 if n is even. Let $f_i(n)$ be the function defined by expression E_i, and define $-f_i(n)$ by

$$-f_i(n) = 0 \quad \text{if} \quad f_i(n) = 1$$
$$-f_i(n) = 1 \quad \text{if} \quad f_i(n) = 0$$

and let E' be the expression "`the function` $-f_n(n)$." E' can be expressed in English as "`the function whose value, for any given natural number as argument, is equal to 0 or 1 if the value, for the given natural number as argument, of the function defined by the expression which corresponds to the given natural number in the enumeration of expressions is 1 or 0, respectively`"

In a proof by contradiction, it is assumed that the enumeration $E_1, E_2, E_3, \ldots$ is *complete* in the sense that it lists *all* expressions corresponding to functions, and that, since E' is obtained by a trivial "bit switch" (diagonalization), E' is expressible in the language (wrong). But then, E' has to be somewhere in the enumeration of all expressions $E_1, E_2, E_3, \ldots$. Yet, E' corresponds to the function $-f_n(n)$, which differs from every function in the enumeration at least at $f_n(n)$. Consequently, E' cannot be in the enumeration of expressions $E_1, E_2, E_3, \ldots$. Yet, E' is supposed to be an expression which is trivially obtained from the enumeration $E_1, E_2, E_3, \ldots$!

There is no other consistent alternative than to assume that the expression E' cannot be expressed by the language and, since E' is obtained by a trivial "bit switch" [(diagonalization) which is surely expressible in any "reasonably complex" language] from the supposedly complete enumeration of expressions $E_1, E_2, E_3, \ldots$, that there does not exist a complete enumeration of expressions; i.e., no enumeration lists *all* expressions corresponding to functions definable by the language.

The Richard paradox can be utilised to formulate the following theorem (resembling the non-enumerability of the recursively enumerable reals, p. 10).

T 12.2 (Incompleteness of intrinsic theories)
In general, no complete intrinsic theory (algorithmic description) of a universal computable system can be obtained actively, *i.e., by self-examination.*

Proof:

(i) Assume a recursive step-by-step enumeration of a mechanistic system. This recursive enumeration can, for instance, be thought of as an infinite computation which corresponds to the mechanistic system. Any event corresponds to some particular output (cf. 4.3, p. 47). It generates the entire phenomenology. The recursive enumeration is specified by a computable total translation function from the natural numbers onto the computable real numbers $\mathrm{ENUM} : \mathbb{N} \rightarrow \mathbb{R}$. Let us call some t a *name* for p if ENUM derives p from t. (In technical terms, t can be interpreted as the Gödel number or index or code of p.) It is thus possible to enumerate all numbers which occur in $\{p_i\}$ *via*

$$\begin{array}{l} 1 \longrightarrow \mathrm{ENUM}(1) = p_1 = 0.p_{11}p_{12}p_{13}p_{14}\cdots \\ 2 \longrightarrow \mathrm{ENUM}(2) = p_2 = 0.p_{21}p_{22}p_{23}p_{24}\cdots \\ 3 \longrightarrow \mathrm{ENUM}(3) = p_3 = 0.p_{31}p_{32}p_{33}p_{34}\cdots \\ 4 \longrightarrow \mathrm{ENUM}(4) = p_4 = 0.p_{41}p_{42}p_{43}p_{44}\cdots \\ \qquad\qquad \vdots \end{array} \quad . \tag{12.1}$$

where the bits p_{ij} correspond to the codes of elementary events.

A notion of *element of (physical or automaton) reality* is used which refers only to *outcomes of actual measurements*. This concept differs from the EPR-terminology [154], where the existence of "elements of physical reality" is claimed even for experiments which have *not* been performed. The sets $\{p_i\}$ or $\{p_{ij}$ or the matrix $[p]$ defined by $[p]_{ij} = p_{ij}$ correspond to the entirety of elements of (physical or automaton) reality.

(ii) Every *intrinsically definable* theory must be consistently representable *from within the system* and therefore must be associated with some sequence p_i in the enumeration.

(iii) One may ask, *"does there exist a* complete *intrinsic theory T in the sense that the entire phenomenology of the system is derivable from T?"* The formal translation of such a statement is the existence of a name t_T, such that $T = \mathrm{ENUM}(t_T)$ encodes *all* p_i's, including itself. Rather than proving the nonexistence of an intrinsic and complete theory by enumeration of all p_{ij}'s, it suffices to show that no intrinsic theory exists which reproduces the diagonal sequence $0.p_{11}p_{22}p_{33}\cdots$ of outcomes.

It is indeed possible to enumerate all p_{ij}'s in a single "universal" real number u by drawing counterdiagonals from the upper right to the lower left in the matrix $[p]$:

$$[p] = \begin{pmatrix} & \swarrow & & \swarrow & & \swarrow & & \swarrow & \\ p_{11} & & p_{12} & & p_{13} & & p_{14} & & \cdots \\ & \swarrow & & \swarrow & & \swarrow & & \swarrow & \\ p_{21} & & p_{22} & & p_{23} & & p_{24} & & \cdots \\ & \swarrow & & \swarrow & & \swarrow & & \swarrow & \\ p_{31} & & p_{32} & & p_{33} & & p_{34} & & \cdots \\ & \swarrow & & \swarrow & & \swarrow & & \swarrow & \\ p_{41} & & p_{42} & & p_{43} & & p_{44} & & \cdots \\ & \swarrow & & \swarrow & & \swarrow & & \swarrow & \\ \vdots & & \vdots & & \vdots & & \vdots & & \ddots \end{pmatrix} \tag{12.2}$$

In this way one obtains

$$u = 0.p_{11}p_{12}p_{21}p_{13}p_{22}p_{31}p_{14}p_{23}\cdots .$$

It is not difficult to show that, starting after the comma, the diagonal matrix elements p_{nn} are in the $(2n^2-2n+1)$'th place of u.

(iv) [DIAGONALIZATION]: Now consider the real number $d = d_1d_2d_3\cdots$, which is constructed from u by taking the subsequent diagonal elements p_{nn} and switching its bit, i.e., by substituting

$$d_n = \begin{cases} 0 \text{ if } u_{(2n^2-2n+1)} = 1 \\ 1 \text{ if } u_{(2n^2-2n+1)} = 0 \end{cases} . \tag{12.3}$$

Notice that d is obtained by a simple function $g(u)$ which would be computable from within the system. (It has been assumed that the system is capable of universal computation.) Evidently, d is different from every one of the ENUM(t) in at least its diagonal element p_{nn}. This means that *there is no t which is a name for d.*

(v) This fact results in the following alternative. Either one of the following statements is true:

(I) There does not exist any t_T for which ENUM$(t_T) = u$. This has as its immediate consequence the incompleteness of any intrinsic theory T' which is representable by some code $t_{T'}$ *within the system.* This means that there is *no intrinsically definable and complete description* of how the system operates; There is no other consistent choice than *(I)*.

(II) There exists a name for u and thus for d, but contradiction occurs: Recall that the enumeration is complete if u and thus d would have some name, say t_u and t_d, then these names would occur somewhere in the enumeration; assume at the m'th and the n'th position. But there exists at least one digit d_n in the binary expansion of d which is *not* equal to p_{nn}.

12.2 Computation of forecast

Of what use is a passive description? Can a mechanistic system "comprehend" itself completely? Indeed, it may be suspected that a finite intelligence to which is presented an (algorithmic) description of itself will never be able to get a complete "self-comprehension" of itself through that description. In a sense, this is a direct consequence of the recursive unsolvability of the halting problem, stating that, in general, no "speedup" is possible. One may still consider predictions obtained by the step-by-step enumeration of a system.

The following argument resembles Zeno's paradox of "Achilles and the Tortoise" [294]. K. Popper has given a similar account [366], based on what he calls *"paradox of Tristram Shandy."* Think of the attempt of a finitely describable "intelligence" or computing agent to understand itself completely. It might first try to describe itself by printing its initial description. (It has been argued above that there is nothing wrong with this attempt *per se,* and that there indeed exist automata which contain the "blueprint" of themselves.) But then it has to describe itself printing its initial description. Then it has to describe itself printing its printing its initial description. Then it has to describe itself printing its printing its printing its initial description $\cdots$ *ad infinitum.* Any reflection about itself "steers" the computing agent into a never-ending vicious circle. In a sense, "in the limit of an infinity of such circles," the agent has completed the task of complete self-comprehension. Yet, for any finite time, this cannot be achieved.

One may state the above argument in terms of the universal self-reproducing automaton M introduced for an informal proof of the existence of such machines. Iteration [which can be encoded by an algorithm of length $O(1)$] of the process of self-reproduction yields a new machine, which is just the concatenation of the original machine and its copy, yields MM', $MM'(MM')'$, $MM'(MM')'(MM'(MM')')'$, $\cdots$ *ad infimum.* Since no generation is equal to any previous one, the process never stops. If self-reproduction is performed in parallel, each iteration step takes one (discrete) unit of time. Of course, the similarity to Zeno's paradox of Achilles and the turtle is only informal, since one does not have any notion of "distance" between two algorithms. [One may count the *program size*, realising that two successive algorithms increase their sizes by a factor of two. Then, if one associates an algorithmic probability of 2^{-L} to an algorithm of length L, "in the limit of infinitely many self-reproductions," a vanishing algorithmic probability is obtained.] Nevertheless, *after* each reproduction, a new algorithm, incorporating two copies of the original algorithm exists. Therefore, at least for finite effective computations and for finite time, the reproduction never yields a *complete* simulation, even if reproduction is automated: if an recursive algorithm allows a finite code for the automatic reproduction of itself, it would not be able to "catch up" with its present form. Stated pointedly: even if a "mechanistic intelligence" knows all about itself, it would neither be able to comprehend its own present state nor to predict its own future.

Chapter 13

Weak physical chaos

It has been argued that due to the *recursive unsolvability of the halting problem*, due to the *recursive unsolvability of the rule inference problem*, due to *extrinsic & intrinsic indeterminism*, due to *complementarity* and related instances, the feature of *undecidability* has to be introduced in a proper description — extrinsically and intrinsically — even for totally computable systems, i.e., even for systems whose evolution is computable on a step-by-step basis. I would like to call this "mild" forms of undecidability in physics *"weak physical chaos"* to distinguish it from stronger forms of chaos which exhibit truly random behaviour.

Paradoxical constructions and the associated proof by contradiction *via diagonalization* is a main tool for an investigation of undecidability. An attempt has been made to apply the method of diagonalization to physics. Recall the correspondence between physical, algorithmic, mathematical and formal logic, which is summarised in table 2.2, p. 35. If this correspondence is extended to *undecidability*, one arrives at table 13.1.

physics	algorithmics	mathematics	formal logic
intrinsic indeterminism			Richard's paradox
		Zeno paradox	
undecidable increase of entropy measure	Chaitin's bounds on computability of algorithmic information	Berry's paradox	
unpredictable observables in deterministic systems	halting problem	Gödel incompleteness	Epimenides' paradox
no effectively computable inference scheme	rule inference problem		

Table 13.1: Correspondence between undecidability in physics, algorithmics, mathematics and formal logic.

Part III
Randomness

Does not the mind's possibility to make mistakes
result in the chance to think correctly?
from "The First Surrealistic Manifesto" by André Breton

Chapter 14

Randomness in mathematics

Presently, there exist (at least) three generic definitions of randomness:

(i) von Mises "collectives";

(ii) Ville type definitions based on statistical tests;

(iii) definitions based on algorithmic information.

Early, intuitive, mathematical concepts of randomness were either too restrictive, such that no mathematical entity existed which was random, or too wide, such that "too regular" sequences were random. The problem is to define randomness narrowly enough to exclude regular sequences while on the other hand broadly enough to assure the existence of sequences characterised as random.

Instead of a comprehensive historical review, I would like to refer briefly to one of the first intuitive but correct accounts in Philipp Frank's book *Das Kausalgesetz und seine Grenzen* [187]. Frank states that (p. 156,157)

> *Es gibt also nur einen Zufall "in Bezug auf ein bestimmtes Kausalgesetz". Ein Zufall schlechthin, also gewissermaßen ein absoluter Zufall wäre dann ein Ereignis, das in bezug auf alle Kausalgesetze ein Zufall ist, das also nirgends als Glied eines Kausalgesetzes auftritt.*
>
> *Die Aussage aber, daß ein bestimmtes Ereignis A in* keinem *Kausalgesetz als Glied auftritt, hätte offenbar nur dann einen Sinn, wenn man ein Verzeichnis aller Kausalgesetze besäße.*
>
> [English translation:] *Any event can be called random only "with respect to a specific causal law." An event would occur absolutely random if it is random with respect to all causal laws, i.e., if it is not the effect of any causal law.*
>
> *But then, the statement that a certain event is the effect of* no *causal law would make sense only if one could obtain a list of all causal laws.*

In recursion theory one could argue that, by identifying "causal law" with *recursive* or *computable function,* it would indeed be possible to enumerate all "causal laws." However, due to the recursive unsolvability of the inference and the halting problem, it would in general be impossible to prove that an event is the "effect" of a "causal law" corresponding to a recursive function. Unfortunately, as has already been suspected by Frank, a proof of randomness is "too strong" to be decidable by finite computational resources — a formal proof of randomness of an infinite sequence would essentially require knowledge of all true universal theorems of number theory [of the form $\forall n A(n)$, where $A(n)$ is quantor-free].

complexity	static	algorithmic/ program size	Martin-Löf/Solovay/Chaitin randomness
		loop depth	—
	dynamic	computational/ time	T-randomness
		storage size	—

Table 14.1: Complexity classes and their associated types of randomness

The heuristic notion of "lawlessness," or "indeterminism" of an entity has been formalised in the context of algorithmics and algorithmic information theory by P. Martin-Löf [312], C. P. Schnorr [404], G. J. Chaitin [109, 110], A. N. Kolmogorov [271] and L. A. Levin [475], among others. As a brief reminder and outlook, table 14.1 lists the complexity classes with the associated types of randomness.

As for algorithmic information theory, G. Chaitin's approach and terminology is adopted. Again, the notation $U(p, s) = t$ will be used for a computer U with program p, input (string) s and output (string) t. $\emptyset$ denotes the *empty* input or output (string). Furthermore, $U(p, \emptyset) = U(p)$.

Sequences of natural numbers may correspond to the codes of time series, measurement results and so on (cf. chapter 4, p. 45). Without loss of generality, one can consider *binary* sequences containing 0's and 1's. The symbol "ω" stands for the (ordinal) number infinity, and 2^ω represents the set of all infinite binary sequences. Infinite sequences $x = x_1x_2x_3\cdots$ in 2^ω can also be represented as binary reals r in the interval $[0, 1]$ if one identifies x with $r = 0.x_1x_2x_3\ldots$. If one wishes, one can then transform this real into an n-ary (radix n) representation corresponding to an n-ary sequence. This bijective map does not alter the complexity-based definitions of randomness given below. All of these definitions hold with probability one.

14.1 "Lawlessness" = "algorithmic incompressibility"

Consider some arbitrary sequence; e.g., the sequence of pointer readings from some experimental device. One has to make precise what is meant by "law-like." As has been suggested above, a reasonable translation of "law-like" appears to be *effectively computable* or, by the Church-Turing thesis, *recursive*.

But then one has to bear in mind that any sequence $x(n)$ of finite length n *can* be generated by an algorithm; e.g., by the program "`PRINT` $x(n)$; `END`." Therefore, in some strict sense, there are no random sequences of finite length. Any attempt to grasp the intuitive notion of what amounts to a random finite sequence remains ambiguous. [Cf. L. Löfgren [305], who proposed to take $H(x(n))/n$, i.e., a "normalised" algorithmic information content per symbol, as a criterion for randomness of finite sequences. However, since by theorem 9.11, p. 123, in general $H(x(n))$ is uncomputable, this criterion is not generally applicable.]

Hence, randomness in the sense of "lawlessness" is defined for sequences of infinite length only. Physically, such sequences are impossible to generate and process. Therefore, any formal definition of randomness which is necessarily based on infinite sequences is, strictly speaking, not operational.

There are non recursive sequences containing "large junks" of recursive (e.g., constant) subsequences, such as the sequence

$$\underbrace{0\cdots 0}_{1000\text{ times}}\; y_2\; \underbrace{0\cdots 0}_{1000\text{ times}}\; y_3\; \underbrace{0\cdots 0}_{1000\text{ times}}\; y_4\; \underbrace{0\cdots 0}_{1000\text{ times}}\; y_5\; \underbrace{0\cdots 0}_{1000\text{ times}}\; y_6\; \underbrace{0\cdots 0}_{1000\text{ times}}\; y_7 \cdots \quad ,$$

where y_i represents the outcome of the i'th toss of a fair coin (0 for "head," 1 for "tail") or the ith place in the decimal expansion of Ω. If the redundancies are eliminated, one obtains a sequence $y_1 y_2 y_3 \cdots$, which is, in an informal sense, "more random" than the previous one. One might also say that the above sequence can be "compressed;" e.g., by an algorithm which reads the $i \cdot 1001$'th element of the old sequence and writes it to the i'th position of the new sequence.

One arrives at a satisfactory concept for (finite) random sequences by assuring that it is "maximally incompressible" by algorithmic means: A sequence is random if it cannot be generated by any shorter algorithm [419, 271, 110]. Therefore, on the average, any algorithm reproducing a finite initial sequence of a random sequence should be at least of the same length as the sequence itself. I.e., the program amounts to a mere enumeration, at best. The terms "law" and "shorter" are implemented with the help of the algorithmic notions of "algorithm" and "algorithmic information," respectively (cf. chapter 7, p. 83).

D 14.1 (Randomness / Chaitin randomness, version I) *A sequence* $x \in 2^\omega$ *is* weakly random *or* Chaitin random *if the algorithmic information of the initial sequence* $x(n) = x_1 \ldots x_n$ *of length n of the base-two expansion of x does not drop arbitrarily far below n:*

$$\lim inf_{n\rightarrow\infty}[H(x(n)) - n] > -\infty \quad ,$$

or $\exists c \forall n\; [H(x(n)) \geq n - c]$.

D 14.2 (Randomness / Chaitin randomness, version II) *A sequence* $x \in 2^\omega$ *is* random *(CR) if the static complexity of the initial segment* $x(n) = x_1 \ldots x_n$ *of length n of the base-two expansion of x eventually becomes and remains arbitrarily greater than n:*

$$\lim_{n\rightarrow\infty}[H(x(n)) - n] = \infty \quad ,$$

or $\forall k \exists N_k \forall (n \geq N_k)\; [H(x(n)) \geq n + k]$.

Stated differently, in general the average information content per unit length of a random sequence cannot be "compressed" into any representation (program code) which is of smaller length than the original sequence itself. As has been shown by G. Chaitin [110], the notions of weak randomness and randomness are equivalent.

Since by theorem 9.11, p. 123, in general $H(x(n))$ is uncomputable, this definition cannot be readily operationalised. In some particular cases, the definition can be directly applied. I.e., if it "appears evident" that a sequence has low algorithmic information, such as with very regular, e.g., periodic, sequences, then randomness can be excluded. However, one must bear in mind that there exist Chaitin random sequences with initial sequences "looking regular," and that there exist very "irregular looking" sequences with extremely low complexity.

14.1.1 Normalised random sequences

As has been pointed out before, in general, uncomputability and randomness are not equivalent. Therefore, randomness implies indeterminism, but not *vice versa:*

$$\text{Martin-Löf/Solovay/Chaitin random} \underset{\not\Leftarrow}{\Rightarrow} \text{uncomputable}$$

The slightest possibility of computing an infinite part of a given sequence makes that sequence non random. For instance, if one can compute an infinity of entries in a sequence, no matter in which order, then the sequence is non random [79]. For a much more detailed account, see C. Calude and I. Chiţescu [69], as well as C. Calude [78]. The distinction between randomness and non recursive enumerability could be perceived as artificial. V. M. Alekseev & M. V. Yakobson and A. A. Brudno [9] proposed to define a sequence x to be random iff its *normalised algorithmic information* $K(x)$ is positive and nonzero:

D 14.3 (Normalised algorithmic information, normalised randomness)
The algorithmic information $K(x)$ *of an infinite sequence* $x \in 2^\omega$ *is defined by*

$$K(x) = \lim inf_{n\rightarrow\infty}[H(x(n))/n] \quad . \tag{14.1}$$

A sequence $x \in 2^\omega$ *is called* normalised random, *iff*

$$K(x) = \lim inf_{n\rightarrow\infty}[H(x(n))/n] > 0.$$

For $K(x) = 0$ *the sequence is non normalised random.*

Notice that not all non recursively enumerable sequences are normalised random. Consider for, instance, the sequence y generated from a random sequence $x = x_1x_2x_3\ldots$ by "padding" it with an ever-increasing density of zeros, i.e.,

$$y = x_1 0 x_2 00 x_3 000 x_4 0000 \ldots x_i \underbrace{0\ldots 0}_{i \text{ times}} \ldots \; .$$

The sequence y is non recursive but has zero limiting density of algorithmic information; i.e., $\lim\inf_{n\rightarrow\infty} H(y(n))/n = 0$. However, due to its equivalence to the metric entropy measure (cf. p. 240), the normalised algorithmic complexity K has very important physical applications.

14.1.2 Halting probability

An illustrative example for a random sequence is G. Chaitin's halting probability Ω [109, 110], which has been defined in 7.8, p. 92. Informally speaking, Ω_U is a measure of the "probability" that an infinite binary string which one inputs in a (not necessarily universal) computer U will eventually halt. It is not difficult to see that Ω can be obtained from an *effectively computable algorithm* in the limit of infinite computing time. The following effective computation, has been implemented by G. Chaitin [110]: Recall than a prefix-free program p which halts contributes $2^{-|p|}$ to Ω. Ω can be obtained in the limit from below

$$\omega_1 \leq \omega_2 \leq \omega_3 \leq \cdots \leq \omega_n \leq \omega_{n+1} \overset{n\rightarrow\infty}{\longrightarrow} \Omega \tag{14.2}$$

by running all prefix-free programs up to k bits in size for k cycles on computer U. G. Chaitin [109, 110] has shown that this algorithm is "very weakly converging." Indeed, the radius of convergence decreases *slower than any recursive function.* Only in this sense it is possible to "compute in the limit of infinite time" sequences of unbounded complexity, in particular random sequences.

One may speculate that the binomial distribution of a fair coin, generated by $2^{-N}(1+u)^N = 2^{-N}\sum_{i=1}^{N}\binom{N}{i}u^N$, for $u=1$ and successive $N = 1, 2, 3, \ldots$, contains all binomial coefficients, and therefore represents all the necessary information for the encoding of Chaitin's Ω (cf. p. 196) into a diophantine equation according to the scheme used by Jones & Matijasevič [247]; at the same time, this distribution contains all information necessary to specify a fair coin.

14.2 "Computational irreducibility"

Although Chaitin randomness is the formal analogue of intuitive notions of "lawlessness" or indeterminism, there may well be other reasonable definitions of randomness employing different types of complexities. In particular one could define a T-randomness from computational complexity (cf. section 8.1, p. 99 and St. Wolfram [453]).

In view of the fact that the class P of polynomial-time algorithms is invariant under the variation of "reasonable" machine models ("reasonable" machine models are circularly defined by the the requirement of invariance), one could define a problem of the order of N to be T-random as follows:

D 14.4 (T-randomness)
Given a problem of the order of N. Its solution $x(N)$ is T-random *if the computational complexity of its solution is not polynomially bounded, i.e., if it is not in P.*

14.3 von Mises collectives

R. von Mises' approach to randomness dates back to the early 20th century. (For an interesting historical account, see the introductory chapters in M. van Lambalgen's dissertation, reference [284, 285].) It is based on a fundamental entity, which is called *collective* [324] and which stands for an ensemble or collection of a very large number of observations. A *collective* may, for instance, be a sequence of events produced by some "random" process, such as coin tossing. It is defined by the following properties.

D 14.5 (Collective) *A* collective *(in the sense of random sequence) is an infinite binary sequence $x \in 2^\omega$ ($2 = \{0, 1\}$; χ denotes the characteristic set function) such that:*

(i) $P(0) = \lim_{n\rightarrow\infty}\frac{1}{n}\sum_{k=1}^{n}\chi_0(x_k)$ and $P(1)$ exists, and

(ii) let ϕ be an "admissible place selection" of x such that a certain infinite partial sequence of x is chosen without making use of the value x_k of the k'th place. *Then (i) holds for the partial sequence ϕx as well.*

The function P is called the probability distribution determined by the collective x.

Remarks:

(i) The notion of an "admissible place selection" is somewhat ambiguous and difficult to implement. The German original [324], p. 12, states:

> *Aus einer unendlichen Folge von Beobachtungen wird durch "Stellenauswahl" eine Teilfolge gebildet, indem man eine Vorschrift angibt, durch die über die Zugehörigkeit und Nichtzugehörigkeit der n^{ten} Beobachtung ($n = 1, 2, \ldots$) zur Teilfolge unabhängig von dem Ergebnis dieser n^{ten} Beobachtung und höchstens unter Benutzung der Kenntnis der vorangegangenen Beobachtungsergebnisse entschieden wird.*

Von Mises wanted this selection to be as general as possible, thereby not restricting the ϕ's. Intuitively, ϕ would correspond to a system of a gambler who does not believe in probability theory. Von Mises' approach proposes that such a system of gambling is of no avail.

(ii) In Richard von Mises vision, the collective comes first, then comes probability (*"erst das Kollektiv, dann die Wahrscheinlichkeit"*): The *probability* of an attribute in a collective equals the *relative frequency* of that attribute within the collective. Thus, in von Mises' approach, randomness of collectives is primary to probability. The collective defines the probability function P. This is in contradistinction to the usual approach to probability.

(iii) For a detailed discussion, see M. van Lambalgen [285, 284, 286, 288] and C. P. Schnorr [404].

Examples

In what follows, four attempts to explicate collectives with the help of special types of place selections are enumerated.

14.3.1 Bernoulli sequences

An example for collectives with a very restricted place selection rule is that of *Bernoulli sequences* or *Bernoulli randomness* (BR). For the explicit construction of the place selection, see M. van Lambalgen [285], p. 729. BR can be defined as follows.

D 14.6 (Bernoulli sequence) *A k-digit word $w_k = y_1 \ldots y_k$ is a string of k integers. A sequence $x \in 2^\omega$ is said to be k-distributed if the probability P that an arbitrary substring $x_n \ldots x_{n+k-1}$ of x is identical to w_k is 2^{-k}; i.e.,*

$$P(x_n \ldots x_{n+k-1} = y_1 \ldots y_k) = \frac{1}{2^k}$$

for arbitrary words w_k.

A sequence is ∞-distributed if it is k-distributed for all positive integers k. An ∞-distributed sequence is called a Bernoulli sequence.

Heuristically, this amounts to counting the relative frequency of an arbitrary word in a given sequence, which should be identical with the probability of the word itself. In the spirit of von Mises' definition, these place selections are too weak a criterion for randomness. For instance, it has been shown by D. G. Champernowne [115] that the enumeration of the set of all finite "decimal" words, i.e., words using the alphabet $0, 1, 2, 3, \cdots 9$, in lexicographic order ($9 > 8 > 7 > \cdots > 1 > 0$) "0 1 2 3 4 5 6 7 8 9 10 11 12 13 ..." is BR although it is based on a "very simple" construction principle.

Bernoulli randomness corresponds to the property of Borel normality; G. J. Chaitin has first investigated [97] this property of Ω; C. Calude has proved [74] that all random sequences have this property. In this context it is interesting to note the example due to von Mises [79] take an arbitrary binary sequence $x(n) = x_1x_2 \ldots x_n \ldots$ and construct the ternary (radix-3) sequence (over the alphabet $\{0, 1, 2\}$) $y(n) = y_1y_2 \ldots y_n \ldots$, where $y_1 = x_1, y_n = x_{n-1} + x_n, n \geq 2$. Then, $y(n)$ is never Bernoulli random because the patterns "02" and "20" do not appear in it. (Cf. C. Calude and I. Chiţescu [69], where this question is discussed within the Chaitin-Martin-Löf theory of randomness.)

14.3.2 Church random sequences

Church proposed to allow only *computable place selections*, corresponding to *recursive functions* ϕ_r in the definition of von Mises.

This definition of Church neglects all non recursive place selections ϕ_{nr}. Since the Bernoulli place selection is recursively enumerable, Church randomness (CHR) implies Bernoulli randomness. Again in the spirit of von Mises' definition recursive enumerability is too restrictive a criterion.

14.3.3 Wald random sequences

Wald proposed to allow an arbitrary but *denumerable set of place selections*. He could show that the set of collectives satisfying this condition has the cardinality of the continuum.

14.3.4 Axiomatisation of randomness

M. van Lambalgen has introduced another approach [288, 289] based, intuitively speaking, on the axiomatisation of the "relative independence" of sequences. If $x, y \in 2^\omega$, then one can use the i'th digit of y to "select" or "unselect" the digit at the i'th place of the binary enumeration of x. For further information, see M. van Lambalgen's articles.

14.4 Statistical-based randomness

J. Ville suggested an approach which is, in a sense, *dual* to the notion of collectives: *A random sequence should satisfy 'all' properties of probability one.* These properties of probability one are *probability laws* of the form "$\mu(\{x \in 2^\omega \mid A(x)\}) = 1$," where μ is a (normalised) measure and A is a formula. Just as for "allowed place selections" in von Mises' approach, the problem is to specify the term "all properties A of probability 1."

For details, the reader is referred to P. Martin-Löf's original work [312], G. Chaitin's book *Algorithmic Information Theory* [110], C. P. Schnorr's book *Zufälligkeit und Wahrscheinlichkeit* [404] and M. van Lambalgen's dissertation [284, 285], among others.

Again, some definitions require the construction of a real binary number $r = 0.x_1x_2x_3 \ldots \in [0,1]$ from the sequence $x \in 2^\omega$.

14.4.1 Martin-Löf random sequences

D 14.7 (Martin-Löf randomness) *A real r is* Martin-Löf random *(MLR) if it is not contained in any set of an recursively enumerable infinite sequence A_i of sets of intervals, such that the measure $\mu(A_i)$ (no double-counting) is always less than or equal to 2^{-i}, i.e., $\mu(A_i) \leq 2^{-i}$. I.e., r is MLR if*

$$\forall i[\mu(A_i) \leq 2^{-i}] \Rightarrow \neg \forall i[r \in A_i] \quad .$$

Remarks:

(i) Any recursive evaluation of sequences of sets A_i corresponds to a statistical test with increasing significance levels.

(ii) The choice of the radius of convergence 2^{-i} corresponding to the levels of significance is arbitrary. By substituting $2^{-f(i)}$ for 2^{-i}, any computable, non decreasing "regulator of convergence" $f(i) \stackrel{i\to\infty}{\longrightarrow} \infty$ for i would work just as well.

14.4.2 Solovay random sequences

The following definition is not based upon any "regulator of convergence" or "level of significance" as Martin-Löf randomness.

D 14.8 (Solovay randomness) *A real r is* Solovay random *(SR) if for any recursively enumerable infinite sequence A_i of sets of intervals with the property that the sum of the measures of A_i converges, i.e.,*

$$\sum_i \mu(A_i) < \infty \quad ,$$

r is contained in at most finitely many of the A_i's. I.e.,

$$\sum_i \mu(A_i) < \infty \Rightarrow \exists N \forall (i > N)[r \notin A_i] \quad .$$

14.5 Equivalencies and comparisons

Whereas von Mises considers the notion of a collective as the primary concept and statistical tests as secondary, the situation is reversed in Ville's and in particular Martin-Löf's approach: there, random sequences are defined by requiring that they satisfy all computable statistical tests with probability one. This is guaranteed if such sequences cannot be algorithmically compressed into a shorter form. This intuitive argument suggests that statistical and complexity based definitions of randomness are equivalent. (By requiring "algorithmic incompressibility," Chaitin randomness requires *more* than uncomputability.) The following equivalencies will be stated without proof (cf. [110]).

T 14.9 ((Martin-Löf, Solovay, Schnorr, Chaitin))

A real number (an infinite sequence) is Martin-Löf random $\Longleftrightarrow$ it is Solovay random $\Longleftrightarrow$ it is Chaitin random $\Longleftrightarrow$ it is weakly Chaitin random

T 14.10 *With probability one, a real number is Martin-Löf/Solovay/Chaitin random.*

Furthermore,

T 14.11 (Ville, van Lambalgen [284, 285])

There exist uncountable many Church random sequences which are not *Martin-Löf/Solovay/Chaitin random. I.e., for suitable measures μ, the set of Church random sequences $\{CHR\}$ and the set of Martin-Löf random sequences $\{MLR\}$ satisfy*

$$\mu(\{\mathrm{CHR}\} - \{\mathrm{MLR}\}) = 1 \quad .$$

Chapter 15

Random fractals and $1/f$ noise

15.1 Self-similarity

Self-similar or *fractal* [307] patterns "look identical" when viewed at different resolution scales. Informally stated, if one looks at such a pattern "a little closer" and with "higher resolution," the pattern appears similar. Probably the most simple among all fractal creatures is the *Cantor set.* It is obtained by starting with the interval $[0, 1]$. In the first construction step, the middle third interval, i.e., $(\frac{1}{3}, \frac{2}{3})$, is cut out from the original interval, leaving two disjoint subsets $[0, \frac{1}{3}]$ and $[\frac{2}{3}, 1]$. In the next construction step, each of these two remaining intervals is treated in the same manner as the original interval. I.e., the middle thirds of the remaining intervals are cut out, leaving four disjoint subintervals. In the n'th construction step there are $N(n) = 2^n$ intervals of length $\delta(n) = 3^{-n}$. An infinite recursion of this process yields the Cantor set. It is drawn in Fig. 15.1.

An alternative construction technique uses the *random iterated algorithm* [241, 23, 163]. It uses the iterated function system $S_1(x) = \frac{1}{3}x$, $S_2(x) = \frac{1}{3}x + \frac{2}{3}$ with probabilities $p_1 = p_2 = \frac{1}{2}$ (other values $\neq 0$ would do just as well, but less efficiently). This iterated function system has been obtained by identifying the point 0 in the zeroth construction step with the points 0 and $\frac{2}{3}$ in the first construction step, and by identifying the point 1 in the zeroth construction step with the points $\frac{1}{3}$ and 1 in the first construction step. By starting with an arbitrary "seed" $0 \leq x \leq 1$, the successive iteration of either S_1 and S_2 chosen at random yields the Cantor set as an "attractor" or "invariant set" of the iterated function system: Let $\mathfrak{C}$ stand for the Cantor set; then

$$\mathfrak{C} = \bigcup_{i=1}^{2} S_i(\mathfrak{C}) = S_1(\mathfrak{C}) \cup S_2(\mathfrak{C}) \quad . \tag{15.1}$$

A *Mathematica* program forthe generation of $\mathfrak{C}$ is listed on page 264.

The Cantor set can be brought into a one-to-one correspondence with the binary (base-2) interval $[0, 1]$ by associating the numbers "0" and "1" with the remaining left and right subintervals in each construction step, respectively. The n'th construction step corresponds to the n'th position in the code of the binary real. The prefix "0." is associated with the zeroth step. Yet, the Cantor set has Lebesgue measure zero, since $\lim_{n\to\infty}\left(\frac{2}{3}\right)^n = 0$. Moreover, when measured at different resolutions $\delta(n) = 3^{-n}$, the Cantor set has different length $\left(\frac{2}{3}\right)^n$ between 1 (corresponding to $n = 0$) and 0 (corresponding to $n = \infty$). It is possible to overcome the "paradoxical" feature of a resolution-dependent length by using a "fractal" measure $\lim_{n\to\infty} 2^n \left(\frac{1}{3^n}\right)^D$, the price being the introduction of a non integer

Figure 15.1: The Cantor set.

fractal or *similarity* dimension D. More generally, if one requires scale invariance of the measure, then

$$N(n)\delta(n)^D = N(n+1)\delta(n+1)^D \tag{15.2}$$

and thus the "fractal" (box-counting) dimension is given by

$$D \approx \frac{\log[N(n)/N(n+1)]}{\log[\delta(n+1)/\delta(n)]} \quad . \tag{15.3}$$

For the Cantor set, equation (15.3) yields $D = \frac{\log 2}{\log 3}$.

There are several alternative definitions of dimensions; detailed discussions and comparisons can be found in the books of K. J. Falconer [162, 163] and M. Barnsley [23] and in an article by J. D. Farmer, E. Ott and J. A. Yorke [165]. By setting $N(n+1) = \delta(n+1) = 1$, by defining $N(n) = N$, $\delta(n) = \delta$ and by taking the limit of infinite resolution, one obtains

$$D = \lim_{\delta \to 0} \frac{\log N}{\log(1/\delta)} \quad . \tag{15.4}$$

For actual physical measurements, a limit corresponding to infinite resolution $\delta = 0$ is unattainable. Instead, one usually has a parameter $\delta_0 > 0$ characterising either the *maximal experimental resolution* or the *scale of breakdown of self-similarity*, and consequently one has to resort to a limit $\delta \to \delta_0$. Additionally, there is a trivial bound δ_∞ from above on the resolution by the overall system size or a largest resolution scale. E.g., for the Cantor set, $\delta_\infty = 1$; for an experimental time series, δ_∞ is the overall observation time *et cetera.* I.e., in summary,

$$\delta_0 \leq \delta \leq \delta_\infty \quad . \tag{15.5}$$

For a much more detailed discussion of operational definitions of dimension, the reader is referred to B. B. Mandelbrot's *Fractals: Form, Chance and Dimension* [307], K. J. Falconer's books *The Geometry of Fractal Sets* and *Fractal Geometry* [162, 163], or K. Svozil's and A. Zeilinger's articles [469, 430, 435], among others.

The number of elements at the $n+1$'st construction (observation) level with resolution $\delta(n+1)$ relative to the number of elements at the n'th construction (observation) level with resolution $\delta(n)$ is given by

$$\frac{N(n+1)}{N(n)} = \left[\frac{\delta(n)}{\delta(n+1)}\right]^D \quad , \tag{15.6}$$

where equation (15.2) has been used. Let E be the integer dimension of the smallest space $\mathbb{R}^E$ in which the fractal object can be embedded. Then, the number of cells of diameter $\delta(n+1)$ which fit into a cell of diameter $\delta(n)$ is given by

$$\left[\frac{\delta(n)}{\delta(n+1)}\right]^E \quad . \tag{15.7}$$

Combining equations (15.6) and (15.7) yields a "relative filling factor" $r(D)$ of covered cells at dimension D by

$$r(D) = \frac{\frac{N(n+1)}{N(n)}}{\left[\frac{\delta(n)}{\delta(n+1)}\right]^E} = \left[\frac{\delta(n)}{\delta(n+1)}\right]^{D-E} \quad . \tag{15.8}$$

In the case of the Cantor set, which can be embedded in $\mathbb{R}^1$, $E = 1$ and $r(D) = 2/3$. I.e., the Cantor set can be constructed in a step-by-step fashion recursively by taking the latest "active" intervals and cutting $1 - r = 1/3$ out of the middle of them.

15.2 Random fractal construction

The self-similarity of the cantor set is *exact:* every part of the pattern contains an *identical* replica of itself. This exact self-similarity is seldom encountered in physics. Natural self-similarities are *statistical* ones. I.e., enlargements of parts of the pattern have the same *statistical* distribution as the whole pattern. Such structures will be called *random fractals*

A similar approach as for the construction of the Cantor set can be applied for the construction of *random* fractals. Random fractals of dimension D can be recursively defined by successively cutting out a fraction of $1-r(D)$ elements of length $\delta(n+1)$ [corresponding to the $n+1$'st construction (observation) level] from an interval of length $\delta(n)$ *at random* (see also [163], chapter 15, p. 224). A *Mathematica* program for a construction of such structures is listed in the appendix, p. A.4. Fig. 15.2 shows random fractals of various dimensions between 0 and 1. For alternative constructions and discussions of random fractals, see K. J. Falconer [162, 164], R. D. Mauldin and S. C. Williams [315] and U. Zähle [466], among others.

The *Morse-Thue sequence* featurs a different type of "self-similarity." The sequence can be constructed by several methods (cf. M. R. Schroeder [405], p. 316). A step-by-step enumeration is obtained by taking the sum modulo 2 of the digits of successive integers, written in binary notation; i.e.,

enumeration of integers	0	1	2	3	4	5	6	7	8	9	...
binary representation	0	1	10	11	100	101	110	111	1000	1001	...
sum of digits modulo 2	0	1	1	0	1	0	0	1	1	0	...

The first 300 bits of the Morse-Thue sequence are drawn in Fig. 15.3. The sequence can also be constructed by appending to each subsequence its complementary sequence (demonstrating its aperiodicity). Still another iteration procedure yielding the Morse-Thue sequence: starting from the "seed" 0, at any stage of the construction, one adds the complementary symbol to the right (demonstrating that there are as many 0's as there are

Figure 15.2: Random fractal patterns.

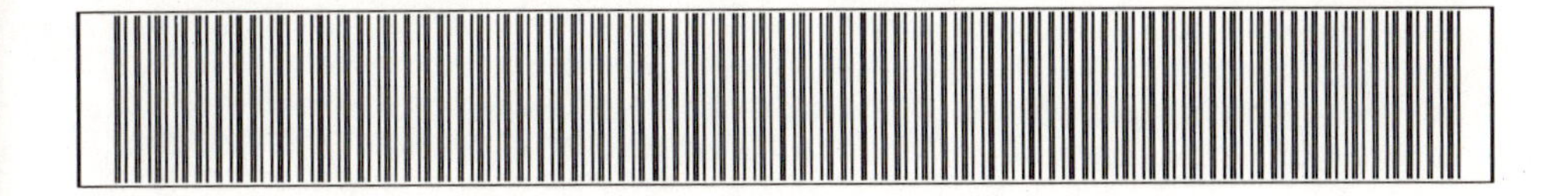

Figure 15.3: Plot of the Morse-Thue sequence up to 300 places.

1's). This explains also another one of its features: by discarding every second place, the original Morse-Thue sequence is recovered. This kind of "self-similarity" manifests itself in the frequency plot drawn in Fig. 15.5(c) below.

15.3 1/f spectral density

One may investigate the *spectral density* $S_V(f)$ of a fractal random signal $V(t)$ by [383]

$$S_V(f) \propto |V(f)|^2 \quad , \tag{15.9}$$

where f is the frequency and $V(f)$ is the Fourier transform of $V(t)$. A signal is called a $1/f$-signal if $S_V(f) \propto 1/f^\beta$ with $0.5 \leq \beta \leq 1.5$. For a review as well as for some explanations of a $1/f$-spectrum, see P. Dutta & P. M. Horn [143], A. van der Ziel [472], among others. For the connection between $1/f$-signals and fractal geometry, see R. F. Voss [450] and K. J. Falconer [163], p. 158.

Roughly speaking, the graph (not the set of the pulses!) of a signal with a power spectral density $S_V(f) \propto 1/f^\beta$ has a fractal (box-counting) dimension of

$$D \approx E + \frac{3-\beta}{2} \quad , \tag{15.10}$$

where again E is the integer dimension of the smallest space $\mathbb{R}^E$ in which the fractal object can be embedded [450, 163].

Fig. 15.4 shows plots of $\log(S_V(f))$ as a function of $\log(f)$ for the random fractals in Fig. 15.2, revealing typical signatures of a $1/f$ noise. This compares to the spectral density of white noise, which is a constant function of the frequency; see Fig. 15.5(a). The spectral analysis of the Cantor set (Fig. 15.1) is drawn in Fig. 15.5(b). The spectral analysis of the Morse-Thue sequence is drawn in Fig. 15.5(c).

15.4 Signal amplification *via* multiple channels

The standard model of signal transmission in information theory is drawn in Fig. 15.6. For certain types of signals it is useful to consider signal transmission of a more general type. In what follows, random fractal signals are considered which are transmitted *via*

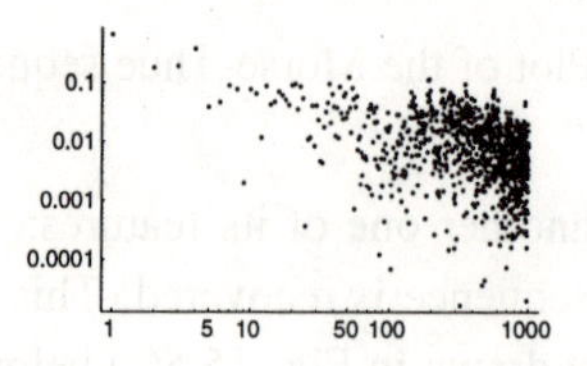

$D = 0.4$

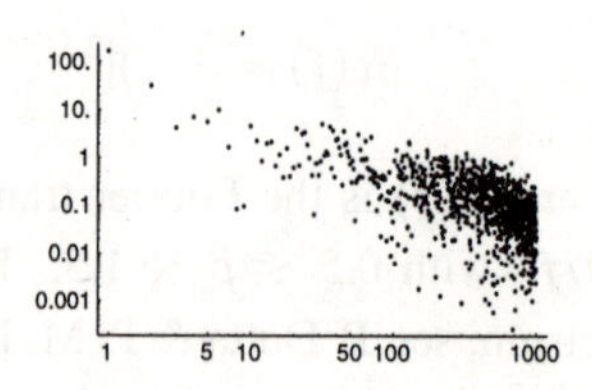

$D = 0.8$

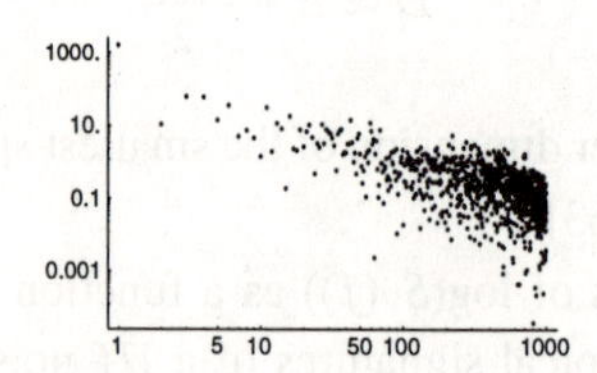

$D = 0.9$

Figure 15.4: $\log S_V(f)$ against $\log f$ for random fractal patterns.

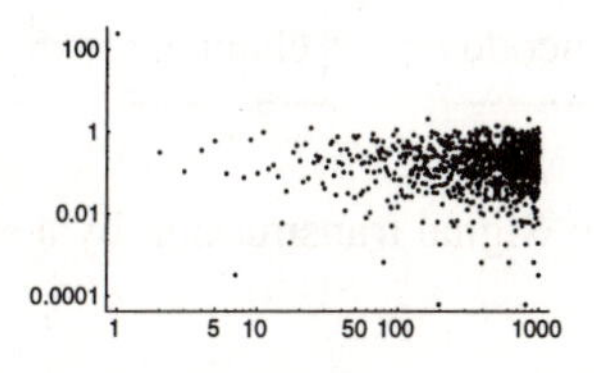

(a)

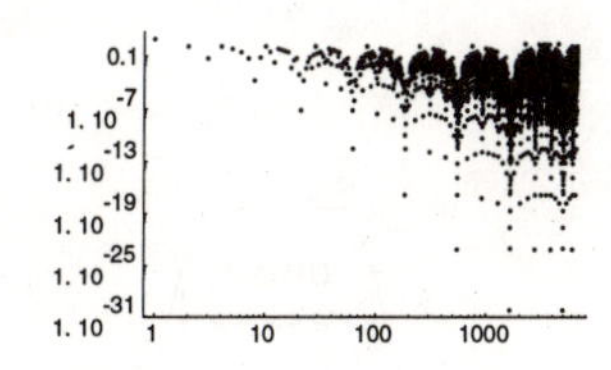

(b)

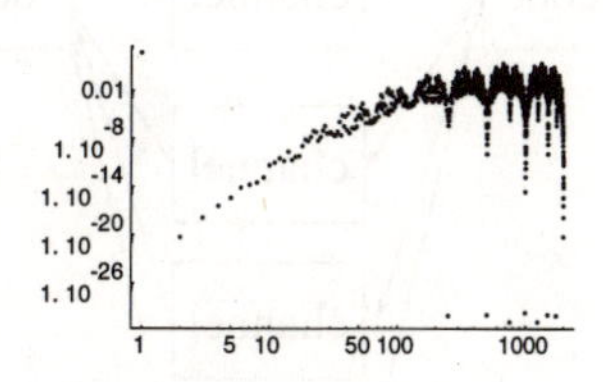

(c)

Figure 15.5: $\log S_V(f)$ against $\log f$ (a) for white noise, (b) for the Cantor set, (c) for the Morse-Thue sequence.

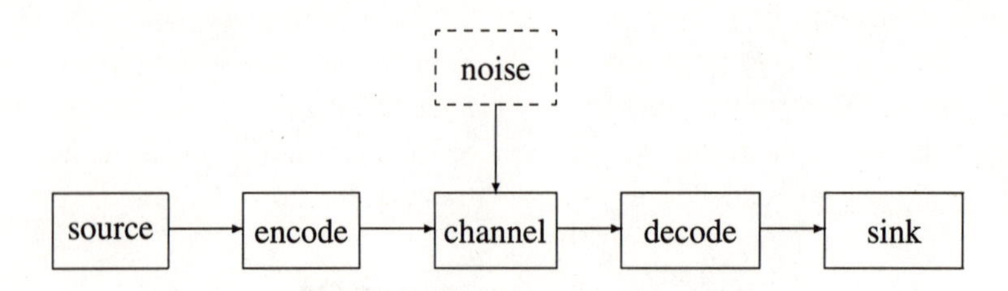

Figure 15.6: Signal transmission by a single channel.

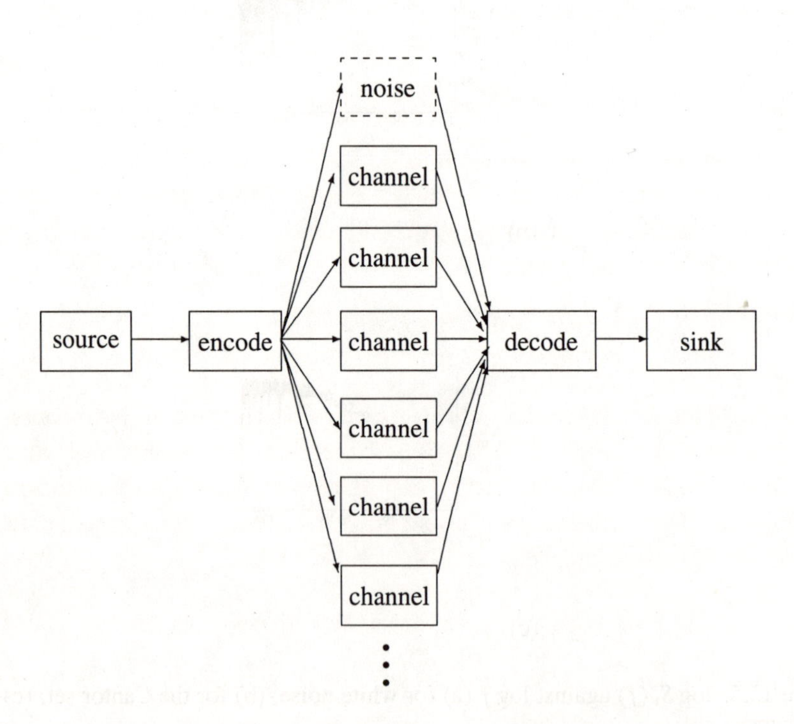

Figure 15.7: Signal transmission *via* multiple channels.

multiple channels. This setup is drawn in Fig. 15.7. (Indeed, adding noise in the standard setup amounts to adding a parallel channel with a noise signal.)

Such signals can be considered as sets of points or sequences of symbols; e.g., of binary symbols. Special attention will be given to the *intersection* of random fractal signals. As has been pointed out by K. J. Falconer [163], theorem 8.2, p. 102, 103, under certain "mild side conditions," the intersection of two random fractals A_1 and A_2 which can be minimally embedded in $\mathbb{R}^E$ is again a random fractal with dimension

$$D(A_1 \cap \sigma(A_2)) \leq \max\{0, D(A_1) + D(A_2) - E\} \quad . \tag{15.11}$$

"Often" ("often" means "for a set of σ positive measure"),

$$D(A_1 \cap \sigma(A_2)) \geq D(A_1) + D(A_2) - E \quad . \tag{15.12}$$

$\sigma \in G$ stands for a set of motions of positive measure, where G is some group of transformations, e.g., similarities, rigid motions *et cetera.* See K. J. Falconer for a proof and for an exact statement of the theorem. For integer-dimensional objects, equation (15.11) can be readily verified: the intersection of a plain ($D = 2, E = 2$) and a line ($D = 1$) yields $D \leq 1$; the intersection of a cube ($D = 3, E = 3$) and a line ($D = 1$) yields $D \leq 1$; *et cetera.*

By induction, this result generalises to the intersection of an arbitrary number of random fractal sets. The dimension of the intersection of n random fractals is "frequently" given by

$$D^{\cap}(\{A_i\}) = D(\bigcap_{i=1}^{n} A_i) \approx \sum_{i=1}^{n} D(A_i) - E(n-1) \quad . \tag{15.13}$$

The embedding dimension of random fractal signals is $E = 1$.

Consider the coding of a signal by random fractals by their *dimension* parameter. An example is the case of just two source symbols s_1 and s_2 (cf. 4.3, p. 47) encoded by (*RFP* stands for "random fractal pattern")

$$\#(s_i) = \begin{cases} RFP \quad \text{with} \quad 0 \leq D(RFP) < 0.5 \quad \text{if} \quad s_i = s_1 \\ RFP \quad \text{with} \quad 0.5 \leq D(RFP) \leq 1 \quad \text{if} \quad s_i = s_2 \end{cases} \quad . \tag{15.14}$$

Let us call the A_i's the *primary signals* (primary sources), and the intersection $\bigcap_{i=1}^{n} A_i$ of the primary signals as the *secondary signal* (secondary source). We shall study interesting special cases of equation (15.13). The addition of white noise to a random fractal signal, denoted by $\mathbb{I}$ with $D(\mathbb{I}) \approx 1$, results in the recovery of the original fractal signal with the original dimension; i.e.,

$$D^{\cap}(A, \mathbb{I}) \approx D(A) + D(\mathbb{I}) - 1 \approx D(A) \quad . \tag{15.15}$$

By assuming that all random fractals have equal dimensions, i.e., $D(A_i) = D$ and $D^{\cap} \geq 0$, equation (15.13) reduces to

$$D^{\cap} \approx \max\{0, n(D-1) + 1\} \quad . \tag{15.16}$$

In Fig. 15.8, the dimension $D^{\cap}(n)$ of the secondary signal is drawn as a function of the number of channels $1 \leq n \leq 20$ for various values of the dimension D of the primary

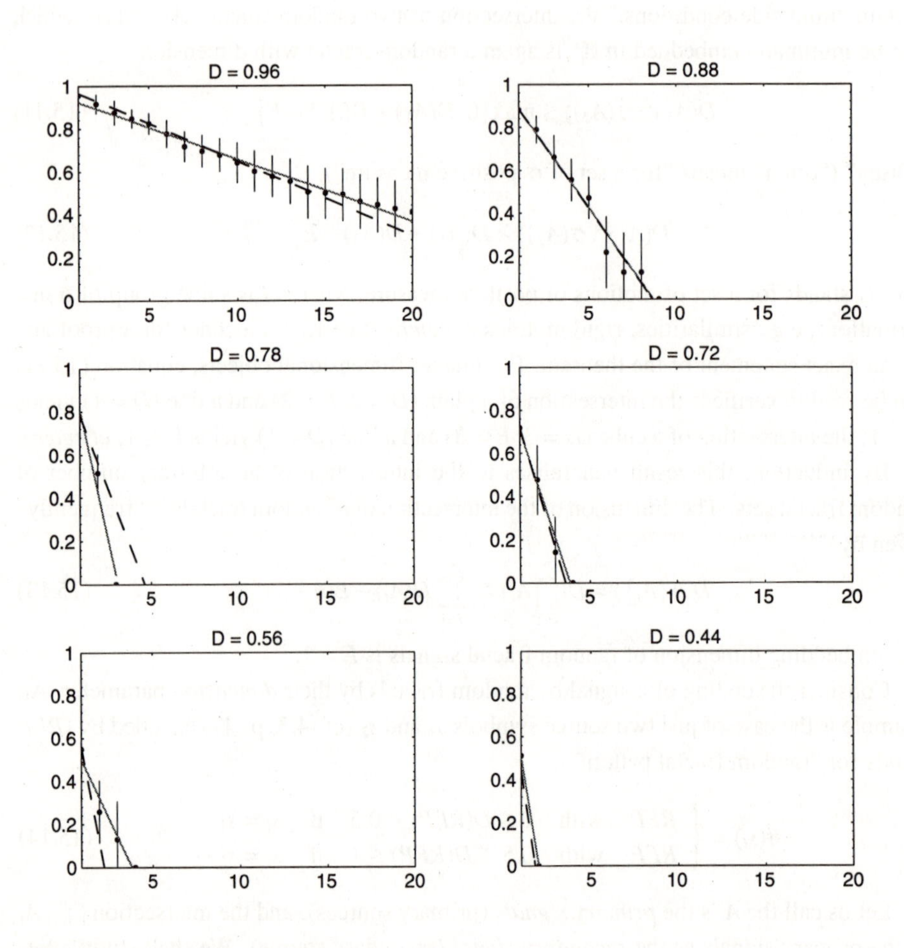

Figure 15.8: $D^{\cap}(n)$ versus n for various values of dimensions D. The points indicate data points, generated from computer experiments. The solid grey line indicates the quadratic fit to a linear function. The dashed line indicates the theoretical prediction.

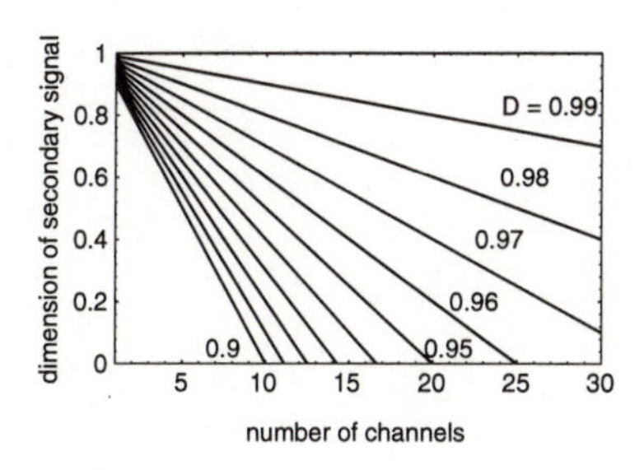

Figure 15.9: Theoretical prediction of $D^{\cap}(n)$ versus n for various values of dimensions D.

signal. The primary signals were generated numerically; the secondary signals were measured by box-counting methods (cf. programs `RandomBurstsList` and `Dim`, p. 265). Fig. 15.9 shows the theoretical prediction of $D^{\cap}(n)$ versus n for various values of dimensions D. From these figures it can be seen that, intuitively speaking, the larger the number of channels, the higher the dimension of the primary signal has to be in order to obtain a secondary signal of nonzero dimension. Fig. 15.10 shows the theoretical prediction of the critical number of channels as a function of the dimension of the primary signal.

An immediate consequence of (15.16) is that, for truely fractal signals ($D < 1$), any variation of the fractal dimension of the secondary signal $D^{\cap}$ is directly proportional to the number n of the primary signals; i.e.,

$$\Delta D^{\cap} \approx n \Delta D \quad \text{for} \quad D \neq 1 \quad . \tag{15.17}$$

Stated pointedly: the more channels there are, the more will the dimension of the secondary source vary in response to variations of the primary source; there is an "amplification" of any change in the primary signal. The case $D = 1$ corresponds to white noise. Since in this case, $d D^{\cap}/d D = 0$, there is no such "amplification" and the effect vanishes.

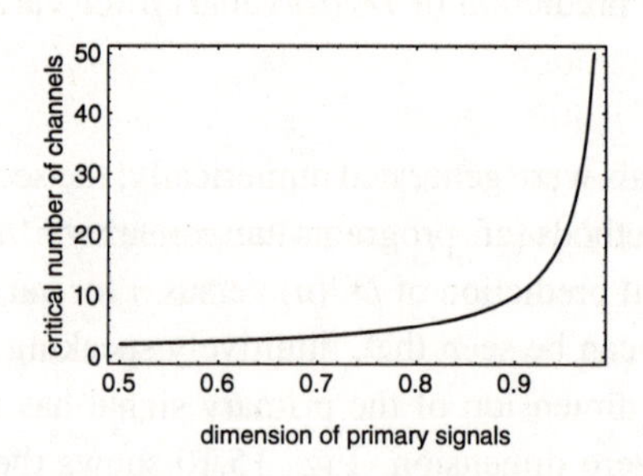

Figure 15.10: Theoretical prediction of the critical number of channels as a function of the dimension of the primary signal.

Chapter 16

Chaotic systems are optimal analogues of themselves

Consider, for the moment, as a working hypothesis the following conjecture:

> "Chaos *in physics corresponds to* randomness *in mathematics (Chaos = randomness)."*

With this in mind, it is of greatest physical relevance which concept of randomness is envisaged and if this abstraction is appropriate for the perception of chaotic motion. One could, for instance, call a number "random" if it is irrational (i.e., if it has no periodic representation), if it is Chaitin random, or if it is contained in some other set with non vanishing measure. Another possibility would be to call a number (representing the evolution of a system) "weakly random" if it is impossible to infer from previous places the future ones. The complexity-based approach to randomness, which is mainly pursued here, suggests the following statement (T. Klein's *"every system is a perfect analogue of itself"* has been communicated to me by A. Zeilinger.)

> *"With respect to computational resources, every chaotic system is an* optimal *analogue of itself; i.e., one cannot simulate a chaotic system with less than its own resources."*

Depending on the type of computational resources (i.e., algorithmic information or computational complexity), one can identify at least four classes of chaos, which will be discussed below. [From a purely algorithmic point of view, classes I ("deterministic chaos"), II and III are equivalent, because there is no fundamental difference between the "program" and the "input" (code).]

16.1 Chaos I — "deterministic chaos"

Chaos I — "deterministic chaos" is characterised by two [three] criteria:

(i) effectively computable/recursive/deterministic evolution;

(ii) property of the (non-linear) system evolution to separate initially close trajectories exponentially fast in time [positive Lyapunov exponent(s)]; unfolding of a unknown (random?) initial value throughout evolution;

[*(iii)*] random initial value. — In the literature, randomness is usually specified by Martin-Löf/Solovay/Chaitin randomness [180]. If the initial value is element of the continuum, the probability that it is Martin-Löf/Solovay/Chaitin random is one, i.e., "almost all" initial values are random reals. (That statement is, of course, also true for weaker definitions of randomness, e.g., for the set of Bernoulli random numbers, for the irrationals, for the transcendentals *et cetera.*)

Stated pointedly, in "deterministic chaos," the randomness or, in a weaker *dictum*, the incomplete information of the initial value "unfolds" throughout evolution. A criterion for *"deterministic chaos"* therefore is a "suitable" evolution function capable of "unfolding" the information of a random real associated with the "true" but unknown initial value x_0. I.e., either the uncertainty δx_0 of the initial value or a corresponding variation of the initial value increases with time. The *Lyapunov exponent* λ can be introduced as a measure of the separation of two distinct initial values. Consider a discrete time evolution of the form $x_{n+1} = f(x_n)$ and an uncertainty interval $(x_0, x_0 + \varepsilon)$ of measure ε which, after n iterations, becomes $(f^{(n)}(x_0), f^{(n)}(x_0 + \varepsilon))$, which is of measure $\varepsilon \exp\{n\lambda(x_0)\}$. $f^{(n)}$ stands for the n-fold iteration of f. The *Lyapunov exponent* $\lambda(x_0)$ is defined for $\varepsilon \to 0$ and $n \to \infty$ as

$$
\begin{aligned}
\lambda(x_0) &= \lim_{n\to\infty}\lim_{\varepsilon\to 0}\frac{1}{n}\log\left|\frac{f^{(n)}(x_0+\varepsilon)-f^{(n)}(x_0)}{\varepsilon}\right| \\
&= \lim_{n\to\infty}\frac{1}{n}\log\left|\frac{d}{dx}f^{(n)}(x)\bigg|_{x=x_0}\right| \\
&= \lim_{n\to\infty}\frac{1}{n}\log\left|\prod_{i=0}^{n-1}f'(x)\bigg|_{x=x_i}\right| \\
&= \lim_{n\to\infty}\frac{1}{n}\sum_{i=0}^{n-1}\log\left|f'(x)|_{x=x_i}\right| \quad ,
\end{aligned}
\tag{16.1}
$$

where the chain rule

$$
\frac{d}{dx}f^{(n)}(x)\bigg|_{x=x_0} = \frac{d}{dx}f(f(\cdots f(x)\cdots))\bigg|_{x=x_0} = f'(x_{n-1})f'(x_{n-2})\cdots f'(x_1)f'(x_0)
$$

and $|a_1a_2a_3\cdots a_n| = |a_1||a_2||a_3|\cdots|a_n|$ for $a_i \in \mathbb{R}$, $i = 1, \ldots n$, has been used (for more details, see, for instance, R. Shaw [411] and H. G. Schuster [408]). For $\lambda > 0$, the initial uncertainty increases exponentially with time; see Fig. 16.1. Since the natural unit in algorithmics is bit, it is more appropriate to define the Lyapunov exponent in terms of the basis 2 (instead of e); all logarithms are then *binary* logarithms $\log_2$, yielding Lyapunov exponents which are by a factor of $1/\log_e 2$ greater then the ones defined in terms of the basis e.

For continuous maps $G(x) = dx/dt$, one obtains a change of uncertainty δx by

$$
\frac{d(\delta x)}{dt} \approx \frac{\partial G(x)}{\partial x}\delta x \quad , \tag{16.2}
$$

with $\lambda(x_0) = \partial G/\partial x\,|_{x_0}$. For constant λ the solution of (16.2) can be obtained by integration: $\delta x(t - t_0) = \delta x_0 \exp[\lambda(t - t_0)]$.

For $\lambda > 0$, a *linear* increase in the precision precision of the initial value δx_0 renders merely a *logarithmic* increase in the accuracy of the prediction. The above scenario is precisely the signature for *"deterministic chaos"* in classical, deterministic continuum mechanics: the evolution by "suitable" (positive Lyapunov exponents) recursive / effectively computable functions whose "random" arguments (i.e., initial values) are elements

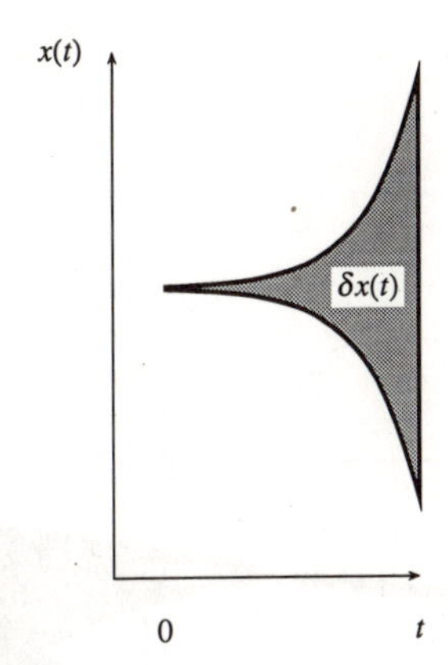

Figure 16.1: For positive Lyaponov exponent, an initial uncertainty increases exponentially with time.

of a continuous spectrum. Such continua serve as a kind of "pool of random reals". Therefore, the *"randomness" of classical "deterministic chaos" resides in its initial configuration.* (See also J. Ford [180, 181] for a very clear and illuminating discussion of this topic.) In classical physics, continua are modelled by $\mathbb{R}^n$ or $\mathbb{C}^n$; and "randomness" translates into Martin-Löf/Solovay/Chaitin randomness. If these classical continua turn out to be appropriate for physical theories remains to be seen.

16.2 Verhulst/Feigenbaum scenario

For a much more detailed review as well as for alternative passes to "deterministic chaos," see H. G. Schuster [408], J.-P. Eckmann [148], P. Cvitanović [130], Hao Bai-Lin [224], K. J. Falconer [163] and others. The symbolic dynamics aspects of nonlinear maps have been worked out in detail by P. Grassberger [209] and Ch. D. Moore [326] and will not be reviewed here.

The starting point of the *Verhulst/Feigenbaum scenario* [169, 170] to "deterministic chaos" is the observation that many nonlinear systems behave generically: there exist "tuning parameters" α which determine the periodicity or stochasticity of the state evolution. For example, the logistic map

$$x_{n+1} = f(x_n) = \alpha x_n(1 - x_n)$$

has several regimes in the interval $\alpha \in [0, \infty)$. See Fig. 16.2 for iterates of this map and the Lyapunov exponents, which were computed numerically from equation (16.1) as a function of α.

Periodic regime

(i) for $\alpha \in [0, a_1]$, there exists one stable fixed point $x_1^* = f(x_1^*)$ and a system converges and remains at x_1^*;

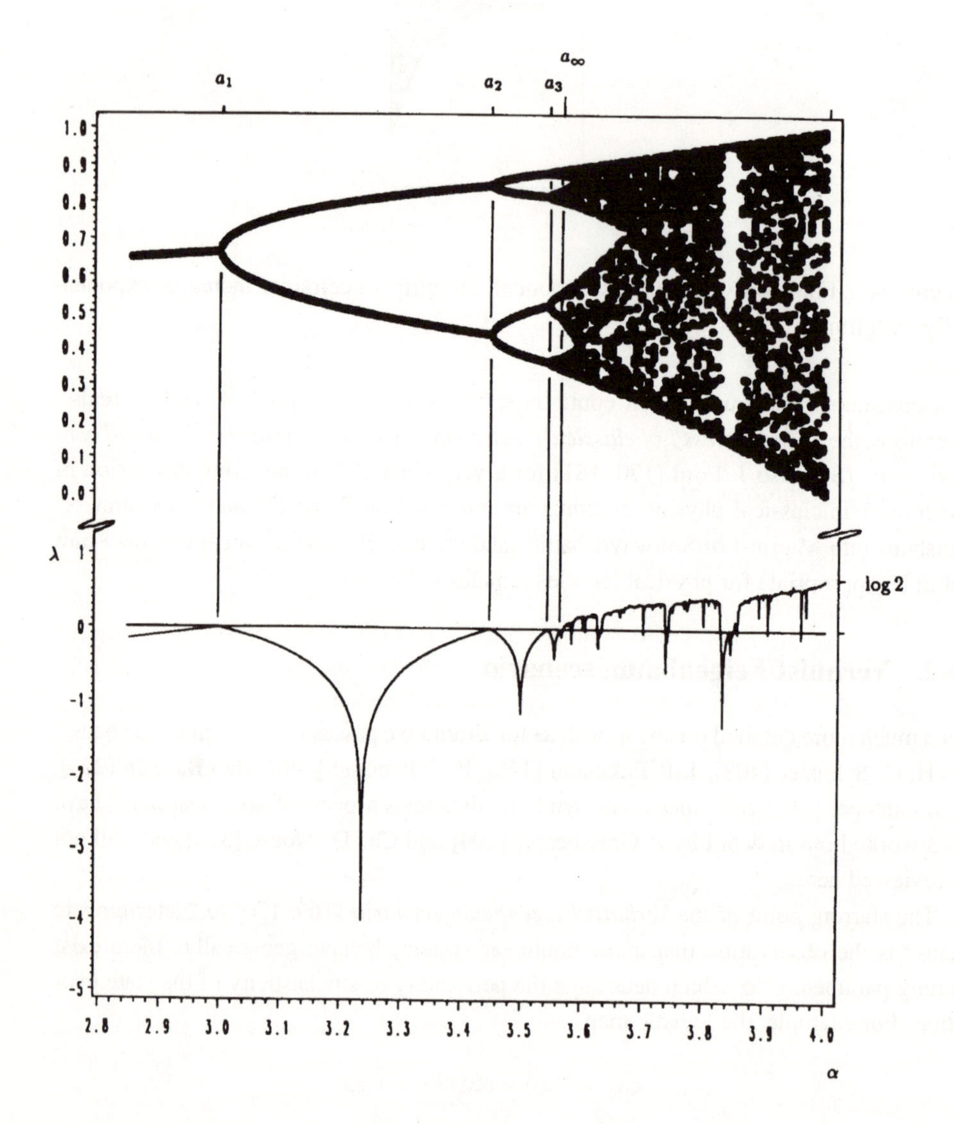

Figure 16.2: Study of the iterates of the logistic map $x_{n+1} = f(x_n) = \alpha x_n(1 - x_n)$ and the associated Lyapunov exponents as a function of the parameter α.

(ii) for $\alpha \in (a_1, a_\infty)$ there is, depending on the parameter α, a hierarchy of fixed points and associated periodic trajectories. By varying α one notices a succession of fixed point instabilities accompanied by bifurcations at a_N: if an N'th order fixed point x_N^* is defined by its recurrence after N computing steps (and not before), that is after N iterations of f,

$$x_N^* = f^{(N)}(x_N^*) = \underbrace{f(f(\cdots f(x_N^*)\cdots))}_{N \text{ times}} \quad ,$$

then x_N^* characterises a periodic evolution in state space (for the logistic equation, $a_1 \approx 3.00$ and $a_\infty \approx 3.57$ [316]). for large N, the following scaling laws are universal: for adjacent fixed points,

$$F_1 = \lim_{N\to\infty} \frac{a_N - a_{N-1}}{a_{N+1} - a_N} = 4.6692\cdots, \quad F_2 = \lim_{N\to\infty} \frac{x_N^* - x_{N-1}^*}{x_{N+1}^* - x_N^*} = 2.5029\cdots \quad ;$$

"Chaotic" regime

(iii) for $\alpha \in [a_\infty, 4)$ aperiodicity sets in, followed by a fine-structure which is not explained here. In this regime the Lyapunov exponent is mostly positive, the unfolding of the algorithmic information of the initial value and thus chaos I or IV,1 sets in;

(iv) for $\alpha = 4$ and after the variable transformation $x_n = \sin^2(\pi X_n)$ one obtains a map $f : X_n \to X_{n+1} = 2X_n \pmod 1$, where (mod 1) means that one has to drop the integer part of $2X_n$. By assuming a starting value X_0, the formal solution to n iterations is $f^{(n)}(X_0) = X_n = 2^n X_0 \pmod 1$. f is easily computable: if X_0 is in binary representation, $f^{(n)}$ is just n times a left shift of the digits of X_0, followed by a left truncation before the decimal point (see Fig. 16.3). Now assume $X_0 \in (0, 1)$ is Martin-Löf/Solovay/Chaitin random. Then the computable function $f^{(n)}(X_0)$ yields a Martin-Löf/Solovay/Chaitin random evolution. It should be stressed again that in *"deterministic chaos,"* the evolution function f itself is computable / recursive, X_0 is random, and f "unfolds" the "information" contained in X_0 in time. (For $\alpha \in (4, \infty)$ the evolution for most points $x_0 \in (0, 1)$ diverges.)

One may criticise this scenario by its assumption of a *continuum*. Indeed, there is no effective computation capable of simulating such a "chaotic" evolution exactly. The evolution requires oracle capacity. If the Verhulst/Feigenbaum scenario as well as other types of "deterministic chaos" is simulated by a universal or (worse) finite computer such as (insert your favourite brand here:) "...," in finite time one never obtains a Martin-Löf/Solovay/Chaitin random evolution, i.e., chaos I, but just chaos of type IV (see below). As von Neumann put it [339], *"Any one who considers arithmetical methods of producing random digits is, of course, in a state of sin."* One might add though, *"if he thinks he can do so in finite time,"* because computation of Chaitin's Ω (cf. 14.1.2 p. 196 and Chaitin's books [109, 110]) is such a sinful endeavour.

Despite the fact that classical mechanics and electrodynamics (as well as quantum theory) are continuum theories, such an assumption cannot be physically operationalised. One of the greatest advantages of the continuum assumption is, in fact, not physically motivated, but rather stems from the formal convenience by which techniques of calculus can be developed and applied. However, as has been shown by E. Bishop and D.

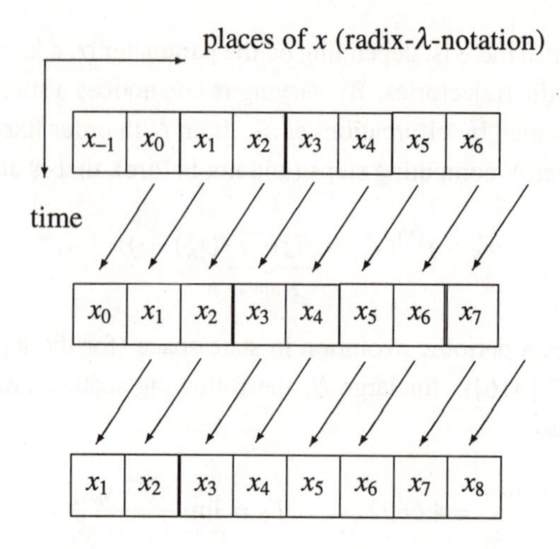

Figure 16.3: Left shift, accompanied by left truncation.

Bridges in *Constructive Analysis* [44], it is possible to develop a "reasonably" comprehensive analysis based on constructive concepts, which are, to a certain extend, related to recursion theory (see also *Computable Analysis* by O. Aberth [2] and *Computability in Analysis and Physics* by M. B. Pour-El and J. I. Richards [373]).

One may, of course, retreat from the postulate of the actual existence of continua by claiming that the arbitrariness residing in the continuum assumption simply reflects the fact that *we do not know the exact initial state of the system.* In this pragmatic "weak chaos" approach, some random element of the continuum is substituted for an undecidable initial value (cf. chapter 13, p. 189).

16.3 Chaos II

Chaos II is generated by the uncomputable evolution of a system with computable initial values. It is the uncomputable equivalent of Chaos IV,2 (cf. below), and it operates with computable initial values and non recursive evolution laws. For an introduction to stochastic Lyapunov exponents, see ref. [12].

16.4 Chaos III

Chaos III is generated by the uncomputable evolution of a system with uncomputable initial values.

Chaos II or chaos III may correspond to the undecidability of single events on the quantum scale. (One must not confuse the undecidability of single events on the quantum scale with the evolution of the linear time-dependent Schrödinger equation.) Both routes are driven by indeterministic evolutions such as Chaitin's diophantine equation for Ω. They require oracle capacity.

	deterministic	indeterministic
computable initial values	**chaos IV** *T-random* *(Chaitin random)* effective computation	**chaos II** *Chaitin random* oracle
uncomputable initial values	**chaos I** *Chaitin random* oracle	**chaos III** *Chaitin random* oracle

Table 16.1: Chaos classes I-IV.

Whereas chaos of class I, II and III supports static or Martin-Löf/Solovay/Chaitin randomness, it requires not only infinite means, but very strong forms of non recursivity. The following chaos class can, for finite times, only support T-randomness. It has the advantage of requiring only recursive (for finite strings merely finite) resources.

16.5 Chaos IV

Chaos IV is generated by the computable evolution of a system with computable initial values. Chaos IV could be characterised by T-randomness, i.e., by computational irreducibility. Chaos IV can then be divided into two subclasses.

Chaos IV,1 is characterised by an unfolding of the T-random initial value X_0. The computational complexity of X_0 unfolds in time, as is the case for the algorithmic complexity of Chaos I. A suitable dynamical evolution function is characterised by positive Lyapunov exponents.

Chaos IV, 2 is characterised by a non random initial value X_0, say a finite number. Its computational complexity resides in the dynamical law governing the system.

In the infinite time limit, chaos IV may be capable of becoming Martin-Löf/Solovay/-Chaitin random. A constructive example for this claim is the program for calculating the Chaitin random real Ω in the limit from below of a computable sequence of rational numbers. However, there is no computable regulator of convergence; i.e., one never knows how far to go to get Ω within a given accuracy. In fact, convergence is slower than any computable function.

There are some speculations that biological evolution resulting in the DNA sequence of nucleic acids as radix-4 symbols may become Chaitin random as well [109]. This scenario is driven by the "creation" of algorithmic information in the course of evolution modelled by an non halting computable process.

In table 16.1 the various aspects of the four classes of chaos are represented schematically. From a purely algorithmic point of view, classes I ("deterministic chaos"), II and III are equivalent, because there is no fundamental difference between a "program" and its "input."

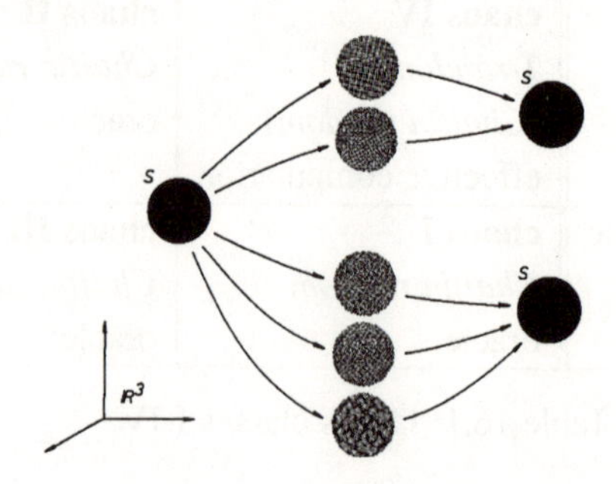

Figure 16.4: Ball doubling in $\mathbb{R}^3$ can be performed using only five (non recursively enumerable, unmeasurable) pieces and distance preserving maps.

16.6 Chaos *via* isometries

The following speculations should point to the possibility of a route to chaos *via* isometric, i.e., distance-preserving, maps such as rotations and translations. This is somewhat different from the usual affine and/or mixing tranformations, which increase the distance of two points on an attractor in time. The characteristics of isometric chaos is *(i)* an isometric evolution function and *(ii)* a non measure-preserving evolution of phase space. If one assumes the *axiom of choice*, then an increase in the separation of two arbitrary points may be possible even for *finite* partitioning of phase space if the dimension of space is ≥ 3. This is due to the Banach-Tarski paradox, reviewed by St. Wagon [451]; cf. Fig. 16.4.

Chapter 17

Quantum chaos

In what follows the term "quantum chaos" will be very widely understood as chaos in the context of quantum mechanics. This chapter has been written from a very subjective point of view. No attempt has been made to review the present discussion of quantum chaos. For more details, the reader is referred to M. Berry [40], B. V. Chirikov, F. M. Izrailev and D. L. Shepelyansky [117], G. M. Zaslavskii [467] and A. J. Lichtenberg and M. A. Lieberman [302], among others.

The discussion of quantum chaos will be divided into two distinct sections: *(i)* Chaotic phenomena may originate from the evolution of the wave function Ψ. The wave function evolves according to the time-dependent linear Schrödinger equation $i\hbar\frac{\partial}{\partial t}\Psi = \hat{H}\Psi$; *(ii)* Processes related to state preparations and measurements may cause an irreversible and sometimes unpredictable "reduction" of the wave function. Not very much attention is paid to the question of how "classical chaos" originates in "quantum chaos."

17.1 Evolution of the wave function

One of the first and most important discoveries which lead to today's quantum formalism was M. Planck's and A. Einstein's derivation of the blackbody radiation from the quantization of the radiation field into "lumps" (later called "quanta") of energy $n\hbar\omega$, $n = 0, 1, 2, \ldots$ per field mode with frequency ω. Action can only be transferred in discrete units $h, 2h, 3h, \ldots$ — This is what is generally meant when it is stated that quantum phase space is discrete [248].

Classically, the partitioning of phase space is in general not invariant with respect to the time flow of a system. One is tempted to speculate that, rather than the classical continuous phase space spanned by position/momentum variables, a space of action variables is an appropriate concept for a "it-from-bit"-reconstruction of quantum mechanics.

It is interesting to note the following remark by A. Einstein [153], p. 163,

> *"There are good reasons to assume that nature cannot be represented by a continuous field. From quantum theory it could be inferred with certainty that a finite system with finite energy can be completely described by a finite number of (quantum) numbers. This seems not in accordance with continuum theory and has to render trials to describe reality with purely algebraic means. However, nobody has any idea how one can find the basis of such a theory."*

17.1.1 Finite automaton model and quantum recurrence

The case of a bounded quantized system with a computable evolution function can be modelled by a finite automaton model or a finite digital computer with finite precision, say m, in a binary (radix 2) representation, corresponding to $M = 2^m$ different discrete states.

This scenario is different from the *busy beaver* scenario which yields a maximal recurrence time of $\Sigma(m) \gg M \gg 2^m$ because it is assumed that the internal storage capacity of a quantized system is bounded by m bits. If we do not restrict the internal storage resources, then the characteristic time scale of the quasi random regime may become "very large," in fact, $\Sigma(M)$ (cf. 8.7, p. 104). The time evolution is also discrete. (In contradistinction, the Schrödinger theory uses continuous space and time parameters as well as continuous expansion coefficients of the eigenstates.) If one cycle time is normalised by $\Delta t = 1$, the maximal time for a single state to reach the initial state is $T_{MAX} = M$, but on the average, $\langle T \rangle = M^{1/2}$. This is the time when (due to the computability of the deterministic evolution) on the average a new period starts.

The above consideration shows that there are essentially two time scales, separated by $\langle T \rangle$, associated with quasi random and periodic evolution: *(i)* For times $t \ll \langle T \rangle$, the complexity of the system evolution can be of the order of the number of cycles (which is identical to t). Hence this starting segment may be random (not absolutely random, but in the sense of "finite random sequence"), but not necessarily so, because aperiodicity is no sufficient criterion for randomness; *(ii)* when the periodic regime is reached, the system can no longer be random.

The above *finite automaton model* for bounded discrete quantum systems [117, 118, 119] suggests that the time evolution of a quantized system can be divided into two distinct regimes: in the *pseudo random* regime the system evolves according to the Schrödinger theory until it reaches a state arbitrary close to its initial configuration; afterwards the system is in a Poincaré-recurrent regime.

(i) Quasi random regime: The quasi random behaviour after a state preparation (during measurement) can be motivated by heuristic arguments as follows: the average time period is estimated by $\langle T \rangle \approx \hbar\varepsilon^{-1}$, where ε is the level spacing. If ΔE defines the energy uncertainty, $t \ll \langle T \rangle$, and $\Delta E t \approx \hbar$. For $\hbar(\Delta E)^{-1} \approx t \ll \langle T \rangle \approx \hbar\varepsilon^{-1}$ and thus $\Delta E \gg \varepsilon$, ΔE exceeds the average energy level spacing, yielding a kind of "quasi continuum" of the energy spectrum associated with quasi randomness.

(ii) Poincaré-recurrent regime: The following theorem has first been discussed by N. S. Krylov [280] and independently by S. Ono [343] and by P. Bocchieri & A. Loinger [47]. For more detailed discussions see, among others, L. S. Schulman [407], B. Chirikov *et al.*, [117, 118, 119], T. Hogg & B. A. Huberman [238], A. Peres [353] and A. Peres & L. S. Schulman [356].

T 17.1 (Quantum recurrence theorem) *Let $\Psi(0)$ be a wave function evolving in time under the Hamiltonian H which has only discrete eigenvalues E_n, $n \in \mathbb{N}$. Then, for each $\delta > 0$ and for each t, there is a $T > t$ such that ($\| \cdot \|$ is the norm)*

$$\| \Psi(T) - \Psi(0) \| < \delta \quad .$$

Classically, Poincaré-recurrence is characterised by probability one for finding the trajectories within an arbitrary neighbourhood $\varepsilon > 0$ of its initial value. I.e., with respect to a probability measure *almost all* trajectories get arbitrary close to their initial values. This is not the case for quantum recurrence, where *all* microstates return to their initial values.

17.1.2 Nonlinear Schrödinger equation

There are some reasons to perceive the linearity of the wave function in the Schrödinger equation as an *artefact* of the bare theory without any interaction, in particular self-interaction. For instance, taking into account the nonlinear self-energy terms $R[\Psi]$, the Schrödinger equation is no longer linear [25],

$$i\hbar\frac{\partial}{\partial t}\Psi = \hat{H}_0\Psi + R[\Psi] \quad .$$

For nonrelativistic quantum mechanics, a nonlinear model for the renormalised theory was indeed proposed by E. Fermi [172] to account for radiation reactions ($\hat{H}_0 = \hat{p}^2/2m + V$)

$$R[\Psi] = -\alpha\,\frac{2\hbar c}{3}\,\Psi\vec{y}\cdot\frac{d^3}{dt^3}\int \Psi^*\vec{x}\Psi\, d^3x \quad .$$

With $K(m, \alpha; \Psi) = -i(\hat{H}_0\Psi + R[\Psi])/\hbar$, the nonlinear Schrödinger equation $\partial/\partial t\Psi = K$ may well be able to produce chaos *via* the Feigenbaum scenario, with the mass m and the coupling strength α as "tuning" parameters. A test of this scenario, however, is rather demanding. In principle one can vary both α and m, for instance $\alpha(Q^2)$ in deep inelastic scattering [Q^2 stands for the energy-momentum transfer at the vertex], or by "sandwiching" a charged particle of mass $m(a)$ between two parallel conducting plates at distance a apart. For the time being, the associated time scales and mass changes (e.g. $\Delta m(1\text{ cm})/m \approx 10^{-12}$) are prohibitively small.

I shall briefly review speculations suggesting that problems to represent the so-called "collapse of the wave function" within the framework of the Schrödinger theory could be resolved by the introduction of a nonlinear Schrödinger equation [350]. The idea is that nonlinear dynamics gives rise to a reduction of the state function, very much like Feigenbaum's scenario reduces a system's motion to points, limit cycles or strange attractors by restricting its effective degrees of freedom. These nonlinearities may originate in radiative corrections to the bare theory [26, 172], as discussed above, or may have different physical origins [350]. The rôle of a macroscopic measuring apparatus is the input of large fluctuations. In the language of complexity theory, this is equivalent to saying that due to the enormous (static or dynamic) complexities of macroscopic devices, they disturb the regular deterministic motion (according to the Schrödinger theory) of quantized systems by introducing an evolution which is effectively random (i.e., white noise).

This argument requires essentially two distinct elements: *(i)* reduction of the state vector due to a reduction of the effective degrees of freedom by attractors in generic *nonlinear* evolutions (Feigenbaum scenario); and *(ii)* undecidability of single events on the quantum scale by the interaction with macroscopic measuring apparata and their associated high complexity measures. A deterministic evolution of both the quantum system and the measuring device is not excluded. The argument states that large fluctuations prevent predictions of single events.

17.2 Single events on the quantum scale

17.2.1 General framework

For convenience, quantized systems with only finite-dimensional (N-dimensional) Hilbert spaces are considered. The basis vectors (pure states) Ψ^n are denumerated by whole numbers $\#(\Psi^n) = n \leq N < \infty$. The symbols $\#(\cdot)$ stand for some code.

Suppose m repeated experiments of the following form: in the ith experiment the system is prepared to be in an identical initial state Ψ_0. Then it evolves a time t_i without any external measurement or state preparation. At t_i it is measured to be in a state Ψ_i. Suppose further the times are ordered such that $t_1 < t_2 < \cdots < t_i < \cdots < t_m$, then one obtains a radix-$N < \infty$ sequence of length $m+1$: $\psi(m) = \#(\Psi_0)\#(\Psi_1)\cdots\#(\Psi_m)$. This series should not be confused with the time flow of the system, represented by $\Psi(t) = \exp(-it\hat{H}/\hbar)\Psi_0 = \sum_{n\leq N<\infty} c_n(t)\Psi_n$.

I shall concentrate on the relationship between the time-ordered actual measurement results in $\psi(m)$ and the time-dependent Schrödinger function $\Psi(t)$ next. Since the positions $\#(\Psi_i)$ of $\psi(m)$ represent pure states, $\langle\Psi(t_i)|\Psi_i\rangle = c_{n_i}(t_i)$. Hence, at a time t_i the system is in a state Ψ_i with probability $|c_{n_i}(t_i)|^2$. As long as $c(t)$ is a nonsingular distribution, the probabilistic interpretation of $|c|^2$ allows for arbitrary Martin-Löf/Solovay/Chaitin random sequences $\psi = \lim_{m\rightarrow\infty}\psi(m)$, even with computable coefficients c. An example, the "quantum coin toss" is discussed in detail next.

17.2.2 The quantum coin toss

From now on, the coding signs are omitted. We next turn our attention to the generation of suitable sequences of pointer readings $\psi(n) = \psi_0\cdots\psi_{n-1}$ from "quantum coin tosses." These can then be subject to statistical and complexity tests, as suggested below. T. Erber and S. Putterman [160] have put forward very similar considerations in the context of the "telegraph signals" generated by the on/off time of a single atom's fluorescence observed by W. Nagourney, J. Sandberg and H. Dehmelt [137].

For any test of quantum mechanical undecidability it is essential to use signals with no (extrinsic) noise from a controllable source of very low extrinsic complexity. To the author's knowledge the optimal realisation of such a source is a laser emitting coherent and linearly polarised light. All emitted quanta from such a source are in an identical state. The polarised laser light is then directed towards a material with anomalous refraction, such as a $CaCO_3$ crystal, which is capable of separating light of different polarisations. Its separation axis should be arranged at ±45° with respect to the direction of polarisation of the incident laser beam. Then each of the two resulting beams, denoted by 0 and 1, respectively, has a polarisation direction ±45° from the original beam polarisation. A detector is in each of the beam passes (see Fig. 17.1). For an ideal anomalous refractor, the probability that a light quantum from the polarised source will be in either one of the two beams is 1/2.

A binary sequence $\psi(n)$ can be generated by the time-ordered observation of subsequent quanta. Whenever the quantum is detected in beam 0 or 1, a corresponding digit 0

Figure 17.1: Measurement configuration for the quantum coin toss.

or 1 is written in the next position of $\psi(n)$, producing $\psi(n+1)$. In this way, n observations generate a sequence $\psi(n)$.

One perception of this process is the amplification of noise from the vacuum fluctuations of the photon field (cf. R. Glauber [202]). If, for any reason, this noise would exhibit *regular non random* characteristics [rendering, for instance, amplitude oscillations $|\psi_t\rangle = \sin(\omega t)\,|0\rangle + \cos(\omega t)\,|1\rangle$ with constant frequency ω], one could detect these regularities and find discrepancies with the postulate of microscopic randomness.

It is suggested [470] that such a sequence is published and suitably distributed (e.g. by electronic mail) by a *bureau of standards.* This sequence could then be taken as a reference for statistical tests, some of which are suggested below, and more generally, as a standard for a generic random sequence. Of course, as has been pointed out by C. Calude [79], there is no guarantee that such an initial sequence (or any other sequence) of finite length cannot be algorithmically compressed substantially; this comes from the fact that, for example, a sequence of a thousand 0's should occur with equal probability as a particular "irregular (i.e., algorithmically incompressible) sequence" containing a thousand symbols.

Compare ψ to any pseudo random sequence φ, generated by a finite deterministic automaton. Whereas φ could be applicable to a great variety of purposes such as numerical integration or optimisation of database retrieval, it will inevitably fail specific statistical tests. Take for example the statistical test corresponding to the generating algorithm of φ itself — the law which is encoded by this algorithm is *per definitionem* capable of generating ("predicting") all digits of φ. Thus, at least with respect to its own generation law, φ is provable non random.

The postulate of microphysical indeterminism and randomness asserts that there is no such "generating" law and hence no statistical test to "disprove" the randomness property of ψ. Indeed, with this postulate, ψ should pass *all* statistical tests with probability one, at least not for infinite sequences. Thus ψ can serve as generic source for a random bit sequence.

In what follows several statistical and algorithmic tests are suggested which could be applied to $\psi(n)$.

(i) Frequency counting: for $\psi(n)$ to pass this test it has to be proven that any arbitrary sequence of m digits occurs in $\psi(x)$ with a limiting frequency 2^{-m}. In order to obtain a reasonable confidence level (see D. Knuth [268] for details), m has to be smaller than approximately $n-7$. An infinite sequence passing this test for arbitrary m is called *Bernoulli sequence*. As has already been mentioned, this criterion is rather weak. It is satisfied by the enumeration of the binary words in lexicographic order [115], and, within finite accuracy, by the decimal expansion of π [338, 19]. Actually, in the above experimental setup, the statistics of a 1-digit string ($m = 1$) should be used for calibration of a suitable angle, which is defined by the requirement that 0 and 1 should occur in $\psi(n)$ with frequency 1/2.

(ii) Algorithmic compressibility: $\psi(n)$ could be the input of various compression algorithms (e.g., the Huffman algorithm), which should produce a (compressed) string of length $H_c(n)$ with $H(\psi(n)) \leq H_c(n) \leq n$. On the average, $H_c(n)$ should increase as n increases, i.e., $\langle \Delta H_c(n)/\Delta n \rangle = 1$. Every compression algorithm is a kind of "code breaking device" based upon a hypothesis on "laws" governing sequences. Some of them are used for commercial applications and are readily available.

(iii) Spectral test: This is a critical test at least for linear congruential sequences. For a detailed discussion see D.Knuth [268]. The idea is to investigate the "granular" structure of $\psi(n)$ in D-dimensional space in the following way. Split $\psi(n)$ into $N \equiv n/k$ subsequent partial sequences $\psi(n, i)$ of length k. Generate N binary numbers $0 \leq x_i < 1$ by $x_i \equiv \psi(n, i)/2^k$. For a D-dimensional analysis, arrange subsequent x_i's into $M \equiv N/D$ D-touples X_j. The X_j's could be perceived as points in $\mathbb{R}^D$. Consider further all families of $(D-1)$-dimensional parallel hyperplanes with points X_j. If $1/\nu(D)$ denotes the maximal distance of these hyperplanes, $\nu(D)$ is called the D-dimensional *''accuracy''* of $\psi(n)$. $\nu(D)$ should on the average be independent of the dimension, i.e., $\langle \Delta\nu(D)/\Delta D \rangle = 0$. For statistical reasons, one cannot achieve a D-dimensional accuracy of more than about $2^{k/D}$ and $1/M^D$. Thus the spectral test is reliable only for $\nu(D) < 2^{k/D}$ and sequence length $n > kD(\nu(D))^D$.

(iv) High-dimensional integration: Assume an analytically computable D-dimensional integral $F(D) \equiv \int_0^1 \cdots \int_0^1 dx_1 \cdots dx_D f(x_1, \ldots, x_D)$. Consider again a representation of $\psi(n)$ by $M = n/kD$ points X_j in the D-dimensional unit interval. Define $F'(D) \equiv (1/M) \sum_j f(X_j)$. Then for arbitrary test functions f and with probability 1, the discrepancy $|F(D)-F'(D)| \propto M^{-1/2}$ only depends on the number of points and not on the dimension. In order to obtain accuracies of the order of $M^{-1/2}$ with the Simpson method, one needs at least $M^{D/8}$ points to obtain the same order of discrepancy. There the number of points depends on the dimension.

The proposed tests are not independent. Certain compression algorithms use tables of repeating sequences and are thus connected to frequency counting methods. The spectral test analyses the distribution of points generated from sequences in a unit interval of highdimensional space. It is thus a criterion for the quality of approximation in numerical integration.

There are other fairly strong statistical tests such as the law of the iterated logarithm

[171], but many of them turn out to be not practical for their low confidence levels in applications.

17.2.3 The role of random single events for the "peaceful coexistence" between quantum mechanics and relativity theory

Presently there seems to be an overwhelming evidence for what A. Shimony calls a "peaceful coexistence" of quantum mechanics and relativity theory [412, 414], despite the fact that they evolved from quite different contexts. Although there are difficulties originating in the "collapse of the Schrödinger wave function" [6], and even with sophisticated setups using delayed choice [457], the Aharonov-Bohm effect [5], variations of boundary conditions [213], "haunted" measurements [214] and photon cloning [232, 202], nonrelativistic quantum mechanics remains relativistically causal.

This "peaceful coexistence" is amazing when viewed in the context of *inseparability* or *non-locality* for entangled states [215]: One important feature of quantum inseparability is the "stronger-than-classical" correlation (cf. A. Peres [352]) between events from an entangled state; i.e., the correlations between such states are *higher* than can be accounted for by local classical models representable by Bell-type inequalities [121].

Attempts to construct local classical models which feature these "stronger-than-classical" quantum correlations with abstract set theoretical concepts have been made in the context of *nonpreservation of the probability measure* and the *Banach-Tarski "paradox"*, see I. Pitowsky [363], and in the context of the application of *random ultrafilters* as "hidden variables", see W. Boos [49].

In accordance with the conjecture of "peaceful coexistence", it is commonly accepted that these "stronger-than-classical" quantum correlations *cannot* give rise to any faster-than-light signalling and thus cannot violate relativistic causality [412, 198, 146]. This is ultimately guaranteed by the assumption of unpredictability (or, more strongly: of randomness) of single events, resulting in "uncontrollable non-localities" [412]: inseparability establishes itself only *after* the recollection of entangled events, which, perceived separately, occur randomly. By this bailout, nonrelativistic quantum mechanics violates the locality assumption without violating relativistic causality. To put it pointedly [413]: *quantum theory predicts* event dependence *but* parameter independence *for entangled states.*

Obedience to relativistic causality for nonrelativistic quantum mechanics is amazing enough by itself, and even more so if one recalls that consistency remains only conjectural if "manifestly covariant" terms (such as tensors and spinors) are the entities in which relativistic quantum field theories are expressed [415]. Let me briefly review the folklore belief that such a procedure ensures relativistic causality.

The usual implementation of what is called "local causality" in relativistic quantum field theory requires independence of the field amplitudes at spatially separated points. Local causality is then guaranteed by a proper connection between spin and statistics. For instance, in the case of a massive scalar field, the commutator is given by the Pauli-Jordan function and vanishes for spacelike separations $t^2 - \vec{x}^2 < 0$, i.e., outside the light-cone. [However, this renders causal Green's functions with nonvanishing contributions for spacelike separated points, i.e., $\Delta_c(x) = \langle 0|T\phi(0)\phi(x)|0\rangle \propto \theta(-x^2)(m/|x|)K_1(m|x|)+\cdots$. For the massless case and for small $x^2 = t^2 - \vec{r}^2 \neq 0$ (close to the light-cone), Δ_c can be expanded, yielding $\Delta_c(x) \propto x^2$.] This presupposes the invariance of the speed of light for *arbitrary operational configurations*, in particular in propagation processes in which light quanta are exchanged. Such processes have to be described in the framework of quantum field theory itself. In other words, what is treated as a *prerequisite* here has to be actually an *outcome* of relativistic quantum field theory. But in the spirit of quantum field theory it is not unreasonable to consider the velocity of light inserted in the "bare" theory as a parameter which becomes renormalised *en route* to the full model, very much like mass or charge. This is exactly the theme of an ongoing debate whether for instance quantum electrodynamics may give rise to acausal effects in the regime of finite space-time processes [415, 230, 448], or for negative vacuum energy densities such as in cavity quantum electrodynamics [403, 24], in the charged vacuum state [321] and with wormholes [329]. Possible violations of Lorentz invariance have also been discussed for very short distances [341, 468].

More generally, one could speculate about a breakdown of Lorentz invariance without causality violation in relativistic field theory. As for nonrelativistic quantum mechanics, this yields a scenario of violation of locality by uncontrollable events, associated with the preservation of relativistic causality, specifying a principle of "peaceful coexistence" of relativistic quantum field theory and relativity theory.

Chapter 18
Algorithmic entropy

18.1 Information theory entropy

The information theoretic approach to entropy reviewed in this section has been pioneered, among others, by E. T. Jaynes [246], A. Hobson [236], A. Katz [260] and A. Rieckers and H. Stumpf [385].

18.1.1 Probability

All physical knowledge has to be (re-)constructed from outcomes of a *finite* number of elementary experiments, i.e., from finite sequences of `TRUE-FALSE` alternatives. One of the most important problems in statistics is the problem of the *"best"* representation of an incomplete state of knowledge about a physical system.

Intuitively speaking, the *probability* of an event (i.e., a particular outcome) in an experiment or trial can be interpreted either *(i)* subjectively, as a *rational measure of belief*, expressed before or, if the outcome is not yet known, after the experiment (F. P. Ramsey [380]); or *(ii)* empirically, as *frequency* of occurrence (R. von Mises [324], K. Popper [367]); or *(iii)* by an inductive logic approach (J. M. Keynes [263], R. Carnap [81]). For an interesting discussion of these interpretations with respect to quantum mechanics, see I. Pitowsky [364], p. 182.

Let e_i be an event in a manual $\mathfrak{M}$ and let h stand for some rational hypothesis, then $P(e_i, h)$ is the probability of obtaining the event e_i given the hypothesis h. After an infinite number of experiments, the *inner probability* $P(e_i, h_\infty) = P(s_i) = p_i$ would be the maximum prior knowledge about a system which is otherwise unpredictable.

According to A. N. Kolmogorov [270], one can axiomatise probability theory by requiring that for all $e_i \subset \mathfrak{M}$ and a certain event $\mathbf{1}$,

$$\begin{aligned} 0 \leq P(e_i) &\leq 1 \\ P(\mathbf{1}) &= 1 \\ P(e_1, e_2, \ldots) &= P(e_1) + P(e_2) + \cdots \\ &\qquad \text{for } e_1 \cap e_2 \cap \cdots = \emptyset \quad . \end{aligned}$$

Since infinite experimental series are impossible, one has to rely upon data and probabilities from finite observations. This results in an arbitrariness of the definition of the probability distribution $P(e_i, h_{N<\infty})$, corresponding to choices of different hypothesis. The arbitrariness has to be eliminated by additional constraints on P. These constraints, in particular Jaynes' principle, use the notion of *"Shannon information"* and *"information theory entropy,"* which will be introduced next.

18.1.2 Shannon information and entropy

Assume a *source alphabet* with q symbols, $s_1, s_2, \ldots, s_q$ which occur with probability $P(s_i, \infty) = p_i$. The infinity sign "∞" stands for the assumption that one knows these probabilities *for sure*, i.e., after one has performed "an infinite number of" experiments. The case of *finite* experimental knowledge shall be treated below

One may ask, *"is it possible to quantify information by a function $I(p_i)$ which measures the amount of information in the occurrence of an event with probability p_i?"* It is reasonable to require that I satisfies three properties:

(i) $I(p_i) \geq 0$;

(ii) for independent events, the *additivity* property $I(p_i p_j) = I(p_i) + I(p_j)$ holds;

(iii) $I(p_i)$ is a continuous function in p_i.

From property *(ii)* follows that for n independent events with equal probability p,

$$I(p^n) = nI(p) \quad ,$$

and after redefining $p = y^{m/n}$,

$$I(y^{n/m}) = \frac{n}{m} I(y) \quad .$$

At least for rational numbers α, $I(p^\alpha)$ behaves, at least up to multiplicative constant k, like the logarithm. [From property *(iii)* follows that α may also be irrational.] Therefore one obtains

$$I(p) = C \log_a(p)$$

for some base a. The choice $C = -k$, $k > 0$ is consistent with requirement *(i)*, such that

$$I(p) = -k' \log p = k' \log \frac{1}{p} \quad , \tag{18.1}$$

where k' depends on the base of the logarithm. (Note that $\log_a x = \log_a b \log_b x$ and $\log_a b = 1/\log_b a$.) k (k') is an arbitrary positive constant which is equal to $1/\log 2$ if the algorithmic entropy is measured in bits and equal to Boltzmann's constant for physical units. For a definition of the information theory entropy, k is set to unity.

D 18.1 (Shannon information) *The* Shannon information *of a symbol which occurs with probability p is defined by*

$$I(p) = -\log p \quad . \tag{18.2}$$

A similar argument shows that (18.1) is a unique choice: assume there is another function $g(p)$ satisfying *(i)—(iii)*, then

$$g(p^n) - k \log \frac{1}{p^n} = n \left[g(p) - k \log \frac{1}{p} \right] .$$

For $p_0 \neq 0, 1$, choose $k = g(p_0)/\log(1/p_0)$. Then, with $z = p_0^n$, follows that $g(z) = k \log(1/z)$, which is true for all p's due to the continuity assumption *(iii)*.

Recall that p_i was defined as the probability for the occurrence of the source code symbol s_i and therefore the probability for obtaining the information $I(p_i)$. From this follows that *on the average*, i.e., over the whole alphabet of symbols $S = \{s_i\}$, one obtains an *average information per symbol* of

$$h = -k \sum_{i=1}^{q} p_i \log p_i \quad . \tag{18.3}$$

h is called the *"information theory entropy."*

The following theorem is stated without proof. For details, see R. W. Hamming ([223], page 119).

T 18.2 *For instantaneous (prefix) encoding, the* information theory entropy *yields a bound from below for the average codeword length by*

$$h \leq \sum_{i=1}^{q} p_i l_i \quad , \tag{18.4}$$

where $l_i \in \mathbb{R}^+$ is the codeword of the i'th source symbol s_i.

18.1.3 Jaynes' principle

As mentioned before, after a *finite* number of experiments one obtains only incomplete knowledge of the inner probability. This case will be treated next. The rational measure of "information" and "uncertainty" will now be used in a reformulation expressing the "relative information content" and the "relative uncertainty" of a measurement series. Assume we have performed M experiments and ask ourselves, *"what is the information gain after $N \geq M$ experiments?"* Motivated by equation (18.2), the relative information content can be modelled by $I(M, N) = I(P(s_i, N)) - I(P(s_i, M))$, and a relative Shannon information can be defined as follows.

D 18.3 (Relative Shannon information, uncertainty) *Let*

$$I(N; M) = I(\ldots, P(e_i, h_{N \geq M}), \ldots; \ldots, P(e_i, h_M), \ldots)$$

denote the additional information gain after $N - M \geq 0$ experiments have been performed. Then the relative Shannon information *is given by*

$$I(N; M) = k \sum_i P(e_i, h_N) \log[P(e_i, h_N)/P(e_i, h_M)] \quad , \tag{18.5}$$

where k is an arbitrary positive constant which is equal to $1/\log 2$ if the algorithmic entropy is measured in bits and equal to Boltzmann's constant for physical units. For a definition of the information theory entropy, $k = 1$. h_M denotes the statistical hypothesis after M trials.

The uncertainty *is the missing information relative to the maximum information from ∞ trials:*

$$h(N; M) = I(\infty; M) - I(N, M) \quad . \tag{18.6}$$

If one has not yet performed any experiments and $M = 0$ and/or $N = 0$ but one knows the possible outcomes, then all probabilities $P(e_i, h_0)$ are defined to be equal, i.e., 1/(the number of possible outcomes). This "principle of insufficient reason" guarantees that none of the outcomes is preferred over the other, a preference which would not be justified by the (missing) data. Note that for sufficiently large experimental series N,

$$\begin{aligned} h(N;M) &= I(\infty;M) - I(N,M) \qquad (18.7)\\ &= k\left\{\sum_i [P(i,\infty)(\log P(i,\infty) - \log P(i,M)) - \right.\\ &\quad -P(i,N)(\log P(i,N) - P(i,M))]\} \qquad (18.8)\\ &\approx -k\left\{\sum_i [P(i,N)\log P(i,N) - \right.\\ &\quad -P(i,\infty)\log P(i,\infty)]\} \qquad (18.9)\\ &\approx -k\sum_i P(i,N)\log P(i,N) + O(1), \qquad (18.10)\\ &\approx h(N) \quad , \qquad (18.11) \end{aligned}$$

where $P(i,N) = P(e_i, h_N)$. The uncertainty $h(N)$ is also called *"information theory entropy."* Its relation to thermodynamic entropy has been discussed by E. T. Jaynes [246].

As has been proved in chapter 4, p. 232, the functional form (18.5) of I is uniquely determined by certain "reasonable" requirements, such as continuity, $I(N;N) = 0$ *et cetera.* Reviews of these requirements as well as a proof of the uniqueness of I can be found in R. W. Hamming ([223], p. 103), A. Hobson ([236], p. 35) and A. Katz ([260], p. 14). For a discussion of E. T. Jaynes' maximum entropy principle and its relation to other concepts, such as R. A. Fisher's maximum likelyhood principle, see an article by M. Li and P. M. B. Vitányi [301].

Since infinite experimental series are not realisable, one has to guess the hypothesis h_N with only a finite number N of experiments performed. Such guesses do not yield unique choices. Therefore, one has to assume "reasonable" requirements in order to specify the choice of the hypothesis h_N further.

One such "reasonable" side condition is the requirement that the probability distribution should *not contain any additional "information"* which is not suggested by the experimental data and therefore should *maximise the "amount of uncertainty."* The above considerations can be summarised in

C 18.4 (Jaynes' principle) *Assume the experimental outcomes e_i and functions $f_j(e_i) = f_j(i)$. Given N additional data from experiments. The hypothesis h_N has to be adjusted such that the probabilities $P(e_i, N)$ obey*

$$\sum_i P(i,N) = 1 \quad , \qquad (18.12)$$

$$\langle f_j(N)\rangle = \sum_i P(i,N) f_j(i) \quad , \qquad (18.13)$$

and at the same time maximize *the uncertainty / the information theory entropy (18.11), i.e.,*

$$h(N) = -k\sum_i P(i,N)\log P(i,N) \quad , \qquad (18.14)$$

minimising the Shannon information $I(N, 0)$.

Jayne's principle is similar to an approach by R. A. Fisher [179, 150], who added a second criterion of *efficiency* by requiring that the *variance* of the estimating statistic (at least for large sample spaces) should not exceed that of any other consistent statistic estimating the same parameter. Roughly speaking, whereas with probability e_i is treated as variable and the hypothesis h is constant, with likelihood the hypothesis h is the variable and the events are constant. More precisely, the *likelihood* $L(h, e_i)$ of the hypothesis h given data e_i is defined by

$$L(h, e_i) = kP(e_i, h) \quad ,$$

the constant of proportionality $k = O(1)$ being an arbitrary "weight factor." Fisher's criterion is that the *support function* S defined as the logarithm of the likelihood function L. For a binomial distribution, for which $L \propto P(e_0, h_N)^K P(e_1, h_N)^{N-K}$, the support thus is $S(P) = K \log P(e_0, h_N) + (N - K) \log P(e_1, h_N) + O(1)$. The best-supported probability distribution P, i.e., the distribution P for which $S(P)$ is a maximum, is then taken for the estimate.

Summhammer's principle: Whereas Janes' principle fixes the arbitrariness in the guessing hypothesis, there is yet another undesirable feature of certain statistics. It may happen that *the variance, i.e., the confidence uncertainty, increases with increasing data sample.*

Consider the following example [427]: two physicists perform coin tosses (for quantum ones, cf. p. 226) and have collected a total of $N = 1000$ events, $K_0 = 800$ for "head," coded by #(head) $= 0$ and $K_1 = 200$ for "tail," coded by #(tail) $= 1$. They hypothesise that the distribution is binomial and

$$\begin{aligned} P(0, 1000) &= K_0/N = 0.8 \quad , \\ \Delta P(1000) &= \{P(0, K_0)P(1, K_0)/N\}^{1/2} \approx 0.01265 \quad . \end{aligned} \tag{18.15}$$

One experimenter goes home, the other one performs 20 additional coin tosses, only four of these are "heads." Hence,

$$P(0, 1020) \approx 0.7882, \quad \Delta P(1020) \approx 0.01279 \quad .$$

I.e., for these particular observations, $\Delta P(1020) > \Delta P(1000)$, and one has *lost* confidence in determining the inner probability of the coin when performing additional experiments. The lazy experimenter got a better "confidence!"

This somewhat pathological feature of the binomial distribution can be overcome by a transformation [424, 425, 426, 427] $T : P \rightarrow \chi$. χ is called *"phase"* for reasons which will become clearer below. We require that χ only depends on the *number of experiments* N, and *not on the outcomes* K_0, K_1, i.e.,

$$\chi = \chi(N) \quad . \tag{18.16}$$

In terms of $P = P(0, N)$, χ's variance $\Delta\chi$ can be written as

$$\Delta\chi = \left| \frac{d\chi(N)}{dP} \right| \Delta P \quad . \tag{18.17}$$

Substitution of $[P(1-P)/N]^{1/2}$ [equation (18.15)] yields

$$N^{1/2}\Delta\chi = \left|\frac{d\chi(N)}{dP}\right| [P(1-P)]^{1/2} \quad . \tag{18.18}$$

Assume the law of large numbers. Then, for large N, $P = P(0,N)$ varies "very slowly" compared to N; furthermore by equation (18.16), χ should solely depend on the sample size N. Therefore both sides of equation (18.18) can be approximated by a constant c_1. Upon integration and one obtains

$$\chi = c_2 + c_1 \arcsin(2P-1) \quad , \tag{18.19}$$

$$\Delta\chi = c_1/N^{1/2} \quad , \tag{18.20}$$

where c_1, c_2 are two constants. By requiring $\chi(P=0)=0$ and $\chi(P=1)=1$, one obtains $c_1 = 1/\pi$ and $c_2 = 1/2$, and

$$P = \sin^2(\pi\chi/2), \quad \chi = \frac{2}{\pi}\arcsin(P^{1/2}) \quad . \tag{18.21}$$

This is the *arcsine distribution*; for details see, e.g., W. Feller [171], p. 79. For complex P and real χ a similar argument yields

$$P = \frac{1}{2}[1-\exp(-i\pi\chi)] = i|\sin(\pi\chi/2)|\exp(-i\pi\chi/2) \quad . \tag{18.22}$$

Note that the property $\chi(N)$ is unique modulo a linear transformation [424, 425, 426, 427] $T' : \chi \rightarrow a\chi + b$.

One generalisation of J. Summhammer's approach is the requirement that, besides the variance, *all moments* of a distribution should depend only on the number of experiments, i.e., the size of the sample space.

C 18.5 (Generalised Summhammer principle) *Let $P(e_i, h_N)$ be a probability distribution. A new distribution χ is obtained by requiring that* all moments about the mean $\langle P(h_N)\rangle = \sum_i P(e_i, h_N)$ *are independent of P and only dependent on the size of the sample space N, i.e., for $r > 1$,*

$$\left[\sum_i (P(e_i,h_N) - \langle P(h_N)\rangle)^r\right] /N = m_r(N) \quad . \tag{18.23}$$

By this procedure it is assured *in advance* that (for sufficiently large sample sizes) the *accuracy* of the estimate of the probability distribution χ as well as other characteristics such as *skewness* or *kurtosis* [422] only depend on the number of experiments and not on the particular measurement series.

18.1.4 Information theory based scenarios for entropy increase

An increase of information theory entropy may be due to a lack of information about the initial value of the system. In this scenario, throughout evolution, the state of the system becomes increasingly sensitive with respect to variations of initial values; a lack of knowledge of these drives the system description towards higher values of information theory entropy; even if the system evolution is reversible. "Deterministic chaos" is such a scenario. There exist nice Cellular Automaton models demonstrating this feature [186].

18.2 Algorithmic probability

If the computer U is perceived as an information source, then the algorithmic probability $P(s)$ can be identified with the probability of the occurrence of an object s from that source. As has been shown by G. Chaitin [100, 101], there is a connection between the algorithmic information $H(s)$ and its algorithmic probability $P(s)$ of an object such as a symbol or a sequence of symbols s [a notable exception are infinite computations; cf. equation (7.36), p. 97]:

$$H(s) = -\log_2 P(s) + O(1) \quad ;$$

cf. equation (7.32), p. 95. The halting probability $P(s)$ has been defined in (7.8), p. 92 by the probability that a universal computer U will calculate an object s: if a prefix-free program p of length $|p|$ is assigned a measure $2^{-|p|}$, then $P(s) = \sum_{U(p)=s} 2^{-|p|}$. R. J. Solomonoff [419] also suggested an algorithmic probability measure which is induced by the output strings of a universal Turing machine. Because of the absence of self-delimiting programs, i.e., prefix coding, two problems occurred: *(i)* the sum for $P(s)$ diverged in the absence of the Kraft inequality (7.1), as well as *(ii)* the missing subadditivity property (7.27). D. G. Willis [461] modified this definition by considering only finite-time processes which are terminated by a *deus ex machina.* L. A. Levin [475, 298] proposed a semicomputable probability semi-measure. For each one of these alternative definitions of $P(s)$, a relation of the form "$H(s) = -\log P(s) + O(1)$" is satisfied. For detailed discussions, see S. K. Leung-Yan-Cheong & T. M. Cover [297] and R. J. Solomonoff [420].

At this point technical obstacles appear.

(I) One did not require the most efficient encoding, for which the Kraft sum in (7.1) holds with equality. Moreover, for universal machines, not all allowed (instantaneous) codes halt. Since $\Omega = \sum_s P(s)$ has been defined to be the sum of the halting probabilities $P(s)$, for prefix code, $P(\emptyset, 1, 2, 3, \ldots) = \Omega < 1$, and there is no unit element $\mathbf{1}$ such that $P(\mathbf{1}) = 1$. I.e., one cannot directly associate an interpretation as probability measure to P, as would be necessary for identifying P with the probability in the Shannon information.

There is yet another difficulty, since there exist programs which halt but output nothing, i.e., the empty list $\emptyset$. This problem can be circumvented by identifying no event from the information source with some additional or existing source symbol $\#(\emptyset)$.

A probability measure interpretation for P could be restored at least in two ways.

(i) include a not-halting element "NH," then define $P(NH) = 1 - \Omega$ and $\mathbf{1} = \{NH, \emptyset, 1, 2, \ldots\}$;

(ii) divide $P(i)$ by Ω, such that $P'(i) = P(i)/\Omega$ and thus $\mathbf{1} = \{\emptyset, 1, 2, 3, \ldots\}$.

At first glance, *(ii)* looks more appealing for physical applications, since a program which does not halt and outputs nothing corresponds to no event. It seems strange to ascribe a probability to non-occurrence, as suggested by *(i)*. A second thought, however, might support *(i)*, since this view reflects a physical system's inability to perform a given task, say, to produce a specific object i, and/or its redundancy in producing one and the same object by one or more processes.

As is often done in information theory (cf. R. W. Hamming [223], p. 119), one may assume *(ii)* and normalise $P(s)$ by the Kraft sum, i.e.,

$$P'(i) = \frac{P(i)}{\Omega} \quad . \tag{18.24}$$

Remarks:

(i) The important identity "$H = -\log P + O(1)$," relating the algorithmic information content H of a single object to the probability of its occurrence is *not* changed by transformation (18.24), since the additional constant $-\log\Omega$ can be absorbed into $O(1)$, i.e., $H(x) = -\log_2(\Omega P'(x)) + O(1) = -\log_2 P'(x) - \log_2 \Omega + O(1) = -\log_2 P'(x) + O(1)$.

(ii) Unlike for instantaneous encoding of source information which contribute *with certainty* to the Kraft sum, only the halting programs and therefore *not all* allowed codes contribute to the halting probabilities $P(s)$. Substitution (18.24) therefore does not only normalise an ineffective encoding, but also normalises the capacity of the computer (the physical system) to produce output (states).

(iii) $\sum_{i=\emptyset,1,2,\dots} P(i) = 1$ can no longer be calculated in the limit from below such as Ω, but may oscillate [114] if it is computed by Chaitin's techniques; cf. equation (14.2), p. 196.

(II) The algorithmic information $H(x(n))$ of "almost all" sequences $x(n)$ grows "maximally" in the sense that [cf. equation (7.30), p. 95]

$$\exists m \forall k \exists n \geq k [H(x(n)) > n + H(n) - m] \quad .$$

According to G. Chaitin [109], p. 45, this statement is due to R. M. Solovay; for a proof, see M. van Lambalgen [284], p. 144. I.e., with probability one, the algorithmically defined entropy grows (by an additional logarithmic term) *faster* than the number of (experimental) bits collected. This problem can be overcome by "normalising" the entropy by dividing it through the number of experiments; see below.

(III) As has been pointed out already in section 7.2.2, p. 92, a redefinition of P makes it non subadditive.

18.3 Symbolic dynamics and metric entropy

The following nomenclature is taken mainly from V. M. Alekseev & M. V. Yakobson [9], K. Petersen, *Ergodic Theory* [361], M. van Lambalgen [285, 284] and A. I. Khinchin [264]; see also chapter 4, p. 45.

D 18.6 (Dynamical system) *Let* (X, f) *denote a* dynamical system. *X is a manifold which can be interpreted as a* generalised phase space *(with points $x \in X$) and f is a measurable transformation on X, representing a* discrete evolution *of the system.*

Consider some measurable regions $E_i \subset X$ of X, which, intuitively speaking, contain all the regions which can be resolved by measurements. The higher the measurement resolution, the smaller the diameter of the regions and the more such regions will be necessary to cover all of X. $X = \bigcup_{i=1}^{q} E_i$ defines a *covering* $\xi = \{E_1, \dots, E_q\}$ of X. The *constituents* E_i need not be disjoint, i.e. $E_i \cap E_j \neq \emptyset$. If the covering is disjoint, we call it a *partition*.

D 18.7 (Evolution, orbit) *The* map $f : X \to X$ *represents a* discrete time evolution. *Let $t = 0$ be some time at which the system is in some $x_0 \in X$, and let "$f^{(n)}$" denote the n-fold application of f. If f is invertible, then f generates a* cascade *or* orbit *x of x_0 by*

$$x = \dots f^{(-2)}(x_0)f^{(-1)}(x_0)x_0f(x_0)f^{(2)}(x_0)\dots = \dots x_{-2}x_{-1}x_0x_1x_2\dots \in X^{\mathbb{Z}} \quad . \tag{18.25}$$

The orbit of x represents the combined history and future (with respect to t=0) of the system.

We next turn to the *source coding* of (X, f).

D 18.8 (Source coding) *To every measurement region $E_i \in \xi$ associate a* pointer reading *s^i. The set of pointer readings $S = \{s^1, \dots s^q\}$ constitutes a source alphabet with q symbols. Let S be represented by numbers, encoded by the numerals* $0, 1, 2, \dots, \#(s^q)$ *of basis q (radix-q). (For simplicity, we shall restrict ourselves to $q = 2$ below.)*

Let

$$\Psi : X \to q^{\omega} \tag{18.26}$$

denote a map from the initial value $x_0 \in X$ into the sequence of pointer readings, *representing a* sequence of experimental data, *where the j'th position Ψ_j is given by*

$$\Psi_j(x_0) = s^i \text{ if } f^{(j-1)}(x_0) = x_{j-1} \in E_i \quad . \tag{18.27}$$

Define the initial sequence *of $\Psi(x_0)$ by*

$$\Psi(x_0, n) = s_0 \dots s_{n-1} \quad . \tag{18.28}$$

Remarks:

(i) Superscripts are used to denote the *i*'the symbol s^i, whereas subscripts are used to denote the place of the symbol in the sequence of pointer readings Ψ. For example, if the system is at $t = 0$ in $x_0 \in E_2$, at $t = 1$ in $x_1 \in E_4$, at $t = 2$ in $x_2 \in E_1$, at $t = 3$ in $x_3 \in E_4$, ..., then $\Psi(x_0) = \dots s^2s^4s^1s^4 \dots$.

(ii) f induces a *left shift* T of one time unit on q^{ω}, such that

$$T\Psi = \Psi' = \dots s'_{-2}s'_{-1}s'_0s'_1s'_2 \dots \quad , \tag{18.29}$$

where $s'_i = s_{i+1}$ $(-\infty \leq i \leq +\infty)$. T symbolises the time evolution of the system. q^{ω}, the set of all infinite radix-q sequences, is closed under the action of T, i.e., $Tq^{\omega} = q^{\omega}$.

(iii) Ψ can be perceived as a *word* composed from a finite sequence of characters of an alphabet $s_i \in S$. Define $\xi^n = \{E_{s_0\cdots s_{n-1}} \mid E_{s_0\cdots s_{n-1}} = E_{s_0} \cap f^{-1}E_{s_1} \cap \cdots \cap f^{(-n+1)}E_{s_{n-1}} = \bigcap_{i=1}^{n} f^{(-i+1)}E_{s_{i-1}}\}$. The word is *admissible*, if it corresponds to at least one $x_0 \in \xi^n$. In this case there exists a Ψ, which can be perceived as a possible sequence of pointer readings, or a *symbolic* cascade, representing the "history" of the system from an initial value x_0.

(iv) From now on, we shall disregard the past history of the system, and consider only future events symbolised by $\Psi(x_0, n)$, $n \in \mathbb{N}$.

If (X, f) is equipped with a *probability* $P(\Psi)$, it induces a measure μ on ξ^n by

$$\mu(E_{s_0\dots s_{n-1}}) = P(s_0 \dots s_{n-1}) \quad . \tag{18.30}$$

A measure μ is *ergodic*, if

$$f^{-1}E = E \Rightarrow \mu(E) = 1 \text{ or } 0 \quad . \tag{18.31}$$

Furthermore, only stationary measures with

$$\mu T^{-1}A = \mu A \tag{18.32}$$

will be considered.

In the following we shall use a *binary* source alphabet (i.e., q=2) with the two symbols 0 and 1.

D 18.9 (Metric entropy) *The* metric entropy $\bar{h}$ *is defined by*

$$\bar{h} = \lim_{n\rightarrow\infty} -\frac{1}{n} \sum_{w\in 2^n} \mu(w) \log_2 \mu(w) \quad . \tag{18.33}$$

A similar quantity, called *thermodynamic depth*, has been introduced by S. Lloyd & H. Pagels [348].

Recall that, if p_i is the probability of occurrence of symbol $s_i \in S$, in (18.3), page 233, $h = -\sum_{i=1}^{q} p_i \log p_i + O(1)$ has been interpreted as the *average amount of information gain* per symbol. Thus, interpreted in terms of the Shannon information, the metric entropy is the *average amount of information gained per experiment* if one performs infinitely many experiments. A positive value of $\bar{h}$ indicates that each experiment contributes a "substantial" amount of information. In this sense, $\bar{h}$ reflects a system's unpredictability and randomness.

T 18.10 ((van Lambalgen [284, 285])) *Let μ be an ergodic computable measure on 2^ω with metric entropy $\bar{h}$. Then for μ-almost all $x_0 \in 2^\omega$, the* normalised algorithmic information of a single trajectory *is identical to the* metric entropy

$$\bar{h} = K(x_0) = \lim_{n\rightarrow\infty} [H(\Psi(x_0, n))/n] \quad . \tag{18.34}$$

18.4 Metric entropy and Lyapunov exponents

In what follows I state some theorems relating Lyapunov exponents defined in (16.1) on page 216 to entropy measures (see also L.-S. Young [465]).

Let C^n stands for n times continuously differentiable; let diffeomorphisms be differentiable functions with a differentiable inverse.

T 18.11 ((Pesin [359])) *For C^2-diffeomorphisms f preserving an ergodic measure μ which is equivalent to Lebesgue, the metric entropy measure equals the sum over positive Lyapunov exponents,*

$$\bar{h} = \sum_{\lambda_i^+ > 0} \lambda_i^+ . \tag{18.35}$$

T 18.12 ((Ruelle [399])) *For C^1-diffeomorphisms f preserving an ergodic measure μ the metric entropy measure is smaller than or equal to the sum over positive Lyapunov exponents,*

$$\bar{h} \leq \sum_{\lambda_i^+ > 0} \lambda_i^+ . \tag{18.36}$$

T 18.13 ((Young [464, 465])) *For C^2-diffeomorphisms $f:\mathbb{R}^2\rightarrow\mathbb{R}^2$ preserving an ergodic measure, the* Hausdorff dimension *of the attracting set can be related to the metric entropy measure and the reciprocal sum over non vanishing Lyapunov exponents by*

$$D=\bar{h}\left(\frac{1}{|\lambda_1|}+\frac{1}{|\lambda_2|}\right) \quad . \tag{18.37}$$

18.5 Algorithmic information based scenarios for entropy increase

The following scenarios are a subjective selection of the author. For other discussions, see E. T. Jaynes' *Information Theory and Statistical Mechanics* [246], *Complexity, Entropy and the Physics of Information*, ed. by W. H. Zurek, and R. Schack & C. M. Caves [401]. For a discussion on the thermodynamics of computation, see R. Landauer [291, 292], C. H. Bennett [34], E. Fredkin [185] and T. Toffoli [444]. (Cf. the authors of *Physical Review Letters* **53**, 1202-1206 (1984) for an interesting discussion.)

18.5.1 "Creation" of algorithmic information by infinite computations

As has been pointed out by G. Chaitin [114], W. H. Zurek [474] and the author [437, 438], there exists the seemingly paradoxical ability of computable processes to produce objects with higher algorithmic information content than themselves. Indeed, as W. H. Zurek points out, a program which, for example, counts and outputs integers in successive order will eventually end up with an integer of higher algorithmic information content than its length.

This scenario features computable initial states and a computable evolution. Examples for physical systems which realise this scenario are computers which count. For more details, see chapter 7, p. 97, as well as chapter 9, p. 124.

18.5.2 "Creation" of algorithmic information by uncomputable processes

This scenario features computable initial states and an uncomputable evolution. The entropy increase originates from an indeterministic system evolution.

A slight variant of this scenario is the assumption that entropy is increased by elementary events in the quantum domain, some of which occur randomly and contribute to the algorithmic information increase [456].

18.5.3 "Unfolding" of the algorithmic information content of the initial value

This scenario includes processes termed "deterministic chaos." It features a computable evolution and random initial values. The entropy increase is due to the *unfolding of the algorithmic information of the initial value* [180, 408, 9].

18.5.4 Uncomputability of algorithmic information theory entropy

As has been pointed out by G. Chaitin, it is impossible to exactly measure large values of algorithmic information. Any approximation of this measure yields a bound from above

process type/initial value	computable	uncomputable
computable	"creative" processes & uncomputability of H	oracles
uncomputable	"deterministic chaos"	all other entries combined

Table 18.1: Schematic representation of scenarios for algorithmic information theory entropy increase.

on the algorithmic information, since it relies on suboptimal program code, i.e.,

$$H(s) \leq H_{\text{approx}}(s) \quad . \tag{18.38}$$

Hence it is possible that the algorithmic information of the system H remains constant while its approximation H_{approx} increases. This is related to the fact that while the entropy of the microstate is constant, the entropy of *observation levels*, which use less observables as the microstate and which are thus approximations, increases [409, 176]. Compare also chapter 13, p. 189.

For a summary of the discussed scenarios for entropy increase, see table 18.1.

Epilogue: Afterthoughts, speculations & metaphysics

A *caveat:* This final chapter contains some afterthoughts, speculations and metaphysics. Readers who hate speculations and metaphysics are strongly discouraged to read it. Readers who would like to know what this book is all about will not find a brief answer here. This is neither a summary, nor can the material be considered scientific in the usual understanding.

The *"horror vacui"* of self-referential perception

Computer scientists as well as artists creating virtual realities and physicists interested in epistemology find themselves confronted with very similar questions:

"How does the (virtual) world look like from within?"

"Which properties has the (virtual) interface?"

"How does an (virtual) intelligence make sense of its (virtual) environment?"

"Is there any way to distinguish between a 'virtual' reality and a 'real' reality?"

"How can (virtual) theories be created by mere self-referential methods?"

"What kind of 'truth' can be ascribed to such (virtual) theories?"

and so on.

It is not unreasonable to suspect that many of those asking these questions would like to base their investigations on what they consider as "solid foundations;" i.e., some "final" methods which remain unaffected as time goes by and science progresses. This includes the possession of some "Archimedean," external, point, from which a universe could be described without changing it. They may desperately grab for something more real than paradigms which are constantly revised.

This turns out to be impossible. At least in our own universe, all scientific entities — phenomena, theories, methods *et cetera* — are intrinsic by definition. Indeed, as we cannot "step outside" of our own universe, all our perceptions *are* self-referential. — As *The Doors* put it, *"no one here gets out alive."*

Acceptance of the non existence of Archimedean points (extrinsic descriptions) results in the suspicion that a theoretical modelling which is solely based on self-referential methods is weakly founded. Our primary interface seems to be our body and its senses, which can be connected or related to secondary interfaces, such as measurement apparata or virtual reality "eye phones;" also theoretical models to organise phenomena. If we change the interface, we create a different view of the world, which is sometimes incongruent with earlier views. Plato's cave metaphor sounds a little overoptimistic: we not merely perceive shadows by an interface, but shadows created maybe by some light which we may have created by ourselves. Shakespeare's "*... like the baseless fabric of this vision ... We are such stuff / As dreams are made of ...*" suits this picture pretty well. I would like to call this feeling of suspense, followed by the tendency to turn to absoluteness, the *"horror vacui"* of self-referential perception.

Yet, *a priori,* there is nothing wrong with self-referential perception. A typical example of a formal self-referential statement is "`this statement contains five words`." Some self-referential statements are not well-founded, e.g., "`this statement is true`," and others may be self-contradictory, e.g., "`this statement is false`." Could something be learned from inconsistencies and paradoxa? The message of metamathematics and recursion theory seems to be that paradoxical statements can be utilised: they can be reformulated to express the limits of self-referential methods.

Virtual backflow interaction

In *Reason, Truth and History*, H. Putnam [377] argues that there is now way how a virtual creature may be able to refer to "the real world." It has already been argued (cf. chapter 6, p. 74) though that there *is* a difference between a self-perception based on the acceptance of virtuality or the pretension of realness *per se*. The former one may give rise to new phenomenology, in very much the same way in which it is possible for a computer virus to damage the hardware.

Virtual programming

In the spirit of algorithmic physics — the perception of the world as a (universal) computer — any preparation or manipulation of a (physical) system can be perceived as a programming task. Stated differently, experimental acts by an observer might be considered as the self-programming of a pseudo-autonomous agent from within the system. In this sense, such acts or interventions may be viewed as the "creation" of potential phenomena. Cf. I. Hacking's remarks in *Representing and Intervening* ([222], p. 226-229):

> ··· *I suggest, Hall's effect did not exist until, with great ingenuity, he had discovered how to isolate, purify it, create it in the laboratory.*
>
> ...
>
> *But the phenomena of physics — the Faraday effect, the Hall effect, the Josephson effect — are the keys that unlock the universe. People made the keys — and perhaps the locks in which they turn.*
>
> ...
>
> *Talk about creating phenomena is perhaps made most powerful when the phenomenon precedes any articulated theory, but that is not necessary. Many phenomena are created after theory.*

One may even speculate that — as we are living in a universal computational surrounding — any tasks which could be performed on a universal computer could be translatable into physical phenomena and *vice versa.*

Unreasonable effectiveness of mathematics

A. Einstein has expressed the following opinion [152]:

> *Insofar mathematical theorems refer to reality, they are not sure, and insofar they are sure, they do not refer to reality.*

In a talk entitled *"The Unreasonable Effectiveness of Mathematics in the Natural Sciences"* [463], Eugene P. Wigner has stated a somewhat different view:

> *... the enormous usefulness of mathematics in the natural sciences is something bordering on the mysterious and that there is no rational explanation for it.*

Probably it will be difficult to specify exactly what the term "rational" means in this context. Nevertheless, one explanation for the enormous usefulness of mathematics in the natural sciences might be the assumption that *nature "actually" is a machine!* For then, our formal machine concept can be adapted to perfectly suit nature; the only problem remaining is to find the machine specifications, which are interpreted as "natural laws" in physics. In this view, the arena of natural phenomena is nothing but an "inner view" of a complex automaton. One could even argue that, by the nature-as-machine metaphor, E. P. Wigner's statement becomes a tautology: mathematical modelling, such as the concept of what actually is mechanically computable / recursive, has been tailored to fit the natural sciences. Cf. Oswald Wiener's remark ([462], p. 631),

> *Introspection clearly supports the hypothesis that understanding natural processes amounts to being capable of simulating them by some "internal" Turing machines—the clearest indication might perhaps be seen in our arrival of the Turing machine concept itself.*

Hilbert's formalist program translated to physics

Unlike in mathematics whose domain is not finitely axiomatisable [204, 205, 206, 442, 446, 109, 112], in physics one could still attempt to argue that although we presently do not know every natural law, there is only a finite number of them. Yet, once we have found them all, we may just stop experimenting because we have found the ultimate & true laws of nature, and there is nothing left but deriving their consequences, which is a purely syntactic task of deduction. Thus in physics one could still cling to the idea of ultimate finite truth, the ultimate theory of everything, a Hilbertean utopia which is not conceivable for mathematics ever more. — Indeed, people, among them many scientist, who can barely bear the painful recognition of the temporal, historic status of physical modelling tend to believe that indeed we are almost at the very edge of such a theory of everything! This "over-stretch" of scientific methods, which are applied to extreme (space-time-matter) domains characterises the beginning of the creation of "scientific fairy tales" which share a huge publicity; in particular when the public relation campaign manages to convey a mysterious image of both author & subject.

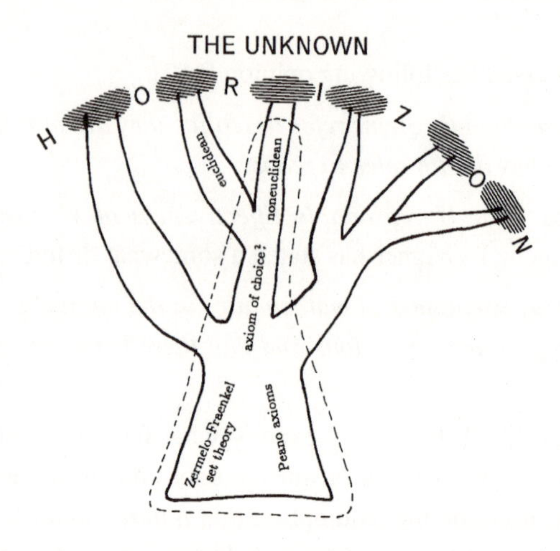

Figure 18.1: The tree of mathematics. Its branches are distinct mathematical universes. The specification of the physical universe is surrounded by dashes.

Diversity of mathematical universes *versus* one physical universe?

How many possible types of phenomena are there? Is there an infinite number of types of phenomena which await discovery-creation (cf. I. Hacking in *The creation of phenomena* [222], p. 220), or is the phenomenology of the physical universe restricted to a finite number of types? Connected to this question is the concept of the "great beyond:" is it infinitely deep or finitely shallow; i.e., does it contain an infinite richness or no more surprises?

Mathematics seems to contain an infinity of possible universes, most of which (e.g., those containing arithmetic) are too rich in structure to be finitely axiomatisable. The physical laws specify the mathematical universe we live in. I.e., whereas there is no *a priori* criterion to select one universe over the other, there is only one physical universe; see Fig. 18.1, based on the *alternative set theory* view of mathematics [449]. This seems strange.

One way to avoid this arbitrariness in the selection of one physical universe is to assume that there are an (uncountable?) infinity of parallel universes; all based on different mathematical models. This is somewhat related to the Everett interpretation of quantum mechanics.

It is as if some external minds were shopping in a huge fair, an exhibition of all possible (virtual) universes, much like J. L. Borges' *Library of Babel.* The mind chooses to get "a taste" of some particular universe by "entering" it: by being born there without memory of its past, nor knowledge of its future. (Otherwise the scenery would be only half the fun.)

Another way of eliminating the arbitrariness is the following. One cannot be sure whether the multitudes of mathematical universes are consistent. Probably there is only one consistent mathematical universe, and this is the same universe in which we physically live in.

Consistency

Why require consistency for physical phenomenology? — There exist reasonable stable systems such as databases or expert systems which are inconsistent; i.e., which yield different, even contradicting statements if consulted with different inputs. This is different from formal logic, where a formal system "blows up" if a theorem and its negation can be derived.

Continuum

Why require the continuum? There seems to be no physical phenomenon for which this assumption is necessary. In a sense, quantum mechanics "stops half-way" between a continuous and a discrete theory; cf. A. Einstein's remark, p. 223.

It it not entirely unreasonable to speculate that the continuity of space and/or time or some action variable is an illusion. Almost all color prints consist of "microscopic" pixels or color dots. To an "uninformed audience," images in cinema, in television or on computer screens, although build out of discrete cadres of events, appear continuous — in fact, they have been constructed to be perceived in that way. It is only from some context-abnormal experience, e.g., waving one's hand in front of a television screen *et cetera*, that discreteness reveals itself. Even the brain appears to organise the seemingly continuous flow of consciousness into discrete time frames of approximately 70 ms [127].

An alternative to the Turing test

An alternative to the Turing test [447] is D. Greenberger's criterion for (artificial) intelligence ([212], after the author's recollection): assume an artificial world and/or an artificial creature. Assume further a "superselection" rule which should not be broken by any circumstances; e.g., eating from the Tree of the Genesis. This creature should be termed "intelligent" if it breaks its "superselection" rule. In this approach it is evident that uncontrollability is a trade-off for intelligence and that it might be impossible to create a machine which is both intelligent and a reliable server.

Appendix A

Program code

Mathematica [455] is an explicit model of universal computation. — It is possible to simulate universal computers such as Cellular Automata, Turing machines *et cetera* in it. Moreover, it is convenient to use and widely available. Thus *Mathematica* will be used as our preferred universal model of computation.

A.1 Cellular Automata simulation

The following *Mathematica* programs simulate the evolution of Cellular Automata used in chapter 3, p. 37.

Arrow motion by right shift:

```
<<cellular.m;
a=Flatten[Append[{"-","-","-",">"},Table[_,{i,60}]]];
TableForm[EvolveCA[a,{{l_,_,_}->l},19],
        TableSpacing->{0,0},
        TableDirections->{Column}]
```

Synchronisation:

```
<<cellular.m;
a={"_","_","_",I,">","_","_","_","_","_","_",I,"_","_","_"};
Print[TableForm[EvolveCA[a,
                              {
                               {">","_",_} -> ">",
                               {_,"_","<"} -> "<",
                               {"_","_","_"} -> "_",
                               {_,"_",">"} -> "_",
                               {"<","_",_} -> "_",
                               {"_",">","_"} -> "_",
                               {"_","<","_"} -> "_",
```

```
                                   {"_",">",I} -> "<",
                                   {I,"<","_"} -> ">",
                                   {_,I,_} -> I,
                                   {"_","_",I} -> "_",
                                   {I,"_","_"} -> "_",
                                   {I,">","_"} -> "_",
                                   {"_","<",I} -> "_"
                                                }, 
     19], TableSpacing->{0,0},TableDirections->Column]];
```

A.2 $\log_2^*$ & $\lg^*$

A *Mathematica* definition of $\log_2^*$ and $\lg^*$ is

```
LogS[x_] := Module[{z = x,y,t = 0},
           While[ (y = N[Log[2,z]]) > 0, t=t+y
             ;z=y]
           ;t]
LgS[x_] := Module[{z = x,y,t = 0},
           While[ (y = Floor[N[Log[2,z]]]) > 0,
                              t=t+Floor[N[Log[2,z]]]
             ;z=y]
           ;t]
(* Plot[{LgS[x],LogS[x]},{x,0.01,20}] *)
```

A.3 Automaton analysis

A very elegant implementation of functions for automaton analysis has been created by Ch. F. Strnadl. It makes use of the package "`DiscreteMath'Combinatorica'`" which is contained in *Mathematica, version 2.* "`DiscreteMath'Combinatorica'`" is documented in *Implementing Discrete Mathematics* by St. Skiena [416], its creator. Caution: althought "`DiscreteMath'Combinatorica'`" is ingenious in many other respects, the function `HasseDiagram` contains a bug and has to be modified. The corrected function, called `HasseD` is contained in the automaton package. A copy of the automaton package can be obtained *via* anonymous ftp: `ftp.univie.ac.at`, directory `packages/mathematica`, file `automata.m`; or from the author by e-mail: `e1360dab@awiuni11.bitnet`, or `e1360dab@awiuni11.edvz.univie.ac.at`, or by sending in a formatted MS-DOS diskette.

```
(*
 * Package Automata.m
 *
 * Mathematica 2.x version
 *
 *
 * Package implementing the Automaton Propositional Calculus
 * for Moore and Mealy automata.
 *
 * Literature: K. SVOZIL, Randomness and Undecidability in Physics,
 *             1992, p137-150
 *
 *
 * Christoph F. Strnadl   (strnadl@tph01.tuwien.ac.at)
 * Insitute for Theoretical Physics, TU Vienna
 * Wiedner Hauptstrasse 8-10, A-1040 AUSTRIA
 *
 * 25-Sep-92 Started implementing Automaton Propositional Calculus
 *          functions.
 * 29-Sep-92 Changed from the Xxxx[AM] functions to generic version
 *          which differentiate between the automata via the
 *          Automat[[1]] entry.
 * 30-Sep-92 Finished the implementation of the generic version.
 *          A final test with a (532) Moore automaton succeeded (in
 *          comparison with the original Moore.m package).
 * 02-Oct-92 A test with Mathematica v2.0 on a 386-PC showed, that
 *          Wolfram has once again changed the order of context
 *          search rules. Automata.m won't work. We have naming
 *          conflicts with Path[] and Partitions[]. These two
 *          have been renamed to PathA[] and PartitionsA[] to re-
 *          flect the Automaton-nature inherent to them :-)
 * 08-Oct-92 K.Svozil suggested the possibility of an error in the
 *          construction of the Hasse-Diagram as implemented in
 *          St. SKIENA's HasseDiagram[]. I confirmed that error,
 *          which actually lies in the fact, that one cannot use
 *          RankGraph[] to construct the rank-ing of the vertices.
 *          An overworked HassD[] function is developed in this
 *          packages. But now we face the additional difficulty of
 *          having connected vertices where no edges are defined --
 *          this being due to a special geometrical aliasing problem.
 *          With HasseD[g,factor] we are able to deal with such types
 *          of graphs.
 * 09-Oct-92 Finished implementation of HasseD[]. And again we had to
 *          face some difficulties regarding the $ContextPath de-
 *          pendencies. Changed the functionality of PropCalc[]: It now
```

```
 *          generates the Graph[] object of the Propositional Calculus.
 *          To display the Hasse-Diagram one now uses ShowPropCalc[]. This
 *          behaviour is in accordance with the big classification task
 *          undertaken by K.SVOZIL, who needs the Graph[] object and not
 *          the picture :-)
 *)

BeginPackage["Automata`","DiscreteMath`Combinatorica`"]

Automat::usage :=
  "Automat[ type, transition-table, output-table,intern, {i1,i2,..} ]
   is the generic form how an automaton is stored.
   <type> = Moore | Mealy   for a Moore or Mealy automaton
   <transition-table> = a list of rules in the form
      { {state1,input1} -> next1 , {state2,input2} -> next2 ,... }
      which specifies the transition of the automaton from state1 into
      state next1 upon reading input-symbol input1 and so on.
   <output-table>  is different for the Moore and Mealy automata:
      Moore: { i1 -> o1, i2 -> o2, ... }  a list of rules specifying
           for each internal state (e.g. i1) the corresponding output
           symbol (e.g. o1).
      Mealy: { {i1,s1} -> o1, {i2,s2} -> o2 ,...} a list of rules specifying
           for each internal state i1  *and* input symbol s1 the
           corresponding output symbol o1.
   <intern> = the number of internal states of the automaton
   {i1,i2,...} = the list of all recognized input symbols of the automaton."

aMoore::usage := "The most well known (422) Moore automaton."

aCounter::usage :=
  "A Moore automaton whose automaton propositional calculus is no lattice."

aMealy::usage :=
   "aMealy is a generic (332) Mealy automaton."

MooreQ::usage :=
  "MooreQ[aut] is True if automaton aut is a Moore automaton, False
   otherwise."

MealyQ::usage :=
  "MealyQ[aut] is True if automaton aut is a Mealy automaton, False
   otherwise."

Feed::usage :=
  "Feed[a,e,{s1,s2,...}] returns the list {i1, i2, ...} of internal
   states automaton a is in when reading the input string {s1, s2,..}
```

```
   from initial internal state e."

PathA::usage :=
  "PathA[a,e,{s1,s2,...}] returns the list {o1, o2,...} of output symbols
   automaton a produces when reading the input string {s1,s2,...} from
   initial state e.
   PathA[a,{e1,e2,...},{s1,s2,..}] returns a list of the output-lists
   automaton a produces when reading input string {s1, s2,..} from initial
   states {e1, e2,...}."

StateFromInput::usage :=
  "StateFromInput[a,{s1,s2,...}] returns the partition v({s1,s2,..})
   of all input states which can be identified by automaton a upon feeding a
   with input string {s1,s2,..}."

PropsFromInput::usage :=
  "PropsFromInput[a,{s1,s2,..}] returns a list of all propositions which
   can be identified by feeding automaton a with input string {s1, s2,..}."

StateFromLevel::usage :=
  "StateFromLevel[aut,l] computes all possible partitions of input states
   which can be identified with automaton aut upon performing an experiment
   with length l on it."

PartitionsA::usage :=
  "PartitionsA[aut,l] computes all possible partitions of input states
   which can be identified by automaton aut, when performing all possible
   experiments of input strings up to (and including) the length l."

Propositions::usage :=
  "Propositions[aut,l] generates the set of all propositions which can
   be identified by automaton aut upon performing experiments up to the
   length l with it.
   Propositions[partlist] generates the set of all propositions for the
   given partition partlist. partlist may be generated by PartitionsA[]."

PropCalc::usage :=
  "PropCalc[v,len] computes the Graph[] object for the Propositional
   calculus of the automaton with partition list v. Input strings up to the
   length len are considered. The graph is the simple graph, no
   reduction to a Hasse-Diagram is done!
   PropCalc[aut,len] computes the same as above, but for Automaton aut."

ShowPropCalc::usage :=
  "ShowPropCalc[aut,l] makes the Graph[] of the automaton propositional
   calculus for automaton aut. Only experiments up to length l are considered.
```

```
   The graph itself is rendered as the Hasse Diagram of the partial
   order induced by the automaton propositional calculus.
   ShowPropCalc[aut,l,fac] makes the same as above, but with each point's
   position stretched by a factor a fac according to the stage the point is
   in the hierarchy.
   ShowPropCalc[partlist] renders the Graph[] for the given list of
   partitions.
   ShowPropCalc[partlist,fac] renders the Graph[] for the given
   partition-list,
   the positions of the vertices on the graph is stretched by factor fac
   to avoid geometrical aliasing."

HasseD::usage :=
  "HasseD[g] renders the Hasse-Diagram of graph g.
   HasseD[g,fac] renders the Hasse-Diagram of graph g with each points
   location stretched by a factor fac according to it's stage."

Begin["'Private'"]


(*
 *   aMoore is the MOORE automaton
 *
 *   Syntax is:
 *   Automat[ type, transition-table, output-table, internal-states,
 *           input-symbols ]
 *
 *   type =  Moore | Mealy
 *
 *   For the sake of convenience there is no MakeAutomaton[]
 *   function, so we cannot easily convert between the Global'Moore
 *   (or Global'Mealy) symbols and private Automata'Private'Moore.
 *   So we just use the Global context to keep user efforts to a minimum.
 *)

aMoore := Automat[Global'Moore,
                {{1, 0} -> 4, {1, 1} -> 3, {2, 0} -> 1, {2, 1} -> 3, {3, _} -> 4,
                {4, _} -> 2},
                {1 -> 0, 2 -> 0, 3 -> 0, 4 -> 1},
                4,
                {0,1}  ]

(*
 * aCounter is the automaton of SVOZIL 1992, p150, which has a propositional
 * structure which is NOT a lattice and whose implication is NO
 * partial ordering. It is the so called counter example.
```

```
 *)
aCounter := Automat[Global'Moore,{{1, _} -> 2, {2, 0} -> 3, {2, 1} -> 2,
    {3, _} -> 4, {4, 0} -> 4, {4, 1} -> 1},
    {1 -> 0, 2 -> 0, 3 -> 1, 4 -> 1},
    4, {0,1} ]


(*
 * aMealy is a generic Mealy automaton
 *)
aMealy := Automat[Global'Mealy,
    {{1, _} -> 1, {2, _} -> 2, {3, _} -> 3},
    {{1, 1} -> 1, {1, 2} -> 0, {1, 3} -> 0, {2, 1} -> 0, {2, 2} -> 1,
         {2, 3} -> 0, {3, 1} -> 0, {3, 2} -> 0, {3, 3} -> 1},
    3,
    {1, 2, 3}]



(*
 * MooreQ[aut] is TRUE if aut is a Moore Automaton
 * MealyQ[aut] is TRUE if aut is a Mealy type Automaton
 *
 * If we don't use Global'Moore here, we'd have Automata'Private'Moore,
 * which means, that the user would have to type the full qualified name if
 * HE wants to create an automaton of his/her own!
 *)
MooreQ[a_Automat] := a[[1]]===Global'Moore
MealyQ[a_Automat] := a[[1]]===Global'Mealy

(*
 * Feed[a,e,i] returns the list of internal states the automaton a
 * is in when reading the input string i from initial state e.
 *)
Feed[m_Automat, i_, l_List] :=
  FoldList[{#1, #2} /. m[[2]] & , i, l]

(*
 * PathA[a,e,i] returns the list of output-symbols the automaton a
 * emits, when reading input string input starting at initial state e.
 * PathA[a,{e1,e2,...,i}] returns the list of output-symbols for
 * different initial states e1, e2, ...
 *)
PathA[a_Automat, e_Integer, input_List] :=
  (Feed[a, e, input] /. a[[3]] )   /; MooreQ[a]
```

```
PathA[a_Automat, e_List, i_List] := (PathA[a, #1, i] & ) /@ e

(* There is a bug in the Apollo Domain/OS Mathematica 1.2 implementation,
 * which causes  Inner[ List,{a,b,c},{1,2,3},List] not to correctly work.
 * Instead of { {a,1},{b,2},{c,3} } we get { {a,b,c},{1,2,3} }.
 * This workaround does produce the correct result even on the Apollos.
 *)
PathA[a_Automat, e_Integer, in_List] :=
   ( Apply[
    List,
    Inner[f, Drop[Feed[a, e, in], -1], in,
    List],  2 ] /. a[[3]] )   /; MealyQ[a]

(* Correct and Simpler Version follows:
PathA[a_Automat, e_Integer, in_List] :=
   Inner[ List, Drop[ Feed[a,e,in], -1], in, List] /; MealyQ[a]
 *)

(*
 * StateFromPath[io,z] gives the set of initial states which
 * generate the output-string z. The list io must contain an internal
 * representation of the input/output analysis of an automaton.
 * (actually, the output of FullPath[] will right do it) So,
 * do not use this function from outside. Use StateFromInput[]
 * instead!
 *)
StateFromPath[io_, z_] :=
  (#1[[1]] & ) /@ io[[(#1[[1]] & ) /@ Position[io, z]]]

(*
 * FullPath[m,e,i] returns a list { {e1,e1path},{e2,e2path},...}
 * were together with the initial state ei the corresponding output
 * string eipath for an input list i is displayed. This form is used
 * by StateFromPath[].
 *)
FullPath[m_Automat, e_List, i_List] :=
  Thread[{e, PathA[m, e, i]}]

(*
 * StateFromInput[m,e,z] returns the partition v(z) (cf. SVOZIL 1992,
 * p140) of initial states, which can be identified by automaton m
 * upon feeding m with the input list z.
 *
 * Note: A Mealy automaton produces output only when feeding it with at
 *       least one input symbol.
 *)
```

```
StateFromInput[m_Automat,e_List,{}] := { {} } /; MealyQ[m]

StateFromInput[m_Automat, e_List, z_List] :=
  Block[ {iopath, zpath},
         iopath = FullPath[m, e, z];
         zpath = Union[(#1[[2]] & ) /@ iopath];
         (StateFromPath[iopath, #1] & ) /@ zpath
  ]

(*
 * StateFromInput[m,z] is the same as above, but with implicitely
 * taking all possible input-states of the automaton into account
 *)
StateFromInput[m_Automat, z_List] :=
  StateFromInput[m, Range[ m[[4]] ], z]

(*
 * PropsFromInput[m,e,z] generates the list of all discernible initial
 * states, which can be identified by performing the experiment z
 * (i.e. input string is z) on the automaton m with initial states
 * e. The output is in the form of { {}, {e1}, {e1,e2},...}.
 *)
PropsFromInput/:
 PropsFromInput[m_Automat, e_List, i_List] :=
      PowerSet[StateFromInput[m, e, i]]

(*
 * PropsFromInput[m,i] generates a list of all discernible initial
 * states of automaton m when feeding it input-list i.
 *)
PropsFromInput[m_Automat, i_List] :=
  PropsFromInput[m, Range[ m[[4]] ], i]


(*
 * PowerSet[ { s1, s2, ..}] produces the power set of all combinations
 * of unions of sets s1, s2,.. : { {1},{2},{3,4},.. } ->
 * { {1},{2},{3,4},{1,2},{1,3,4},{2,3,4},{1,2,3,4},... }.
 *)
PowerSet[e_List] := Union[ Apply[Union, Subsets[e], {1}] ]


(*
 * MakeInput[ level, {s1, s2, ...} ] generates all lists of input-
 * sequences with level input-symbols in each list. This function
 * is the same as St. SKIENA's Strings[l,n] function, but implemented
```

```
 * differently.
 *)
MakeInput[0,_] := {}
MakeInput[level_Integer, s_List] :=
   Flatten[Apply[Outer, Join[{List}, Table[s, {i, level}]]], level - 1]

(*
 * StateFromLevel[m,l] computes all different partitions available
 * for any input string of length l of automaton m.
 *)
StateFromLevel[m_Automat,0] :=
   { StateFromInput[m,{}] } /; MooreQ[m]
                              (* consistency with higher StateFromLevel[]
                                 list output needs surrounding braces *)

StateFromLevel[m_Automat,0] := { {} } /; MealyQ[m]

StateFromLevel[m_Automat, l_Integer] :=
   Union[(StateFromInput[m, #1] & ) /@ MakeInput[l, m[[5]] ] ]


(*
 * PartitionsA[m,l] determines all initial state partitions, which
 * can be identified experimentally with input-strings up to length
 * l for automaton m.
 * The difference between the MOORE and MEALY automata is more subtil
 * for this function: A MOORE automat emits an output symbol even when
 * presented with no input (therfore we have an iterator {j,0,l}).
 * A MEALY automat just shows output when eating an input symbol, so
 * the iterator starts at 1: {j,1,l}.
 *
 * The Union[ Sort /@ ... ] construct eliminates identical state partitions
 * via first bringing all the state partitions into canoncial (= ordered)
 * form (Sort[]) and then making a union thereof.
 *)
PartitionsA[m_Automat, l_Integer] :=
   Union [
    Sort /@
     Flatten[
       Union[(StateFromLevel[m, #1] & ) /@ Table[j, {j, 0, l}]], 1 ]
   ] /; MooreQ[m]

PartitionsA[m_Automat, l_Integer] :=
   Union[                                (* sort out multiple entries *)
     Sort /@
     Flatten[ Union[(StateFromLevel[m, #1] & ) /@ Table[j, {j, 1, l}]], 1]
```

```
  ]   /; MealyQ[m]

(*
 * Propositions[v] generates a list of all possible automaton
 * propositions. v is the PartitionList as returned by Partitions[].
 *)
Propositions[v_List] :=
  Union[Map[Sort, Flatten[(Flatten /@ Subsets[#1] & ) /@ v, 1], 1]]

(*
 * Propositions[m,l] generates the automaton propositional calculus
 * for automaton m with a maximum length of input strings of l.
 *)
Propositions[m_Automat, l_Integer] := Propositions[PartitionsA[m, l]]



(*
 * PropCalc[v] renders the Automaton Propositional Calculus for the
 * partition list v. v is in the same format as returned from PartitionsA[].
 *
 *)
PropCalc[v_List] :=
  Block[{vfull, vp},
        vp = Propositions[v];
        vfull = Map[Sort, (Flatten /@ Subsets[#1] & ) /@ v, 2];
        DiscreteMath`Combinatorica`MakeGraph[vp,
                   Intersection[#1, #2] === #1 &&
                   Intersection[first[Position[vfull, #1]],
                            first[Position[vfull, #2]]] =!= {} &
        ]
  ]

PropCalc[m_Automat,i_Integer] := PropCalc[ PartitionsA[m,i] ]

(*
 * ShowPropCalc[] generates the Hasse-Diagram of the Automaton Propositional
 * calculus, which is the displayed. fac is a factor (default = 1), with
 * which each vertex x-position is multiplied according to it's stage to
 * avoid geometrical aliasing.
 *)
ShowPropCalc[v_List,fac_:1] :=
  Block[ {g, vp},
         vp = Propositions[v];   (* need only for labelling the graph *)
         g = PropCalc[v];
         DiscreteMath`Combinatorica`ShowLabeledGraph[ HasseD[g,fac],vp]
  ]
```

```
ShowPropCalc[m_Automat,i_Integer,fac_:1] :=
   ShowPropCalc[ PartitionsA[m,i], fac]


(*
 * first[{ {a,1,..},{b,2,...},{c,...},...}] generates the list
 * {a,b,c,...} consisting of all the first elements of the sublists of
 * list.
 *)
first[l_List] := (#1[[1]] & ) /@ l


(*
 * SetLevel[{p1,p2,...},lvl,rank] sets the positions p1, p2,.. of
 * list rank to the level lvl, if the old entry at that position
 * is less than level.
 *)
SetLevel[l_List,lvl_,rank_List] :=
    Block[ {r=rank},
           If[ r[[#]] < lvl, r[[#]] = lvl ] & /@ l;
           r
    ]


(*
 * MakeLevel[l,level,adjm,rank] constructs recursively the ranks of
 * each vertex according to the adjacency matrix adjm of the graph.
 * rank is the current ranking, level the new level to assign and
 * l = {v1,v2,..} the list of vertices to be set to level.
 *)
MakeLevel[{},_,_,rank_] := rank

MakeLevel[l_List,lvl_,adjm_List,r_List] :=
  Block[ {rank=r, v, lst=l },
        rank = SetLevel[lst,lvl,rank];  (* make this level ready *)
        While[ lst != {},
              v = First[lst];
              rank = MakeLevel[adjm[[v]], lvl+1,adjm,rank];
              lst = Rest[lst];
        ];
        rank
  ]
```

```
(*
 * HasseD[g] renders a graph corresponding to the HasseDiagram of
 * the partial order induced by the directed graph g.
 * HasseD[g,fac] renders the HasseDiagram in which each vertex'
 * position is stretched by factor fac. In each stage that factor
 * is taken to the power of the distance to the 1 element.
 *
 * This function also uses some functions of Combinatorica.
 * Unfortunately, St. SKIENA's implementation HasseDiagram[] is faulty
 * for certain types of posets!
 *)
HasseD[g_,fak_:1] :=
       Block[{r, rank, m, stages,
            freq=Table[0,{DiscreteMath'Combinatorica'V[g]}], adjm, first},
               r = DiscreteMath'Combinatorica'TransitiveReduction[
                      DiscreteMath'Combinatorica'RemoveSelfLoops[g]  ];
           adjm = DiscreteMath'Combinatorica'ToAdjacencyLists[r];
           rank = Table[ 0,{ DiscreteMath'Combinatorica'V[g]} ];
           first = Select[ Range[ DiscreteMath'Combinatorica'V[g]],
                      DiscreteMath'Combinatorica'private'InDegree[r,#]==0& ];
           rank = MakeLevel[ first, 1, adjm, rank];
           first = Max[rank];
           stages = DiscreteMath'Combinatorica'Distribution[ rank ];
           DiscreteMath'Combinatorica'Graph[
                    DiscreteMath'Combinatorica'Edges[r],
                     Table[
                           m = ++ freq[[ rank[[i]] ]];
                       { ((m-1) + (1-stages[[rank[[i]] ]])/2)
                                         fak^(first-rank[[i]]),
                       rank[[i]] },
                           {i, DiscreteMath'Combinatorica'V[g]}
                     ]
               ]
       ] /; DiscreteMath'Combinatorica'AcyclicQ[
                           DiscreteMath'Combinatorica'RemoveSelfLoops[g],
                           DiscreteMath'Combinatorica'Directed ]


End[]            (* 'Private' *)

EndPackage[]      (* Automata.m *)

(* Automata.m -------------------------------------------------------*)
```

Fig. 10.8, p. 157 shows the Hasse diagrams of generic logics for automata up to 4 states. They have been obtained by a similar procedure as before, despite the fact that the set of state partitions has been generated by permutation; e.g., by the following *Mathematica* code.

```
$DefaultFont={"Helvetica",20};
<<\math\packages\Discrete\Combinat.m ;
<<Automata.m;
Print["----"];

Compress[h_]:=
Block[{n,index,iset,i1,in,i,j},
n=Length[h];
hh={h[[1]]};
index={1};
iset={1};
Do[
 If[Intersection[index,{i}]==={i}, index,
  Do[
    If[Intersection[index,{j}]==={j}, index,
      If[i===j,
         index=Append[index,j];
         iset=Append[iset,j];
         hh=Append[hh, h[[j]] ],
         If[
            ii=False; in=Length[iset];
            Do[ii=Or[ii,IsomorphicQ[ h[[j]],hh[[i1]] ]],
              {i1,in}];ii,
              index=Append[index,j], index
           ];
        ];
      ];
  ,{j,i,n}];
 ];
Print[i, " // ",index];
Print[i, "    ",iset];
,{i,n}];
hh]


Print["----"];


a=Map[ToCycles,Permutations[{1,2,3,4}]];
Print[a,Length[a]];
```

```
 a1=Rest[a];a=a1;
 b={};
 Do[If[First[Dimensions[a[[ii]]]]!=1,b1=Append[b,a[[ii]]];b=b1],{ii,
                                                        Length[a]}];
 a=b;
 b=Table[{},{Length[a]}];
 Do[b[[i]]=Map[Sort,a[[i]]],{i,Length[a]}];
 aa=Union[b];
                              (* reduction of trivial state partitions*)
 Print[b];
 sa=Subsets[aa];
 Print["Length=",Length[sa]];
(*
 g = Union[PropCalc[#]& /@ sa];
*)
 g={};
 Do[g=Union[Append[g,PropCalc[sa[[i]]]]],
   {i,Length[sa]}];
                              (* graphs of the propositional calculi *)
Print["(* graphs of the propositional calculi *)"];
 h=Table[HasseD[g[[i]],1.2],{i,2,Length[g]}];
                                   (* construction of Hasse diagrams *)
Print["(* construction of Hasse diagrams *)"];
 hh=Compress[h];
              (* construction of set of nonisomorphic Hasse diagrams *)
hh >> result. ;
hh=Prepend[hh,
Graph[{{0, 1}, {0, 0}}, {{0, 1}, {0, 2}}]
];
 (*
 * Generate the pictorial representation of all nonisomorphic graphs
 *)
hhg=Table[ShowGraph[ShakeGraph[hh[[i]],0.00]],{i,Length[hh]}];
 (*
 * Generate a partition of these pictures with mm numbers of rows
 *)
mm=5;
hhp=Partition[hhg,mm];
 (*
 * Generate an array of graphics objects
 *)
hh1=Append[hhp,Complement[hhg,Flatten[hhp]]];
If[MemberQ[hh1,{}],hh2=Delete[hh1,Length[hh1]],hh2=hh1];
xxa=Show[GraphicsArray[hh2]];
 (*
 * Generate PostScript description of array of graphics objects on file
```

```
 *)
HardcopyF[xxa,"ga.ps"];
```

A.4 Random fractals

The following *Mathematica* program is a random iteration algorithm generating the Cantor set; see chapter 15, p. 203.

```
(*
 random iteration algorithm for the Cantor set
 iteration nit times, starting ''seed'' xin
 *)
cantor[nit_,xin_]:= Block[{i},
(*
 iterated function system data
 *)
a={1/3, 1/3};
b={0, 2/3};
x=xin;
Do[
   ra=Random[Integer,{1,2}];
   (*
    apply affine transformation
    *)
   newx=a[[ra]]*x + b[[ra]];
   x=newx;
,{i,20}];
(*
 iteration starts nit times!
 *)
gr={};
Do[
   ra=Random[Integer,{1,2}];
   (*
    apply affine transformation
    *)
   newx=a[[ra]]*x + b[[ra]];
   x=newx;
   (*
    append point to graphics
    *)
   ge=Append[gr,Graphics[AbsoluteThickness[1],Line[{{x,0},{x,1}}]]]; gr=ge;
```

```
   ge=Flatten[gr];
   ,{i,nit}];
(*
display graphics
*)
Show[ge,Frame->True,FrameTicks->None
(*,PlotLabel->d,
DisplayFunction->Identity*)]
];
```

The following *Mathematica* programs generate random fractal patterns; see chapter 15, p. 203.

```
(*
 *    Bursts.m
 *
 *    Package implements one-dimensional fractals, both randomly
 *    and sequentially generated.
 *
 *    Christoph F. Strnadl    (strnadl@tph01.tuwien.ac.at)
 *    Institute for Theoretical Physics
 *    Wiedner Hauptstrasse 8 - 10, TU Vienna
 *    A-1040 VIENNA,  AUSTRIA / Europe
 *
 * HISTORY:
 *    09-Dec-92 First implementation on Apollo Domain/OS Mma v1.2
 *    10-Dec-92 Added functionality for the generation of
 *              Sequential bursts (like the Cantor Dust).
 *
 *    03-Mar-93 Added functionality for evaluation of the
 *              ''box-counting'' fractal dimension:
 *              function Dim[ ] by Franz Pichler, TU-Wien
 *
 *
 *)

BeginPackage["Bursts`"]


UniqueRandom::usage =
   "UniqueRandom[typ, n, {from,to}] generates a list of n different
    random items of type typ in the range of {from,..., to}.
```

```
   typ is the same as for the built-in Random[Integer,...] etc.
   functions."

RandomBursts::usage =
  "RandomBursts[dim, ninter, npart] generates a one-dimensional random
   burst of (approximate) fractal dimension dim by dividing the unit
   intervall into ninter sub-intervalls and choosing (according to the
   fractal dimension dim) sub-intervalls which are again divided, and
   so on... for npart times.The resulting random fractal is then
   displayed as a Graphic[]s object.
   Default values:  ninter = 10, npart = 3."

RandomBurstsList::usage =
  "RandomBurstsLists[] functions as RandomBursts[] but it returns the
   list of the subintervalls in the last level instead of the Graphics-
   object."

ShowBursts::usage =
  "ShowBursts[{e1, e2,...}] shows the list of values ei = 0 | 1 as a
   Burst-Graph."

SequentialBurstsList::usage =
 "SequentialBurstsList[lst,patt,lvl] makes a list corresponding to a
  sequential burst pattern, which is generated by applying the
  substitution pattern patt to the initial list lst for lvl times.
  SequentialBurstsList[patt] has a default initial list {1} and goes
  for 3 levels deep.
  SequentialBurstsList[patt,lvl] has a default initial list {1}."


SequentialBursts::usage =
 "SequentialBursts[] has the same functionality of SequentialBurstsList[]
  the only difference being the fact that the final list of intervalls
  is displayed as a Graphic[]s object and not outputted in form of a
  list."


Dim::usage=
 "Dim[ ldat ] evaluates the ''box-counting'' fractal dimension of the
  list ldat."

Begin["'Private'"]

(*
 *   UniqueRandom[typ,n,{from,to}] makes a list of n different random
 *   items of type typ in the range from ... to.
```

```
 *
 *   Of course, one could have written UniqueRandom[] more functional
 *   like, but -- again -- the SameTest-Option for FixedPoint[] is
 *   missing in v1.2 :-(
 *)
UniqueRandom[type_,n_Integer,{from_,to_}] :=
      Block[ { r = {} },
            While[ Length[r] =!= n,
                  r = Union[ r, {Random[ type, {from,to} ]} ]
            ];
            r
      ]


(*
 * zoom[lst, n, r] expands the list lst according to the (integer)
 * ratio r into sublists consisting of n elements.
 * Every '0' in lst -- representing a discarded intervall -- is replaced
 * by n empty subintervalls {0,....,0}.
 * Every '1' in lst -- representing an intervall which has been kept --
 * is replaced by a list of subintervalls, of which exactly r of the n
 * intervalls being kept, the other being discarded.
 *)
zoom[l_List,n_Integer,r_Integer] :=
   ( # /. { 0 -> Table[0,{n}],
          1 -> ReplacePart[ Table[0,{n}],   (* list to be replaced *)
                          1,            (* replace with '1' *)
                          {#}& /@ UniqueRandom[Integer, r,{1,n}]
                              (* positions at which to replace *)
                         ]
        } )&  /@  l   //Flatten


(*
 * RandomBursts[dim,n,lvl] generates a random burst of approximate
 * dimension dim in which the unit intervall is divided into n
 * (default 10) subintervalls for lvl (default = 3) times.
 *)
RandomBursts[d_,n_Integer:10,lvl_Integer:3] :=
         ShowBursts[ RandomBurstsList[d,n,lvl] ]


(*
 *   RandomBurstsList[] as RandomBursts[] but returns the list of
 *   intervalls instead of the Graphic[]s object.
 *)
```

```
RandomBurstsList[d_,n_Integer:10,lvl_Integer:3] :=
   Block[ {dim, r, l},
          r = Round[ n^d ];          (* number of substitutions *)
          dim = Log[r] / Log[n];      (* correct dimension *)
          Print["Exact similarity dimension is ",dim//N];
          Print["Substitution ratio r = ",r,":",n," (",r/n//N,")"];
          l = Nest[ zoom[#, n, r]&, {1}, lvl]
   ]

(*
 *  ShowBursts[]
 *
 *  Note, that we must set the PlotRange for the x-axis to be the
 *  full length of the list. Otherwise a lot of trailing (or leading)
 *  empty stripes wouldn't be displyed, which is not what any User
 *  would expect.
 *
 *  Of course, one could have written a much more elegant and stream-
 *  lined functional (tail-recursive ... insert your favorite style here)
 *  version of ShowBursts[] ;-)
 *)
ShowBursts[l_List] :=
 Block[ {gr={}, i},
          Do[  If[ l[[i]]===1,
                   gr = Append[gr,Line[{{i,0},{i,1}}] ]
               ],
               {i, Length[l] }
          ];
          Show[Graphics[ gr ],PlotRange->{{0,Length[l]},Automatic} ]
   ]

(*
 *  szoom[lst,p] expands the list lst in the following way: Each '0'
 *  in lst is replaced by a list {0,0,...,0} of length Length[p],
 *  each '1' is replaced with the pattern p. After these substitutions
 *  the inner lists are Flatten'ed out.
 *
 *  We use a szoom1[] function just for more computational speed, which,
 *  at least in the 2.x versions, could be boosted much more efficiently
 *  by Compiling[] the code. BTW the 's' stands for sequential.
 *)
szoom[lst_List,p_] := szoom1[lst,p,Length[p] ]

szoom1[l_,p_,n_] :=
  ( # /. { 0 -> Table[0,{n}], 1 -> p } )& /@ l    //Flatten
```

```
(*
 * SequentialBursts[lst,patt,lvl] generates a sequential burst pattern
 * out of starting list lst, wherein each '1' is replaced by the
 * pattern patt. This process is repeated for lvl times.
 *)
SequentialBursts[p_,lvl_Integer:3] := SequentialBursts[{1},p,lvl]

SequentialBursts[l_List,p_,lvl_Integer:3] :=
    ShowBursts[  SequentialBurstsList[ l,p,lvl ] ]

SequentialBurstsList[p_,lvl_Integer:3] := SequentialBurstsList[{1},p,lvl]

SequentialBurstsList[l_List,p_,lvl_Integer:3] :=
    Nest[  szoom[#,p]&,  l,  lvl ]


  (*
   * Dim[list,s,name] evaluates the ''box-counting'' fractal dimension
   * of the list 'list', with optional cut-off parameter 's'
   * and plot label 'name'.
   *)

 Dim[list_,s_Real:.3,name_String:"plot"]:=
Module[{n,werte,anz,jmx,jg,delta},
 $DefaultFont={"Helvetica",20};
 n=Length[list];
 If[n<2,Print[ bad data ! ];Abort[] ];
 werte={};
 jmx=Floor[Log[2 n/3]/Log[4/3]]//N;                 (* number of steps *)
 jg=Floor[Log[n/9]/Log[4/3]]//N;    (* parameter for low & high contrast *)
 Do[If[j<=jg,
      delta=Round[E^.4*(4/3)^j],   (* length of the counting-intervalls *)
      delta=Floor[n/(jmx-j)],
     ];
   anz=0;i=1;
   While[i<=n,If[Part[list,i]>s,anz++;i+=delta,i++]];   (* box-counting *)
   AppendTo[werte,{Log[delta],Log[anz]}],    (* plotpoints for graphics *)
   {j,0,jmx-1}
  ];Clear[x];
 gerade=Fit[werte,{1,x},x];                          (* interpolation *)
 dim=-Coefficient[gerade,x,1];
 sum=0;q=Length[werte];
 Do[x=werte[[j,1]]//N;
   y=werte[[j,2]]//N;
   sq=(gerade-y)^2;sum+=sq,                         (* sum of errors^2 *)
```

```
    {j,1,q}
   ];
  sigq=Sqrt[sum/q]*Cos[ArcTan[dim]];         (* confidention-interval  *)
  Print[dim,"+-",sigq];
  grp=ListPlot[werte,DisplayFunction->Identity];
  grg=Plot[gerade,{x,0,Log[n]},DisplayFunction->Identity]
  ;
  Show[grp,grg,
      Graphics[Text[ToString[StringForm[                  (* graphics *)
              "   dim = `` +- ``",dim,sigq]],
              {Log[n]/2,.9*Log[n]}]],
             DisplayFunction->$DisplayFunction,
             AxesLabel->{"log(delta)","log(N)"},
             AspectRatio->Automatic,
             PlotRange->{0,Log[n]},
             PlotLabel->ToString[StringForm["   `` , s = ``",name,s]]
    ]
  ]


End[]             (* Private *)

EndPackage[]      (* Bursts.m *)
```

Bibliography

[1] J. C. Abbott, *Trends in Lattice Theory* (Van Nostrand Reinhold Company, New York, 1970).

[2] O. Aberth, *Computable Analysis* (McGraw-Hill, New York, 1980).

[3] W. Ackermann, *Math. Annalen* **99**, 118 (1928).

[4] L. M. Adleman and M. Blum, *Journal of Symbolic Logic* **56**, 891 (1991).

[5] Y. Aharonov and D. Bohm, *Phys. Rev.* **115**, 485 (1959).

[6] Y. Aharonov and D. Z. Albert, *Phys. Rev.* **D21**, 3316 (1980); *Ibid.* **D24**, 359 (1981).

[7] D. Z. Albert, *Phys. Lett.* **94A**, 249 (1983).

[8] S. Albeverio, J. E. Fenstad, R. Hoegh-Krøhn and T. Lindstrøm, *Nonstandard Methods in Stochastic Analysis and Mathematical Physics* (Academic Press, Orlando, FA, 1986).

[9] V. M. Alekseev and M. V. Yakobson, *Physics Reports* **75**, 287 (1981).

[10] D. Angluin and C. H. Smith, *Computing Surveys* **15**, 237 (1983).

[11] Archimedes (≈ 287-212 B.C.) encountered the mechanical problem of *moving any given weight by any given force (however small)*. His answer to that question was: "give me a place to stand on and I will move the earth" (Δός μοι ποῦ στῶ καὶ κινῶ τὴν γῆν).

[12] L. Arnold and W. Kliemann, *Qualitative theory of stochastic systems*, in *Probabilistic Analysis and Related Topics*, ed. by A. Bharucha-Reid, Vol.3 (Academic Press, New York, 1981).

[13] R. Ash, *Information Theory* (Interscience, New York, 1965).

[14] H. Atmanspacher and H. Scheingraber, *Found. Physics* **17**, 939 (1987).

[15] H. Atmanspacher and G. Dalenoort, *Endo/Exo-Problems in Dynamical Systems*, (Springer, Berlin, *to be published*).

[16] B. W. Augenstein, *International Journal of Theoretical Physics* **23**, 1197 (1984).

[17] M. Baaz, *Über den allgemeinen Gehalt von Beweisen*, in *Contributions to General Algebra 6* (Hölder - Pichler - Tempsky, Vienna, 1988); see also Jan Krajíček and Pavel Pudlák, *Archiv math. Logik und Grundlagen*, in print 1988

[18] M. Baaz, N. Brunner and K. Svozil, *to be published*.

[19] D. H. Bailey, *Mathematics of Computation* **60**, 283 (1988).

[20] L. E. Ballentine, *Rev. Mod. Phys.* **42**, 358 (1970).

[21] H. P. Barendregt, *The Type Free Lambda Calculus*, in *Handbook of Mathematical Logic*, ed. by J. Barwise (North Holland, Amsterdam, 1977), ref. [27], p.1091.

[22] H. P. Barendregt, *The Lambda Calculus (Revised Edition)* (North Holland, Amsterdam, 1984).

[23] M. Barnsley's *Fractals Everywhere* (Academic Press, San Diego, 1988).

[24] G. Barton, *Phys. Lett.* **B237**, 559 (1990).

[25] A. O. Barut, *ICTP preprint, 1987*; *Proc. 2nd Int. Symp. Foundations of Quantum Mechanics* (Tokyo 1986), p.323

[26] A. O. Barut, *Physica Scripta* **35**, 229 (1987).

[27] J. Barwise *et al.*, eds., *Handbook of Mathematical Logic* (North-Holland, Amsterdam, 1977).

[28] W. Brauer, *Automatentheorie* (Teubner, Stuttgart, 1984).

[29] J. S. Bell, *Physics* **1**, 195 (1964).

[30] J. S. Bell, *Speakable and Unspeakable in Quantum Mechanics* (Cambridge University Press, Cambridge, 1987).

[31] P. A. Benioff, *J. Math. Phys.* **17**, 618 (1967); **17**, 629 (1976).

[32] P. A. Benioff, *Phys. Rev.* **D7**, 3603 (1973).

[33] P. A. Benioff, *Annals New York Akademy of Sciences* **480**, 475 (1986).

[34] Ch. H. Bennett, *International Journal of Theoretical Physics* **21**, 905 (1982).

[35] Ch. H. Bennett, *Dissipation, Information, Computational Complexity and the Definition of Organization*, in *Emerging Synthesis in Science*, ed. by D. Pines (Academic Press, New York, 1985).

[36] Ch. L. Bennett, *Phys. Rev* **A35**, 2409 (1987); *Ibid.* **A35**,2420 (1987).

[37] Ch. H. Bennett, *Logical Depth and Physical Complexity*, in *The Universal Turing Machine. A Half-Century Survey*, ed. by R. Herken (Kammerer & Unverzagt, Hamburg, 1988).

[38] E. Berlekamp, J. H. Conway and R. Guy, *Winning Ways* (Academic Press, New York, 1982).

[39] J. Bernstein, *Quantum Profiles* (Princeton University Press, Princeton, 1991), pp. 140-141.

[40] M. Berry, *Some quantum-to-classical asymtotics*, in *Les Houches school on Chaos and Quantum Physics* (North-Holland, Amsterdam, 1991).

[41] E. W. Beth, *The Foundations of Metamathematics* (North-Holland, Amsterdam, 1959).

[42] G. Birkhoff and J. von Neumann, *Annals of Mathematics* **37**, 823 (1936).

[43] G. Birkhoff, *Lattice Theory, Second Edition* (American Mathematical Society, New York, 1948).

[44] E. Bishop and D. S. Bridges, *Constructive Analysis* (Springer, Berlin, 1985).

[45] D. S. Bridges, *Bulletin of the American Mathematical Society* **24**, 216 (1991).

[46] P. van Emde Boas, *Machine Models and Simulations*, in *Algorithms and Complexity, Volume A*, ed. by J. van Leeuwen (Elsevier, Amsterdam and MIT Press, Cambridge, MA., 1990), ref. [296].

[47] P. Bocchieri and A. Loinger, *Phys. Rev.* **107**, 337 (1957).

[48] D. Bohm, *Phys. Rev.* **85**, 166, 180 (1952); D. Bohm and J. Bub, *Rev. Mod. Phys.* **38**, 453 (1966).

[49] W. Boos, *Metamathematical quantum theory: random ultrafilters as ''hidden vaiables''* (preprint, 1990, unpublished).

[50] T. L. Booth, *Sequential Machines and Automata Theory* (John Wiley and Sons, Inc., New York, 1967).

[51] A. H. Brady, *The Busy Beaver Game and the Meaning of Life*, in *The Universal Turing Machine. A Half-Century Survey*, ed. by R. Herken (Kammerer & Unverzagt, Hamburg, 1988), p. 259.

[52] A. Breton, *Manifestes du Surré alisme* (J.-J. Pauvert, Paris).

[53] D. S. Bridges, *Computability --- A Mathematical Sketchbook*, (Springer-Verlag, Berlin, 1993).

[54] L. Brillouin, *Science and Information, Second Edition* (Academic Press, New York, 1962).

[55] L. Brillouin, *Scientific Uncertainty, and Information* (Academic Press, New York, 1964).

[56] J. Brown, *New Scientist* 14 July 1990, p. 37.

[57] A. A. Brudno, *Russian Math. Surveys* **33**, 197 (1978). [*Uspekhi Mat. Nauk* **33**, 207 (1978)].

[58] A. A. Brudno, *Trans. Mosc. Math. Soc.* **44**, 127 (1983).

[59] J. R. Büchi, *Finite Automata, Their Algebras and Grammars* (Springer, New York, 1989).

[60] A. W. Burks, *Behavioral Science* **6**, 5 (1961).

[61] A. W. Burks, ed., *Essays on Cellular Automata* (Illinois University Press, Chicago, 1970).

[62] C. Calude, S. Marcus and I. Tevy, *Hist. Math.* **6**, 380 (1979).

[63] C. Calude and I. Chiţescu, *Internat. J. Comput. Math.* **11**, 43 (1982).

[64] C. Calude, I. Chiţescu, L. Staiger, *Rev. Roumaine Math. Pures Appl.* **30**, 719 (1985).

[65] C. Calude, *Theories of Computational Complexity* (North-Holland, Amsterdam, 1988).

[66] C. Calude and I. Chiţescu, *An. Univ. Bucureşti Mat.-Inf.* **2**, 27 (1988).

[67] C. Calude and I. Chiţescu, *Internat. J. Comput. Math.* **17**, 53 (1988).

[68] C. Calude and I. Chiţescu, *J. Computational Math.* **7**, 61 (1989).

[69] C. Calude and I. Chiţescu, *Boll. Unione Mat. Ital.* (7) 3-B, 229 (1989).

[70] C. Calude and Eva Kurta, *Rev. Roumaine Math. Pures Appl.* **35**, 597 (1990).

[71] C. Calude, G. Istrate, *Theoret. Comput. Sci.* **82**, 151 (1991).

[72] C. Calude, *Theoret. Comput. Sci.* **87**, 347 (1991).

[73] C. Calude, *Elementary Algorithmic Information Theory*, Bucharest University, 1991, 90 pp.

[74] C. Calude. *Borel Normality and Algorithmic Randomness*, Technical Report, The University of Western Ontario, London, December 1992, 19 pp. (with a note by G.J.Chaitin).

[75] C. Calude, *Algorithmic Information Theory. Lecture Notes*, Technical Report, The University of Western Ontario, London, December 1992, 43 pp.

[76] C. Calude and C. Câmpeanu, *Theoret. Comput. Sci.* **112**, 383 (1993).

[77] C. Calude, H. Jürgensen. *Randomness as an Invariant for Number Representation*, Technical Report, The University of Western Ontario, London, 1993, 15 pp.

[78] C. Calude, *Information and Randomness --- An Algorithmic Perspective* (Springer, Berlin, *in preparation*).

[79] C. Calude, *private communication.*

[80] G. Cantor, *Gesammelte Abhandlungen* , eds. A. Fraenkel and E. Zermelo (Springer, Berlin, 1932).

[81] R. Carnap, *Logical Foundations of Probability* (University of Chicago Press, Chicago, 1950).

[82] L. Carroll, *Alice's Adventures in Wonderland* (first published in 1865); *Through the Looking Glass* (first published in 1872).

[83] G. Casati, B. V. Chirikov and D. L. Shepelyanski, *Phys. Rev. Lett.* **53**, 2525 (1984).

[84] G. Casati, B. V. Chirikov, I. Guarneri and D. L. Shepelyanski, *Phys. Rev. Lett.* **57**, 823 (1986).

[85] G. Casati, B. V. Chirikov, I. Guarneri and D. L. Shepelyanski, *Milano university preprint, 1986; Novosibirsk preprint 87-30*

[86] J. L. Casti, *Alternate Realities: Mathematical Models of Nature & Man* (J. Wiley & Sons, New York, 1989).

[87] J. L. Casti, *Beyond Believe: Randomness, Prediction and Explanation in Science* (CRC Press, Boca Raton, Florida, 1990).

[88] J. L. Casti, *Searching for Certainty: How Scientists Predict the Future* (W. Morrow& Company, New York, 1991).

[89] J. L. Casti, *Searching for Certainty: What Scientists Can Know about the Future* (W. Morrow& Company, New York, 1992).

[90] J. L. Casti, *Reality Rules, Vol I: Picturing the World in Mathematics: the Fundamentals* (J. Wiley & Sons, New York, 1992).

[91] J. L. Casti, *Reality Rules, Vol II: Picturing the World in Mathematics: the Frontier* (J. Wiley & Sons, New York, 1992).

[92] V. Černý, *Eur. J. Phys.* **9**, 94 (1988).

[93] V. Černý, *Quantum computers and NP-complete problems*, University of Bratislava preprint, 1990.

[94] G. J. Chaitin, *AMS Notices* **13**, 133 (1966); reprinted in [109].

[95] G. J. Chaitin, *AMS Notices* **13**, 228 (1966); reprinted in [109].

[96] G. J. Chaitin, *Journal of the Assoc. Comput. Mach.* **13**, 547 (1966); reprinted in [109]. (This paper has been received by the journal in October 1965.)

[97] G. J. Chaitin, *J. Assoc. Comput. Mach.* **16**, 145 (1969); reprinted in [109].

[98] G. J. Chaitin, *Notices Amer. Math. Soc.* **17**, 672 (1970). reprinted in [109].

[99] G. J. Chaitin, *Notices Amer. Math. Soc.* **19**, A-712 (1972); reprinted in [109].

[100] G. J. Chaitin, *IBM Research Report* RC-4805, April 1974; reprinted in [109].

[101] G. J. Chaitin, *Journal of the Assoc. Comput. Mach.* **21**, 403 (1974); reprinted in [109].

[102] G. J. Chaitin, *Journal of the Assoc. Comput. Mach.* **22**, 329 (1975); reprinted in [109].

[103] G. J. Chaitin, *Scientific American* **232**, 47 (1975); reprinted in [109].

[104] G. J. Chaitin, *Theoret. Comput. Sci.* **2**, 45 (1976); reprinted in [109].

[105] G. J. Chaitin. *IBM J. Res. Develop.* **21**, 350-359, 496 (1977); reprinted in [109].

[106] G. J. Chaitin, *Int. J. Theoret. Physics* **21**, 941 (1982); reprinted in [109].

[107] G. J. Chaitin, *Computing the Busy Beaver Function*, in *Open Problems in Communication and Computation*, ed. by T. M. Cover and B. Gopinath (Springer, New York, 1987), p. 108; reprinted in [109].

[108] G. J. Chaitin, *Adv. Appl. Math.* **8**, 119 (1987), reprinted in [109].

[109] G. J. Chaitin, *Information, Randomness and Incompleteness, Second edition* (World Scientific, Singapore, 1987,1990), which is a collection of G. Chaitin's publications.

[110] G. J. Chaitin, *Algorithmic Information Theory* (Cambridge University Press, Cambridge, 1987).

[111] G. J. Chaitin, *Scientific American* **259**, 80 (1988); reprinted in [109].

[112] G. J. Chaitin, *Information-Theoretic Incompleteness* (World Scientific, Singapore, 1992).

[113] G. J. Chaitin, *Discussions of Chaitin's Work* (IBM report, 1991); this collection contains more than 60 reviews of G. Chaitin's findings.

[114] G. J. Chaitin, *private communication.*

[115] D. G. Champernowne, *The Journal of London Mathematical Society* **8**, 254 (1933).

[116] compare also G. F. Chew and H. P. Stapp, *Foundations of Physics* **18**, 809 (1988).

[117] B. V. Chirikov, F. M. Izrailev and D. L. Shepelyansky, *Soviet Scientific Review* **2C**, 209 (1981);

[118] B. V. Chirikov, *Dynamical Mechanisms of Noise* (priprint 87-123, Novosibirsk, 1987).

[119] B. V. Chirikov, *Foundations of Physics* **16**, 39 (1986).

[120] B. V. Chirikov, F. M. Izrailev, D. L. Shepelyansky, *Physica* **D33**, 77 (1988).

[121] J. F. Clauser and A. Shimony, *Rep. Prog. Phys.* **41**, 1881 (1978).

[122] E. F. Codd, *Cellular Automata* (Academic Press, New York, 1968).

[123] N. L. Cohen *et al.*, *The New England Journal of Medicine* **328**, 233 (1993).

[124] D. W. Cohen, *An Introduction to Hilbert Space and Quantum Logic* (Springer, New York, 1989).

[125] J. H. Conway, *Regular Algebra and Finite Machines* (Chapman and Hall Ltd., London, 1971).

[126] H. S. M. Coxeter, *Introduction to Geometry* (Wiley, New York, 1961).

[127] F. Crick and Ch. Koch, *Scientific American* **267**, 153 (September 1992).

[128] J. P. Crutchfield and N. H. Packard, *International Journal of Theoretical Physics* **21**, 433 (1982); *Physica* **D7**, 201 (1983).

[129] H. B. Curry and R. Feys, *Combinatory Logic* (North-Holland, Amsterdam, 1958).

[130] P. Cvitanović, *Universality in Chaos* (Hilger, Bristol 1986).

[131] N. C. A. da Costa and F. A. Doria, *Foundations of Physics Letters* **4**, 363 (1991).

[132] N. C. A. da Costa and F. A. Doria, *International Journal of Theoretical Physics* **30**, 1041 (1991).

[133] P. C. W. Davies, in *Quantum Gravity 2*, ed. by C.J.Isham et al. (Claderon Press, Oxford 1981), p.207

[134] M. Davis, H. Putnam and J. Robinson, *Annals of Mathematics* **74**, 425 (1961).

[135] M. Davis, *Computability & Unsolvability* (McGraw-Hill, New York, 1958).

[136] M. Davis, *The Undecidable* (Raven Press, New York, 1965).

[137] W. Nagourney, J. Sandberg and H. Dehmelt, *Phys. Rev. Lett.* **56**, 2797 (1986).

[138] J. P. Delahaye, *La Recherche* **19**, 1492 (1988).

[139] D. Deutsch, *Proc. R. Soc. Lond.* **A 400**, 97 (1985).

[140] D. Deutsch, *Proc. R. Soc. Lond.* **A 425**, 73 (1989).

[141] D. Deutsch and R. Jozsa, *Proc. R. Soc. Lond.* **A 439**, 553 (1992).

[142] A. K. Dewdney, *Scientific American* **251**, 10 (July 1984).

[143] P. Dutta and P. M. Horn, *Rev. Mod. Phys.* **53**, 497 (1981).

[144] A. Dvurečenskij, *Gleason's Theorem and Its Applications* (Kluwer Academic Publishers, Dordrecht, 1983; in co-edition with Ister Science Press, Bratislava, 1993).

[145] P. Eberhard, *Nuovo Cim.* **46B**, 392 (1978).

[146] P. H. Eberhard and R. R. Ross, *Foundations of Physics Letters* **2**, 127 (1989).

[147] J. C. Eccles, *The Mind-Brain Problem Revisited: The Microsite Hypothesis*, in *The Principles of Design and Operation of the Brain*, ed. by J. C. Eccles and O. Creutzfeldt (Springer, Berlin, 1990), p. 549.

[148] J.-P. Eckmann, *Rev. Mod. Phys.* **53**, 643 (1981).

[149] J.-P. Eckmann and D. Ruelle, *Rev. Mod. Phys.* **57**, 617 (1985).

[150] A. W. F. Edwards, *Likelyhood* (Cambridge University Press, Cambridge, 1972).

[151] A. Einstein, *Annalen der Physik* **17**, 891 (1905).

[152] A. Einstein, *Sitzungsberichte der Preußischen Akademie der Wissenschaften* **1**, 123 (1921).

[153] A. Einstein, *Grundzüge der Relativitätstheorie*, 1st edition (Vieweg, Braunschweig 1956).

[154] A. Einstein, B. Podolsky and N. Rosen, *Phys. Rev.* **47**, 777 (1935).

[155] P. Elias, *IEEE Transaction on Information Theory* **IT-21**, 194 (1975).

[156] A very readable introduction into nonstandard analysis and metamathematics for German speaking readers is E. Engeler, *Metamatematik der Elementarmathematik* (Springer, Berlin, 1983).

[157] is E. Engeler, *Jahrbuch 1990 der Kurt-Gödel-Gesellschaft*, 43 (1990).

[158] F. Englert, *Foundations of Physics* **17**, 621 (1987).

[159] F. Englert, J.-M. Frère, M. Rooman, *Nucl. Phys.* **B280**, 147 (1987).

[160] T. Erber and S. Putterman, *Nature* **318**, 41 (1985).

[161] G. L. Eyink, *Quantum field-theory in non-integer dimensions*, dissertation, The Ohio State University

[162] K. J. Falconer, *The Geometry of Fractal Sets* (Cambridge University Press, Cambridge, 1985).

[163] K. J. Falconer, *Fractal Geometry* (John Wiley & Sons, Chichester, 1990).

[164] K. J. Falconer, *Math. Proc. Camb. Phil. Soc.* **100**, 559 (1986); *Ibid.* **103**, 339 (1988); *Ibid.* **111**, 169 (1992).

[165] J. D. Farmer, E. Ott and J. A. Yorke, *Physica* **D7**, 153 (1983).

[166] H. Federer, *Geometric Measure Theory* (Springer, Berlin and New York, 1969), see in particular 3. 2. 19 and 3. 3. 22.

[167] S. Feferman, *Philosophia Naturalis* **21**, 546 (1984).

[168] S. Feferman, *Turing in the Land of Oz*, in *The Universal Turing Machine. A Half-Century Survey*, ed. by R. Herken (Kammerer & Unverzagt, Hamburg, 1988).

[169] M. J. Feigenbaum, *J. Stat. Phys.* **19**, 25 (1978); *Ibid.* **21**, 669 (1979).

[170] M. J. Feigenbaum, *Physica* **D7**, 16 (1983).

[171] W. Feller, *An Introduction to Probability Theory and Its Applications (Vol. 1)* (Wiley, New York 1950).

[172] E. Fermi, *Rendiconti Acad. Lincei* **5**, 795 (1927).

[173] P. K. Feyerabend, *Against Method* (New Left Books, London, 1974); *Science in a Free Society* (London, 1970).

[174] R. P. Feynman, *International Journal of Theoretical Physics* **21**, 467 (1982).

[175] R. P. Feynman, *Opt. News* **11**, 11 (1985).

[176] E. Fick und G. Sauermann, *Quantenstatistik dynamischer Prozesse, Band I* (Verlag Harri Deutsch, Thun, Frankfurt, 1983).

[177] D. Finkelstein, *Holistic Methods in Quantum Logic*, in *Quantum Theory and the Structures of Time and Space, Volume 3*, ed. by L. Castell and C. F. von Weizsäcker (Carl Hanser Verlag, München, 1979), p. 37.

[178] D. Finkelstein and S. R. Finkelstein, *International Journal of Theoretical Physics* **22**, 753 (1983).

[179] R. A. Fisher, *Statistical Methods for the Research Worker* (Oliver and Boyd, Edinburgh, 1950); *Contributions to Mathematical Statistics* (John Wiley & Sons, New York, 1950; *Statistical Methods and Scientific Inference* (Oliver and Boyd, Edinburgh, 1956). *Contributions to Mathematical Statistics* (John Wiley & Sons, New York, 1950.

[180] J. Ford, *Physics Today* **40** (4), 1 (April 1983).

[181] J. Ford, *Chaos: solving the unsolvable, predicting the unpredictable*, in *Chaotic Dynamics and Fractals*, ed. by M. F. Barnsley and S. G. Demko (Akademic Press, New York, 1986).

[182] J. Ford, *Foundations of Physics* **19**, 1275 (1989); *American Scientist* **77**, 601 (1989).

[183] D. J. Foulis and C. H. Randall, *Manuals, Morphisms and Quantum Mechanics*, in *Mathematical Foundations of Quantum Theory*, ed. by A. R. Marlow (Academic Press, New York, 1978), p.105.

[184] A. Fraenkel, Y. Bar Hillel and A. Levy, *Foundations of Set Theory*, 2nd ed. (North-Holland, Amsterdam 1973).

[185] E. Fredkin, *International Journal of Theoretical Physics* **21**, 219 (1982).

[186] E. Fredkin, *Digital Information Mechanics, technical report, August 1989*; Physica **D45**, 254 (1990).

[187] Ph. Frank, *Das Kausalgesetz und seine Grenzen* (Springer, Vienna 1932).

[188] P. Gács, *Soviet Math. Dokl.* **15**, 1477 (1974); correction, *Ibid.* **15**, 1480 (1974).

[189] P. Gács, *Theoret. Comput. Sci.* **22**, 71 (1983).

[190] P. Gács, *Information and Control* **70** , 186 (1986).

[191] P. Gács, *Lecture Notes on Descriptional Complexity and Randomness*, Boston University, 1988, 62 pp.

[192] P. Gács. *The Boltzman Entropy and Randomness Tests*, Manuscript, November 23, 1992, 23 pp.

[193] R. O. Gandy, *Limitations to Mathematical Knowledge*, in *Logic Colloquium '82*, ed. by D. van Dalen, D. Lascar and J. Smiley (North Holland, Amsterdam 1982).

[194] R. O. Gandy, *Church's Thesis and Principles for Mechanics*, in *The Kleene Symposium*, ed. by J. Barwise, H. J. Kreisler and K. Kunen (North Holland, Amsterdam 1980).

[195] M Gardner, *Scientific American* **241 (5)**, 20.

[196] M. R. Garey and D. S. Johnson, *Computers and Intractability, A Guide to the Theory of NP-Completeness* (Freeman, San Francisco, 1979).

[197] K. Gerbel and P. Weibel, *Die Welt von Innen -- Endo & Nano The World from Within -- ENDO & NANO*, ed. by K. Gerbel and P. Weibel (PVS Verleger, Linz, Austria, 1992).

[198] G. C. Ghirardi, A. Rimini and T. Weber, *Lett. Nuovo Cim.* **27**, 263 (1980).

[199] A. Gill, *Information and Control* **4**, 132 (1961).

[200] S. Ginsburg, *Journal of the Association for Computing Machinery* **5**, 266 (1958).

[201] R. Giuntini, *Quantum Logic and Hidden Variables* (BI Wisseschaftsverlag, Mannheim, 1991).

[202] R. J. Glauber, *Amplifyers, Attenuators and the Quantum Theory of Measurement*, in *Frontiers in Quantum Optics*, ed. by E. R. Pikes and S. Sarkar (Adam Hilger, Bristol 1986).

[203] Jean-Luc Godard, *''Alphaville'' (film script)* (Lorrimer Publishing, London, 1984).

[204] K. Gödel, *Monatshefte für Mathematik und Physik* **38**, 173 (1931); English translation in [205] and in Davis, ref. [136].

[205] K. Gödel, *Collected Works, Volume I, Publications 1929-1936*, ed. by S. Feferman, J. W. Dawson, Jr., St. C. Kleene, G. H. Moore, R. M. Solovay, J. van Heijenoort (Oxford University Press, Oxford, 1986).

[206] K. Gödel, *Collected Works, Volume II, Publications 1938-1974*, ed. by S. Feferman, J. W. Dawson, Jr., St. C. Kleene, G. H. Moore, R. M. Solovay, J. van Heijenoort (Oxford University Press, Oxford, 1990).

[207] E. M. Gold, *Information and Control* **10**, 447 (1967).

[208] P. Grassberger, *International Journal of Theoretical Physics* **25**, 907 (1986).

[209] P. Grassberger, *Z. Naturforsch.* **43a**, 671 (1988).

[210] J. R. Greechie, *Journal of Combinatorial Theory* **10**, 119 (1971).

[211] J. R. Greechie, *A non-standard quantum logic with a strong set of states*, in *Current Issues in Quantum Logic*, ed. by E. Beltrametti and Bas C. van Fraassen (Plenum Press, New York, 1981).

[212] D. M. Greenberger, *private communication in Viennese coffee-house, around 1986.*

[213] D. M. Greenberger, *Physica* **B151**, 374 (1988).

[214] D. B. Greenberger and A. YaSin, *Foundation of Physics* **19**, 679 (1989).

[215] D. M. Greenberger, M. Horne and A. Zeilinger, in *Bell's Theorem, Quantum Theory, and Conceptions of the Universe*, ed. by M. Kafatos (Kluwer, Dordrecht, 1989); D. M. Greenberger, M. A. Horne, A. Shimony and A. Zeilinger, *Am J. Phys.* **58**, 1131 (1990).

[216] A. A. Grib and R. R. Zapatrin, *International Journal of Theoretical Physics* **29**, 113 (1990).

[217] A. A. Grib and R. R. Zapatrin, *International Journal of Theoretical Physics* **31**, 1669 (1992).

[218] A. Grünbaum, *Modern Science and Zeno's paradoxes, Second edition* (Allen and Unwin, London, 1968).

[219] A. Grünbaum, *Philosophical Problems of Space of Time, Second, enlarged edition* (D. Reidel, Dordrecht, 1973).

[220] S. P. Gudder, *J. Math. Phys.* **11**, 431 (1970).

[221] H. Gutowitz, ed., *Cellular Automata: Theory and Experiment*, in *Physica* **D45**, 3-483 (1990); previous CA conference proceedings in *International Journal of Theoretical Physics* **21** (1982); *Complex Systems* **2** (1988); *Physica* **D10** (1984).

[222] I. Hacking, *Representing and Intervening* (Cambridge University Press, Cambridge, 1983).

[223] R. W. Hamming, *Coding and Information Theory, Second Edition* (Prentice-Hall, Englewood Cliffs, New Jersey, 1980).

[224] Hao Bai-Lin, *Chaos* (World Scientific, Singapore 1984).

[225] F. Harary, *Graph Theory* (Addison-Wesley, Reading, MA, 1969).

[226] G. H. Hardy and E. M. Wright, *An Introduction to the Theory of Numbers* (3rd edition) (Cambridge University Press, London, 1954).

[227] D. Harel, *Algorithmics, The Spirit of Computing* (Addison-Wesley, Wokingham, England, 1987).

[228] M. A. Harrison, *Canadian J. Math.* **17**, 100 (1965).

[229] J. Hauck, *Zeitschrift für mathematische Logik und Grundlagen der Mathematik* **30**, 561 (1984).

[230] G. C. Hegerfeld, *Phys. Rev. Lett.* **54**, 2395 (1985).

[231] R. Herken, editor, *The Universal Turing Machine. A Half-Century Survey* (Kammerer & Unverzagt, Hamburg, 1988).

[232] N. Herbert, *Foundation of Physics* **12**, 1171 (1982).

[233] D. Hiebeler, *Physica* **D45**, 463 (1990).

[234] D. Hilbert, *Gesammelte Abhandlungen, Volume 3* (Springer, Berlin, 1935).

[235] M. W. Hirsch and St. Smale, *Differential Equations, Dynamical Systems and Linear Algebra* (Academic Press, New York, 1974).

[236] A. Hobson, *Concepts in Statistical Mechanics* (Gordon and Breach, New York, 1971).

[237] D. R. Hofstadter, *Gödel, Escher, Bach: An Eternal Golden Braid* (Random House, New York, 1980).

[238] T. Hogg and B. A. Huberman, *Phys. Rev. Lett.* **48**, 711 (1982); *Phys. Rev.* **A28**, 22 (1983).

[239] J. E. Hopcroft, *Scientific American* **250, # 5**, 70 (1984).

[240] J. E. Hopcroft and J. D. Ullman, *Introduction to Automata Theory, Languages, and Computation* (Addison-Wesley, Reading, MA, 1979).

[241] J. E. Hutchinson, *Indiana University Mathematics Journal* **30**, 713 (1981).

[242] C. J. Isham, in *Quantum Theory of Gravity*, ed. by St. Christensen (Adam Hilger Ltd., Bristol 1982), p.313.

[243] M. Jammer, *The Philosophy of Quantum Mechanics* (John Wiley, New York, 1974).

[244] J. M. Jauch, *Foundations of Quantum Mechanics* (Addison-Wesley, Reading, Massachusetts, 1968).

[245] E. T. Jaynes, *Phys. Rev.* **106**, 620 (1957): *Ibid.* **108**, 171 (1957).

[246] E. T. Jaynes, *Information Theory and Statistical Mechanics*; in *Statistical Physics 3 (Brandeis Summer Institute 1962)*, ed. by K. W. Ford (Benjamin, New York, 1983).

[247] J. P. Jones and Y. V. Matijasevič, *Journal of Symbolic Logic* **49**, 818 (1984).

[248] G. Joos, *Lehrbuch der Theoretischen Physik* (3rd ed.) (Akademische Verlagsgesellschaft, Leipzig 1939), p. 521-523

[249] for a speculative approach, see for instance C. G. Jung, in *Synchronizität als ein Prinzip akausaler Zusammenhänge*, in *Naturerklärung und Psyche*, ed. by C. G. Jung and W. Pauli (Rascher, Zürich, 1952).

[250] G. Kalmbach, *Omologic as a Hilbert Type Calculus*, in *Current Issues in Quantum Logic*, ed. by E. Beltrametti and Bas C. van Fraassen (Plenum Press, New York, 1981), p. 333.

[251] G. Kalmbach, *Orthomodular Lattices* (Academic Press, New York, 1983).

[252] G. Kalmbach, *Measures and Hilbert Lattices* (World Scientific, Singapore, 1986).

[253] G. Kalmbach, *Foundations of Physics* **20**, 801 (1990).

[254] G. Kampis, *Int. J. General Systems* **15**, 75 (1989).

[255] G. Kampis, *Self-Modifying Systems in Biology and Cognitive Science* (Pergamon Press, Oxford, 1991).

[256] G. Kampis, *Process, information theory, and the creation of systems*, in *Evolution of Information Processing* ed. by K. Haefner (Springer, Berlin, 1992), pp. 83-103.

[257] G. Kampis, *Information: course and recourse*, in *Nature and the Evolution of Information Processing*, ed. by K. Haefner (Springer, Berlin, 1992), pp. 49-63.

[258] G. Kampis, *Life-like computing beyond the machine metaphor*, in *Computing with Biological Metaphors*, ed. by R. Paton (Chapman and Hall, London, 1993).

[259] I. Kanter, Phys. Rev. Lett. **64**, 332 (1990).

[260] A. Katz, *Principles of Statisical Mechanics. The Information Theory Approach* (W. H. Freeman and Company, San Francisco, 1967).

[261] H. J. Keisler, *Elementary Calculus (2nd ed.)* (Prindle, Boston, 1986).

[262] H. A. Keller, *Mathematische Zeitschrift* **172**, 41 (1980).

[263] J. M. Keynes, *A Treatise on Probability* (McMillan, London, 1923).

[264] A. V. Khinchin, *Mathematical Foundations of Information Theory* (Dover, New York, 1957).

[265] G. S. Kirk and J. E. Raven, *The Presocratic Philosophers* (Cambridge University Press, Cambridge, 1957).

[266] St. C. Kleene, *Introduction to Metamathematics* (North-Holland, Amsterdam, 1952).

[267] St. C. Kleene, *Turing's Anylysis of Computability, and Mayor Applications of It*, in *The Universal Turing Machine. A Half-Century Survey*, ed. by R. Herken (Kammerer & Unverzagt, Hamburg, 1988).

[268] D. E. Knuth, *The Art of Computer Programming (Vol. 2, 2nd edition)* (Addison-Wesley, Reading 1981).

[269] S. Kochen and E. P. Specker, *Journal of Mathematics and Mechanics* **17**, 59 (1967).

[270] A. N. Kolmogorov, *Grundbegriffe der Wahrscheinlichkeitsrechnung* (Springer, Berlin, 1933).

[271] A. N. Kolmogorov, *Problemy peredači informacii* **1**, 3 (1965).

[272] A. N. Kolmogorov, *IEEE Trans.* **IT-14**, 662 (1968).

[273] A. N. Kolmogorov, V. A. Uspensky, *Theory of Probability and Its Applications* **32**, 389 (1988).

[274] A. Komar, Phys. Rev. **133**, B542 (1964).

[275] L.G.Kraft. *A Device for Quantizing Grouping and Coding Amplitude Modulated Pulses*, (MS Thesis, Electrical Eng. Dept., MIT, Cambridge, Ma., 1949).

[276] G. Kreisel, *Synthese* **29**, 11 (1974).

[277] G. Kreisel, *Biographical memoirs of Fellows of the Royal Society* **26**, 148 (1980); corrections *Ibid.* **27**, 697; **28**, 718.

[278] G. Kreisel, *J. Symb. Log.* **47**, 900 (1982).

[279] L. Kronsjö, *Computational Complexity of Sequential and Parallel Algorithms* (Wiley, Chichester, 1985).

[280] N. S. Krylov, *Works in the Foundations of Statistical Mechanics* (Princeton University Press, Princeton 1979).

[281] T. S. Kuhn, *The Structure of Scientific Revolutions (2nd edition)* (Princeton University Press, Princeton, 1970).

[282] R. Laing, *J. Theor. Biol.* **66**, 437 (1977).

[283] I. Lakatosch, *The Methology of Scientific Research Programmes* (Cambridge University Press, Cambridge, 1978).

[284] M. van Lambalgen, *Random Sequences* (Dissertation, University of Amsterdam, 1987).

[285] M. van Lambalgen, *Journal of Symbolic Logic* **52**, 725 (1987).

[286] M. van Lambalgen, *Complexity, randomness and unpredictability* (Proceedings of the Mack Kac Seminar, Amsterdam, 1988).

[287] M. van Lambalgen, *Journal of Symbolic Logic* **54**, 1389 (1989).

[288] M. van Lambalgen, *Journal of Symbolic Logic* **55**, 1143 (1990).

[289] M. van Lambalgen, *Jahrbuch 1991 der Kurt-Gödel-Gesellschaft*, 65 (1991).

[290] M. van Lambalgen, *private communication*.

[291] R. Landauer, *IBM J. Res. Dev.* **3**, 183 (1961).

[292] R. Landauer, *International Journal of Theoretical Physics* **21**, 283 (1982).

[293] P. S. Laplace, *Théorie analytique des probabiletés. Introduction*, in *Oeuvres de Laplace* **7**, 6 (1847).

[294] H. D. P. Lee, *Zeno of Elea* (Cambridge University Press, Cambridge, 1936; reprinted by Adolf M. Hakkert, Amsterdam, 1967).

[295] K. de Leeuw, E. F. Moore, C. E. Shannon and N. Shapiro, *Computability by probabilistic Machines*, in *Automata Studies*, ed. by C. E. Shannon & J. McCarthy (Princeton University Press, Princeton, 1956).

[296] J. van Leeuwen, ed., *Algorithms and Complexity, Volume A* (Elsevier, Amsterdam and MIT Press, Cambridge, MA., 1990).

[297] S. K. Leung-Yan-Cheong and T. M. Cover, *IEEE Transactions on Information Theory* **IT-24**, 331 (1978).

[298] L. A. Levin, *Dokl. Akad. Nauk SSSR* **212** (1973) [*Soviet Math. Dokl.* **14**, 1413 (1973)].

[299] L. A. Levin, *Problems of Inf. Transmission* **10**, 206 (1974).

[300] M. Li and P. M. B. Vitányi, *Kolmogorov Complexity and its Applications*, in *Handbook of Theoretical Computer Sciences*, ed. by J. van Leeuwen (Elsevier Science Publishers, Amsterdam 1990), ref. [296].

[301] M. Li and P. M. B. Vitányi, *Journal of Computer and System Science* **44**, 343 (1992).

[302] A. J. Lichtenberg and M. A. Lieberman, *Regular and Stochastic Motion* (Springer, New York, 1983).

[303] H. Liermann, *Verbandsstrukturen im Mathematikunterricht* (Diesterweg, Frankfurt, 1971).

[304] E. G. K. López-Escobar, *Jahrbuch 1991 der Kurt-Gödel-Gesellschaft*, 49 (1991).

[305] L. Löfgren, *Complexity of Systems*, in *Systems & Control Encyklopedia*, ed. by M. G. Singh (Pergamon Press, Oxford, 1987), p. 704.

[306] G. W. Mackey, *Amer. Math. Monthly, Supplement* **64**, 45 (1957).

[307] B. B. Mandelbrot, *Fractals: Form, Chance and Dimension* (Freeman, San Francisco, 1977).

[308] N. Margolus, *Physica* **D10**, 81 (1984).

[309] N. Margolus, *Annals New York Akademy of Sciences* **480**, 487 (1986).

[310] The *Bible* contains a passage, which refers to Epimenides, a Crete living in the capital city of Cnossus: *"One of themselves, a prophet of their own, said, 'Cretans are always liars, evil beasts, lazy gluttons.' "* — St. Paul, Epistle to Titus I (12-13). For more details, see A. R. Anderson, *St. Paul's epistle to Titus*, in *The Paradox of the Liar*, ed. by R. L. Martin (Yale University Press, New Haven, 1970).

[311] P. Martin-Löf, *Algorithms and Random Sequences*, Erlangen University, Nürnberg, Erlangen, 1966.

[312] P. Martin-Löf, *Information and Control* **9**, 602 (1966).

[313] P. Martin-Löf, *On the notion of randomness,* in *Intuitionism and Proof Theory,* ed. by A. Kino, J. Myhill and R. E. Vesley (North-Holland, Amsterdam, London, 1970), p. 73.

[314] P. Martin-Löf, *Zeitschrift für Wahrscheinlichkeitstheorie und Verwantdte Gebiete* **19**, 225 (1971).

[315] R. D. Mauldin and S. C. Williams, *Trans. Am. Math. Soc.* **295**, 325 (1986).

[316] R. M. May, *Nature* **261**, 459 (1976), reprinted in [130] and [224].

[317] R. J. McEliece, *The Theory of Information and Coding,* published in the *Encyclopedia of Mathematics and its Applications, Volume 3* (addison-Wesley, London, 1977).

[318] N. D. Mermin, *Phys. Today* **43**, 9 (1990).

[319] A. Messiah, *Quantum Mechanics, Volume I* (North-Holland, Amsterdam, 1961).

[320] P. W. Milonni and Pl. L. Knight, *Phys. Rev.* **A10**, 1096 (1974).

[321] P. Milonni and K. Svozil, *Phys. Lett.* **B248**, 437 (1990).

[322] M. L. Minsky, *Computation: Finite and Infinite Machines* (Prentice Hall, Inc., Englewood Cliffs, New Jersey, 1970).

[323] M. Minsky, *International Journal of Theoretical Physics* **21**, 537 (1982).

[324] R. von Mises, *Probability Statistics and Truth* (Dover, New York, 1957); German original: *Wahrscheinlichkeit, Statistik und Wahrheit (2nd edition)* (Springer, Berlin, 1936); *Wahrscheinlichkeitsrechnung und ihre Anwendung in der Statistik und Theoretischen Physik* (Franz Deuticke, Leipzig, 1931).

[325] *Proceedings of the 1st Symposium on Mathematics in the Alternative Set Theory,* eds. J. Mlček, M. Benešová and B. Vojtášková (JSMF, Bratislava, 1989).

[326] Ch. D. Moore, *Complexity in Dynamical Systems* (Dissertation, Cornell University, 1990).

[327] Ch. D. Moore, *Phys. Rev. Lett.* **64**, 2354 (1990).

[328] E. F. Moore, *Gedanken-Experiments on Sequential Machines,* in *Automata Studies,* ed. by C. E. Shannon & J. McCarthy (Princeton University Press, Princeton, 1956).

[329] M. S. Morris, K. S. Thorne and U. Yurtsever, *Phys. Rev. Lett.* **61**, 1446 (1988).

[330] J. Myhill, *Zeitschrift für mathematische Logik und Grundlagen der Mathematik* **1**, 97 (1955).

[331] E. Nagel and J. R. Newman, *Scientific American* **194**, 71 (June 1956).

[332] E. Nagel and J. R. Newman, *Gödel's Proof* (New York University Press, New York, 1958); also R. Rucker, ref. [397].

[333] M. Navara and V. Rogalewicz, *Math. Nachr.* **154**, 157 (1991).

[334] E. Nelson, *Internal set theory: a new method to nonstandard analysis,* in *Bulletin of the American Mathematical Society* **83**, 1165 (1977).

[335] E. Nelson, *Radically Elementary Probability Theory* (Princeton University Press, Princeton, 1987).

[336] O. Neugebauer, *Vorlesungen über die Geschichte der antiken mathematischen Wissenschaften, 1. Band: Vorgriechische Mathematik* (Springer, Berlin, 1934), p. 172.

[337] J. von Neumann, *Mathematical Foundations of Quantum Mechanics* (Princeton University Press, Princeton, 1955).

[338] N. C. Metropolis, G. Reitweisner and J. von Neumann, *Statistical Treatment of Values of First 2000 decimal digits of e and of π Calculated on the ENIAC, Math. Tables and other Aids to Comp.* **4**, 109 (1950); reprinted in *John von Neumann, Collected Works, (Vol. V)*, ed. by A. H. Traub (MacMillan, New York, 1963), p. 765.

[339] J. von Neumann, *Various Techniques Used in Connection With Random Digits, J. Res. Nat. Bus. Stand. Appl. Math. Series.* **3**, 36 (1951); reprinted in *John von Neumann, Collected Works, (Vol. V)*, ed. by A. H. Traub (MacMillan, New York, 1963), p. 768.

[340] J. von Neumann, *Theory of Self-Reproducing Automata*, ed. by A. W. Burks (University of Illinois Press, Urbana, 1966).

[341] H. B. Nielsen and I. Picek, *Phys. Lett.* **114B**, 141 (1982); *Nucl. Phys.* **211B**, 269 (1983).

[342] P. Odifreddi's *Classical Recursion Theory, Volume 1* (North-Holland, Amsterdam, 1989); *Classical Recursion Theory, Volume 2* (in preparation).

[343] S. Ono, Mem. Fac. Eng. Kyushu Univ. **11**, 125 (1949).

[344] O. Ore, *Theory of Graphs* (American Mathematical Society, New York, 1962).

[345] D. N. Osherson, M. Stob, S. Weinstein, *Systems That Learn* (MIT Press, Cambridge, MA, 1986).

[346] E. Ott, W. D. Withers and J. A. Yorke, *J. Stat. Phys.* **36**, 687 (1984).

[347] D. Page, *Phys. Lett.* **91A**, 57 (1982).

[348] S. Lloyd and H. Pagels, *Annals of Physics (New York)* **188**, 186 (1988).

[349] M. Pavičić, *International Journal of Theoretical Physics* **31**, 373 (1992).

[350] P. Pearle, *Dynamics of the reduction of the statevector*, in *The Wave-Particle Dualism*, ed. by S. Diner *et al.* (Reidel Publishing Company, Amsterdam, 1984), and references cited therein.

[351] R. Penrose, *The Emperor's New Mind: Concerning Computers, Minds, and the Laws of Physics* (Oxford University Press, Oxford, 1990).

[352] A. Peres, *Am. J. Phys.* **46**, 745 (1978).

[353] A. Peres, *Phys. Rev. Lett.* **49**, 1118 (1982).

[354] A. Peres and W. H. Zurek, Am. J. Phys., **50**, 807 (1982).

[355] A. Peres, *Found. Phys.* **15**, 201 (1985).

[356] A. Peres and L. S. Schulman, *J. Phys.* **A21**, 3893 (1988).

[357] A. Peres, *Phys. Lett.* **151**, 107 (1990).

[358] A. Peres, *J. Phys.* **A24**, L175 (1991).

[359] Ya. B. Pesin, *Russian Math. Surveys* **32**, 55 (1977) [*Uspekhi Mat. Nauk* **32**, 55 (1977)].

[360] R. Péter, *Rekursive Funktionen, 2nd Edition* (Verlag der Ungarischen Akademie der Wissenschaften, Budapest, 1957).

[361] K. Petersen, *Ergodic Theory* (Cambridge University Press, Cambridge, 1983).

[362] C. Piron, *Foundations of Quantum Physics* (W. A. Benjamin, Reading, MA, 1976).

[363] I. Pitowsky, *Phys. Rev. Lett.* **48**, 1299 (1982); *Phys. Rev.* **D27**, 2316 (1983); N. D. Mermin, *Phys. Rev. Lett.* **49**, 1214 (1982); A. L. Macdonald, *Ibid.*, 1215 (1982); I. Pitowsky, *Ibid.*, 1216 (1982).

[364] I. Pitowsky, *Quantum Probability --- Quantum Logic* (Springer, Berlin, 1989).

[365] R. Piziak, *Orthomodular lattices and quadratic spaces: a survey* (Baylor University preprint).

[366] K. R. Popper, *The British Journal for the Philosophy of Science* **1**, 117, 173 (1950).

[367] K. R. Popper, *The Logic of Scientific Discovery* (Basic Books, New York, 1959); German original: *Die Logik der Forschung*, 1934.

[368] W. Poundstone, *The Recursive Universe* (Oxford University Press, Oxford, 1987; first published by William Morrow & Company, New York, 1985).

[369] M. B. Pour-El and I. Richards, *Annals of Mathematical Logic* **17**, 61 (1979).

[370] M. B. Pour-El and I. Richards, *Advances in Mathematics* **39**, 215 (1981).

[371] M. B. Pour-El and I. Richards, *Advances in Mathematics* **48**, 44 (1983).

[372] M. B. Pour-El and I. Richards, *Advances in Mathematics* **63**, 1 (1987).

[373] M. B. Pour-El and J. I. Richards, *Computability in Analysis and Physics* (Springer, Berlin, 1989), in which the results of [370, 371, 372] are summarised.

[374] H. Primas, *Time-asymmetric phenomena in Biology* (ETH preprint, 1988).

[375] H. Primas, *Mathematical and philosophical questions in the theory of open and macroscopic quantum systems*, in *Sixty-Two Years of Uncertainty*, ed. by A. I. Miller (Plenum Press, New York, 1990), p. 233.

[376] P. Pták and S. Pulmannová, *Orthomodular Structures as Quantum Logics* (Kluwer Academic Publishers, Dordrecht, 1991).

[377] H. Putnam, *Reason, Truth and History* (Cambridge University Press, Cambridge, 1981).

[378] D. I. Radin and R. D. Nelson, *Foundations of Physics* **19**, 1499 (1989).

[379] T. Rado, *Bell Sys. T.* (May 1962), 877.

[380] F. P. Ramsey, *Truth and Probability*, in *The Foundations of Mathematics*, ed. by R. B. Brathwaite (Routledge and Kegan, London, 1926).

[381] C. H. Randall, *A Mathematical Foundation for Empirical Science --- with special reference to quantum theory, Part I --- A Calculus of Experimental Propositions*, (Knolls Atomic Power Lab. Report KAPL-3147, 1966).

[382] K. Reach, *Journal of Symbolic Logic* **3**, 97 (1938).

[383] F. Reif, *Fundamentals of Statistical and Thermal Physics* (McGraw-Hill, New York, 1965).

[384] J. Richard, *Revue géneérale des sciences pures et appliquées* **17**, 72 (1905); also in *Acta mathematica* **30**, 295 (1906).

[385] A. Rieckers and H Stumpf, *Thermodynamik* (Vieweg, Braunschweig, 1977).

[386] J. Rissanen, *The Annals of Statistics* **11**, 416 (1983).

[387] A. Robert, *Nonstandard Analysis* (Wiley, New York, 1988).

[388] A. Robinson, *Non-standard Analysis (2nd ed.* (American Elsevier, New York, 1974).

[389] R. M. Robinson, *J. Symb. Logic* **16**, 280 (1951).

[390] H. Rogers, *Theory of Recursive Functions and Effective Computability* (MacGraw-Hill, New York 1967).

[391] R. Rosen, *Effective Processes and Natural Law*, in *The Universal Turing Machine. A Half-Century Survey*, ed. by R. Herken (Kammerer & Unverzagt, Hamburg, 1988), p. 523.

[392] O. E. Rössler, *Endophysics*, in *Real Brains, Artificial Minds*, ed. by J. L. Casti and A. Karlquist (North-Holland, New York, 1987), p. 25.

[393] O. E. Rössler, *Endophysics, Die Welt des inneren Beobachters*, ed. by P. Weibel (Merwe Verlag, Berlin, 1992).

[394] O. E. Rössler, *talk at the Endophysics Symposium*, Linz, Austria, June 1992.

[395] O. E. Rössler, *private communication.*

[396] J. Rothstein, International Journal of Theoretical Physics **21**, 327 (1982).

[397] R. Rucker, *Infinity and the Mind* (Birkhäuser, Boston 1982; reprinted by Bantam Books 1986).

[398] R. Rucker, *Rudy Rucker's Cellular Automata Laboratory* (Autodesk, Inc., Sausalito, CA, 1989).

[399] D. Ruelle, *Bol. Soc. Bras. Mat.* **9**, 83 (1978).

[400] B. Russell, *Mathematical logic as based on the theory of types*, in *From Frege to Gödel: A Source Book in Mathematical Logic 1879-1931*, ed. by L. van Heijenoort (Harvard University Press, Cambridge, MA 1967), p. 150.

[401] R. Schack and C. M. Caves, *Phys. Rev. Lett.* **69**, 3413 (1992).

[402] M. Schaller and K. Svozil, *in preparation.*

[403] K. Scharnhorst, *Phys. Lett.* **B236**, 354 (1990).

[404] C. P. Schnorr, *Zufälligkeit und Wahrscheinlichkeit* (Springer Lecture Notes in Mathematics 218, Berlin, 1971).

[405] M. R. Schroeder, *Number Theory in Science and Communication, Second Enlarged Edition* (Springer, Berlin, 1986).

[406] E. Schrödinger, *Naturwissenschaften* **23**, 807; 823; 844 (1935); English translation in *Quantum Theory and Measurement*, ed. by J. A. Wheeler and W. H. Zurek (Princeton University Press, Princeton, 1983), [458], p. 152.

[407] L. S. Schulman, *Phys. Rev.* **A18**, 2379 (1978).

[408] H. G. Schuster, *Deterministic Chaos* (Physik Verlag, Weinheim, 1984).

[409] H. Schwegler, *Z. Naturforschg.* **20a**, 1543 (1965).

[410] C. E. Shannon, *Bell System Tech. J.* **27**, 379, 632 (1948); reprinted in C. E. Shannon and W. Weaver, *The Mathematical Theory of Communication* (University of Illinois Press, Urbana, Illinois, 1949).

[411] R. Shaw, *Z. Naturforsch.* **36a**, 80, 1981.

[412] A. Shimony, *Controllable and uncontrollable non-locality*, in *Proc. Int. Symp. Foundations of Quantum Mechanics*, ed. by S. Kamefuchi *et al.* (Physical Society of Japan, Tokyo, 1984); see also J. Jarrett, *Bell's Theorem, Quantum Mechanics and Local Realism* (Ph. D. thesis, Univ. of Chicago, 1983); *Nous* **18**, 569 (1984).

[413] A. Shimony, *Events and Processes in the Quantum World*, in *Quantum Concepts in Space and Time*, ed. by R. Penrose and C. I. Isham (Clarendon Press, Oxford, 1986). Shimony uses the term "outcome dependence" for "event dependence".

[414] A. Shimony, *Scientific American* **258**, 36 (1988).

[415] M. I. Shirokov, *Sov. Phys. Usp.* **21**, 345 (1979).

[416] St. Skiena, *Implementing Discrete Mathematics* (Addison-Wesley, Redwood City, CA, 1990).

[417] C. Smorynski, *The Incompleteness Theorems*, in *Handbook of Mathematical Logic*, ed. by J. Barwise (North Holland, Amsterdam, 1977), ref. [27], p. 821.

[418] R. Smullyan, *To Mock a Mockingbird and Other Logical Puzzles*, (Knopf, New York, 1985).

[419] R. J. Solomonoff, *Information and Control* **7**, 1 (1964).

[420] R. J. Solomonoff, *IEEE Transactions on Information Theory* **IT-24**, 422 (1978).

[421] R. M. Solovay, *On Random R.E. Sets*, in *Non-Classical Logics, Model Theory, and Computability*, A. I. Arruda, N. C. A. da Costa, and R. Chuaqui (eds.), North-Holland, Amsterdam, 1977, pp. 283–307.

[422] M. R. Spiegel, *Theory and Problems of Statistics*, in *Schaum's outline series* (McGraw-Hill, New York, 1961).

[423] G.Sudan, *Bull. Math. Soc. Roumaine des Sciences* **30**, 11 (1927).

[424] J. Sumhammer, *Phys. Lett.* **A136**, 183 (1989).

[425] J. Sumhammer, *Foundations of Physics Letters* **1**, 113 (1988).

[426] J. Sumhammer, *The Physical Quantities in the Random Data of Neutron Interferometry*, in *The Concept of Probability*, ed. by E. I. Bitsakis and C. A. Nicolaides (Kluwer Akademic Publishers, Amsterdam, 1989).

[427] J. Sumhammer, *Gaining Knowledge from Quantum Experiments*, poster, presented at the *Symposium on the Foundations of Modern Physics 1990*, Joensu, Finland, August 13-17, 1990.

[428] K. Svozil, *On the setting of scales for space and time in arbitrary quantized media* (Lawrence Berkeley Laboratory preprint LBL-16097, May 1983).

[429] K. Svozil, *Europhysics Letters* **2**, 83 (1986).

[430] K. Svozil and A. Zeilinger, *International Journal of Modern Physics* **A1**, 971 (1986).

[431] K. Svozil, *J. Phys.* **A19**, L1125 (1986).

[432] K. Svozil, *Il Nuovo Cimento* **96B**, 127 (1986).

[433] K. Svozil, *J. Phys.* **A20**, 3861 (1987).

[434] K. Svozil, *Jahrbuch 1988 der Kurt-Gödel-Gesellschaft* **1**, 53 (1988).

[435] K. Svozil and A. Zeilinger, *Physica Scripta* **T21**, 122 (1988).

[436] K. Svozil, *Phys. Lett.* **140**, 5 (1989).

[437] K. Svozil, *Physica* **D45**, 420 (1990).

[438] K. Svozil, *Phys. Rev.* **D41**, 1353 (1990).

[439] D. D. Swade, *Scientific American* **268**, 62 (February 1993).

[440] G. Takeuti, *Quantum Set Theory*, in *Current Issues in Quantum Logic*, ed. by E. Beltrametti and Bas C. van Fraassen (Plenum Press, New York, 1981), p. 303.

[441] A. Tarski, *Der Wahrheitsbegriff in den Sprachen der deduktiven Disziplinen*, in *Akademie der Wissenschaften in Wien, Mathematisch-naturwissenschaftliche Klasse, Anzeiger* **69**, 24 (1932).

[442] A. Tarski, *Logic, Semantics and Metamathematics* (Oxford University Press, Oxford, 1956).

[443] R. F. Tichy, *Acta Arithmetica* **48**, 197 (1987); *Manuscripta math.* **54**, 205 (1985).

[444] T. Toffoli, *International Journal of Theoretical Physics* **21**, 177 (1982).

[445] T. Toffoli and N. Margolus, *Cellular Automata Machines* (MIT Press, Cambridge, Massachusetts, 1987).

[446] A. M. Turing, *Proc. London Math. Soc.* **(2), 42**, 230 (1936-7), reprinted in [136].

[447] A. M. Turing, *Mind* **59**, 433 (1950).

[448] A. Valentini, in *New Frontiers of Quantum Electrodynamics and Quantum Optics*, ed. by A. O. Barut (Plenum, New York, 1990).

[449] P. Vopěnka, *Mathematics in the Alternative Set Theory* (Teubner, Leipzig, 1979); see also [325].

[450] R. F. Voss, *Random Fractal Forgeries*, in *NATO ASI Series* **17**, *Fundamental Algorithms fo Computer Graphics*, ed. by R. A. Earnshaw (Springer, Berlin, 1985), p. 805.

[451] St. Wagon, *The Banach-Tarski paradox (2nd printing)* (Cambridge University Press, Cambridge, 1986).

[452] H. Weyl, *Philosophy of Mathematics and Natural Science* (Princeton University Press, Princeton, 1949).

[453] St. Wolfram, *Physical Review Letters* **54**, 735 (1985); *Physica* **D10**, 1 (1984); *Rev. Mod. Phys.* **55**, 601 (1983).

[454] St. Wolfram, *Theory and Application of Cellular Automata* (World Scientific, Singapore, 1986).

[455] St. Wolfram, *Mathematica, A System for Doing Mathematics by Computer, Second Edition* (Addison-Wesley, Redwood City, CA, 1991).

[456] C. H. Woo, *Chaos, ineffectiveness and the contrast between classical and quantal physics* (University of Maryland preprint, 1987); C. H. Woo, *Phys. Rev.* **D39**, 3174 (1989); see also K. Svozil, *Phys. Rev.* **D41**, 1353 (1990); C. H. Woo, *Phys. Rev.* **D41**, 1355 (1990).

[457] J. A. Wheeler, *Law without law*, in *Quantum Theory and Measurement*, ed. by J. A. Wheeler and W. H. Zurek (Princeton University Press, Princeton, 1983). [458], p. 182.

[458] J. A. Wheeler and W. H. Zurek, eds., *Quantum Theory and Measurement* (Princeton University Press, Princeton, 1983).

[459] For a recent review of the steering principle see J. A. Wheeler, *International Journal of Modern Physics* **A3**, 2207 (1988).

[460] W. J. Wilbur, *Transactions of the American Mathematical Society* **233**, 265, 1977.

[461] D. G. Willis, *Journal of the Assoc. Comput. Mach.* **17**, 241 (1970).

[462] O. Wiener, *Form and Content in Thinking Turing Machines*, in *The Universal Turing Machine. A Half-Century Survey*, ed. by R. Herken (Kammerer & Unverzagt, Hamburg, 1988), p. 631.

[463] E. P. Wigner, *''The unreasonable effectiveness of mathematics in the natural sciences''*, Richard Courant Lecture delivered at New York University, May 11, 1959 and published in *Communications on Pure and Applied Mathematics* **13**, 1 (1960).

[464] L.-S. Young, *J. Ergodic Theory and Dyn. Syst.* **2**, 109.

[465] L.-S. Young, *Physica* **A124**, 639 (1984).

[466] U. Zähle, *Math. Nachr.* **116**, 27 (1984).

[467] G. M. Zaslavskii, *Phys. Rep.* **80**, 157 (1981).

[468] A. Zee, *Phys. Rev.* **D25**, 1864 (1982).

[469] A. Zeilinger and K. Svozil, *Phys. Rev. Lett.* **54**, 2553 (1985).

[470] A. Zeilinger, *private communication.*

[471] Ya. B. Zel'dovich and D. D. Sokolov, *Sov. Phys. Usp.* **28**, 608 (1985).

[472] A. van der Ziel, *Adv. Electronics and Electron Phys.* **49**, 225 (1979).

[473] *Complexity, Entropy and the Physics of Information*, ed. by W. H. Zurek (Addison-Wesley, New York, 1990).

[474] W. H. Zurek, *Phys. Rev.* **A40**, 4731 (1989).

[475] A. K. Zvonkin and L. A. Levin, *Russian Mathemathical Surveys* **25**, 83 (1970).

Index